# Digital Literacy and Socio-Cultural Acceptance of ICT in Developing Countries

Emmanuel Eilu • Rehema Baguma
John Soren Pettersson • Ganesh D. Bhutkar
Editors

# Digital Literacy and Socio-Cultural Acceptance of ICT in Developing Countries

Springer

*Editors*
Emmanuel Eilu
Uganda Christian University, Mbale
University College
Mukono, Uganda

Rehema Baguma
School of Computing & Information
Technology
Makerere University
Kampala, Uganda

John Soren Pettersson
Karlstad University
Karlstad, Sweden

Ganesh D. Bhutkar
Vishwakarma Institute of Technology
Pune, Maharashtra, India

ISBN 978-3-030-61088-3 ISBN 978-3-030-61089-0 (eBook)
https://doi.org/10.1007/978-3-030-61089-0



This Springer imprint is published by the registered company Springer Nature Switzerland AG
The registered company address is: Gewerbestrasse 11, 6330 Cham, Switzerland

# Introduction

Compared to developed countries. Digital illiteracy and complex socio-cultural dynamics represent a major concern when it comes to ICT adoption and usage in the developing world, as more than 80% of illiterate adults live in developing countries. As a result, these populations are unable to make use of digital resources. However, the past few years have marked a huge increase in the use of the latest ICT innovations in developing countries, and this has led to an enormous transformation in the health, agriculture, economic, education, and social sectors in these countries. These recent innovations are in the form of mobile payment systems, the Internet of Things, Big Data, Geographic analysis and visualization tools, and other smart technologies. Numerous studies have been conducted on the impact of these innovations on the different economies; however, very few studies have been conducted on the role of HCI design in fostering digital literacy and promoting socio-cultural acceptance and usage of these latest ICT innovations in developing countries.

## Major Audience

(i) Academia: Understanding design dynamics in developing countries
(ii) Business corporations with interests in developing countries: Designing acceptable products and services for the market in developing countries
(iii) Governments: Designing socio-culturally acceptable e-government services

# Contents

# A Bird's Eye View of a Sustainable Learning Game Design

**James Boogere, Benedict Oyo, Emmanuel Eilu, and Paul Birevu Muyinda**

**Abstract** We empirically explore learning games with an aim of empowering learners' lifelong learning skills to enable them deal with complex situations sustainably. Learning is better by doing and the relation between learning and fun in games is essentially much deeper. It is also important for the players to clearly know the game system, which involves goals, entities and rules, in order to improve the vital individual social skills required to succeed. This chapter presents learning designs as a major determinant of motivation towards meaningful learning and challenge learners through continual provision of new aspects to learn, discover and develop oneself. Prior consequences designate that holistic learning game practices are more intense in a home compared to a school context. Moreover, the overall influence of context on perceived learning is significantly smaller than that of other experimental factors such as identification and enjoyment.

**Keywords** Sustainable · Complex situation · Learning games · Context · Learning experience

## 1 Introduction

We present a study in which researchers collaborated with both undergraduate and postgraduate learners in varying environments, in an activity grounded on learning games to generate their own respective lecture slides once the course outlines are given and present them during the lecture. The goal was to address that knowledge gap and deliver a logical description of the lack of widespread self-directed learning

J. Boogere · B. Oyo (✉)
Department of Computer Science, Gulu University, Gulu, Uganda

E. Eilu
Department of Information Technology, Uganda Christian University, Mbale University College, Mukono, Uganda

P. B. Muyinda
Department of Open and Distance Learning, Makerere University, Kampala, Uganda
e-mail: mpbirevu@cees.mak.ac.ug



E. Eilu et al. (eds.), *Digital Literacy and Socio-Cultural Acceptance of ICT in Developing Countries*, https://doi.org/10.1007/978-3-030-61089-0_1

skills envisioning to boost their life-long learning skills. Ontologically, universities are no longer simply viewed as places where individuals just get the qualifications specific to an area of specialization or an occupation. Rather, universities are learning environments where respective learners shape their knowledge, skills and identity. Thus, it is not only demeaning their value but is also a limited and flawed viewpoint to treat universities as just occupational training organizations [1]. In this study, learners worked individually; documented and presented their respective slides into the learning environment, as well as executed the actual game-based activities within or outside the learning environment. This chapter aims to highlight the demands that the implementation and use of an educational game in formal educational settings puts on educators' working processes and skillsets. Thus, we benchmark possibilities for professional development that an educator needs to take on if they are to assimilate games into a curriculum. Game assimilation in the education sector is a demanding one involving examination of roles that educators need to play when assimilating and using learning games in their own environments.

It is important to note that our context is based on rural African perspective and we do not deliberate about learning game effectiveness but rather the roles both learners and their educators play. Ultimately, understanding the constraints under which educators work, the contexts in which games are to be used, and the roles educators play during learning games scenarios, is a necessary foundation for improving the practicability of games. Practical tasks are indispensable when educators institute an environment ready for sustainable gaming activities. The chapter intentionally reports on research aimed at establishing empirical grounds for incorporating SMAR model [2, 3] as a framework within which learning games are evaluated [4].

## *1.1 Classification*

Sustainability as applied in learning would imply quantifiable limits to physical consumption and to material use losses; natural system acceptable damage limits; and that progress must be understood under such conditions [5]. Thus, sustainability is a broad concept involving multiple fields and perspectives extremely difficult to unify but specifically targeting the needs of the present without necessarily denting the future generations' ability to meet their own needs.

A sustainable future requires a right skillset to deal with the complex dynamics and advance the environmental assessment ability and adapting to it. Adapting to an environment demands emotional involvement and commitment, as motivation and fun are fundamental for players' engagement in any game. The key principle to motivate and engage players in learning processes is the learners' ability to control the essential connection between fun and learning. Players are engaged at a cognitive and emotional level, precisely what education for sustainability needs. Learners ought to learn the effects of context on learning game experience.

Complex situation; would inform learners to think more critically, express thoughts more clearly and understand more complex problems. This demands for learning theories that the situated-ness of knowledge and the need to create learning situations in which theoretical ways of knowing were deeply connected to complex practices in which learners are expected and wished to participate.

Learning games; the importance of our findings is in the context of games for learning and where possible commercial-oriented ones. Both learning and commercial-oriented games have been explored, but the use of entirely learning games have certainly reduced participation, as these games are not as widely known and played [6].

Context; Creating and designing a game-based curriculum involves many processes and tasks as well as implementing and using learning game software in a learning environment setting. Each learner ought to acquire specific skillset in order to overcome numerous demands and roles required for effective game-based learning. Thus, integrating games into a formal educational setting is a complex process involving heavy investments in resources and time in order to be considered a feasible part of a learning process.

Learning experience; Learning games covers a broad range of topics featuring different goals and eliciting different types of use. Playing games and learning are, however, context-embedded activities. Hence playing a given game in an educational context may not necessarily be experienced the same way in any other and/or different environment. Whilst the use of learning games has been studied in educational settings, little is known about how these games are experienced when being played in other environments and about the experiential differences between differing settings.

## *1.2 Related Work*

Vast amount of research work outputs in relation to learning games is available more specifically for research conducted in developed countries. Research shows that games can be used to support teaching and learning and eventually improve skills in many different types of areas. Games can be used in areas such as lecture assignments with an aim to improve participation and motivation of the learners.

According to [7–10], games ought not to be restricted to a specific area or locality but should be spatially enabled relative to the learner's everyday life. A restricted game would imply playing it within the boundaries of a university campus, such a game is not spatially pervasive because when the learner leaves the university campus, he or she will no longer be part of the game. There is a need to always emphasize that for a game to be spatially pervasive the learner has to be in the game no matter his location and attend to game-related actions [8].

Younger learners and their teachers in primary school need games that they can easily alter as a means of expressing their creativity to the fullness. Researchers in [11] propose a novel approach that serves as a middle ground between digital and traditional table-top games, grants learners more freedom to express themselves and helps articulate their understanding and difficulties both individually and socially. The approach called card-based model for digital game design is an alternative to the current trend of associating programming with digital creativity. In [11], it is possible to alter a digital game into a trading card game with an aim of exploring the potential of the approach. Results showed that learners' participation in the study shifted between playing and design thinking.

Shifting between playing and design thinking demands motivation, intensive use of resources, and there are few standards established to guide educators through the complex process of integrating games into their working environments. They address the knowledge gap by examining the educators' roles in regard to implementing and using games in their learning environments. Educational effectiveness of the used game is a fundamental part of a curriculum; educators supervise and guide a "game-ready" class as learners engage with the subject through play.

Essentially, the development of sustainable mindset(s) is a complex task and worth investigating [10, 12], and could demand for multi-institutional involvement.

## 1.3 Institutional Perspective

Institutional collaboration is vital especially in the game selection process, which enables learners from various fields such as health care, education and communications to identify and practice their lifelong learning skills throughout their careers. Institutions collaboration at both national and international levels requires communication activities involving international players that aim at managing the international relations and facilitate the assertion of various interests [13–15]. An institution pursues common communication goals jointly with other institutions; a cooperative approach known as public diplomacy. Public diplomacy can be enhanced depending on the nature of games either online or offline games operated in an increasingly interconnected world. For instance, the challenge of an online game-based curriculum does not stop at national borders, but demand transnational cooperation as well as engagement of state and non-state actors. Games serve as meaningful New Learning Environments (NLE) activities which could act as powerful means for improving learners' ability to transfer knowledge, concepts and skills learned in class to real-life contexts hence boosting their respective learning experience.

## 1.4 Game Story

This study focused on game-based contexts to be integrated in a curriculum envisioning the boosting of learners' real-world experiences. The major aim being focusing on design issues that could motivate learners to think critically, sustainably enhance their learning and challenge learners through continual provision of new aspects to learn, discover and develop oneself [5]. Activities included among others content nature, presentation, schedules, formats and styles presented on Playground.tensorflow.org. The game is based very loosely on how humans think; a technique known as neural network for building a computer program that learns from data. First, a collection of software "neurons" is created and connected together, allowing them to send messages to each other. Next, the network is asked to solve a problem, which it attempts to do over and over, each time strengthening the connections that lead to success and diminishing those that lead to failure.

The rest of the chapter is structured as follows: we highlight methods and procedures used (Sect. 2), elaborate on the results and discussion of our findings (Sect. 3) and lastly conclusion (Sect. 4).

## 2 Methods and Procedure

This research never subscribed to any 'right' approach, 'novel' system, or 'true' theory, we were mostly concerned on how our research could advance our ability to solve project and problem-based learning challenges through learning games [10, 16, 17]. We specifically considered observations and thoughts during the implementation of the simulation game (Playground.tensorflow.org.) as a prototype in anticipation that such games would boost learners' critical thinking and knowledge acquisition about how real-world events influence each other in a causal way. Such games are appropriate learning evaluation aspects for a spatially enabled learning design especially during evaluation of concerns such as the following: Which aspects are appropriate for a spatially enabled e-model? Are we addressing the right problems? Are we solving them well? Thus, a combination of approaches was adopted for in-depth understanding of best ways for enhancing learner skills hence rendering staff development opportunities around various learning technologies. Staff development opportunities may be pervasive with the application of the same pattern applied in this study to various sectors [18–20].

## 3 Results

We used the game as an artefactual benchmark for the visualization of cognition of complex data relationships. Learners were engaged at a cognitive and emotional level, precisely what education for sustainability needs. The engagement boosted learners to learn the effects of context on learning game experience. They visualized the application of such neural network tools in areas such as speech recognition, character recognition, signature verification application and/or human face recognition. Below are some of the snapshots and corresponding descriptions learners gave:

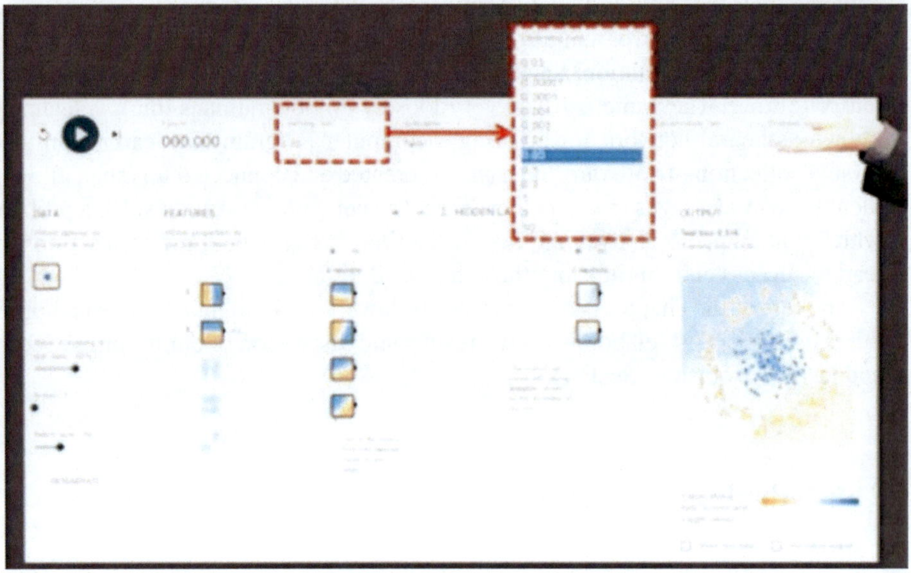


**Fig. 1** Learning rate

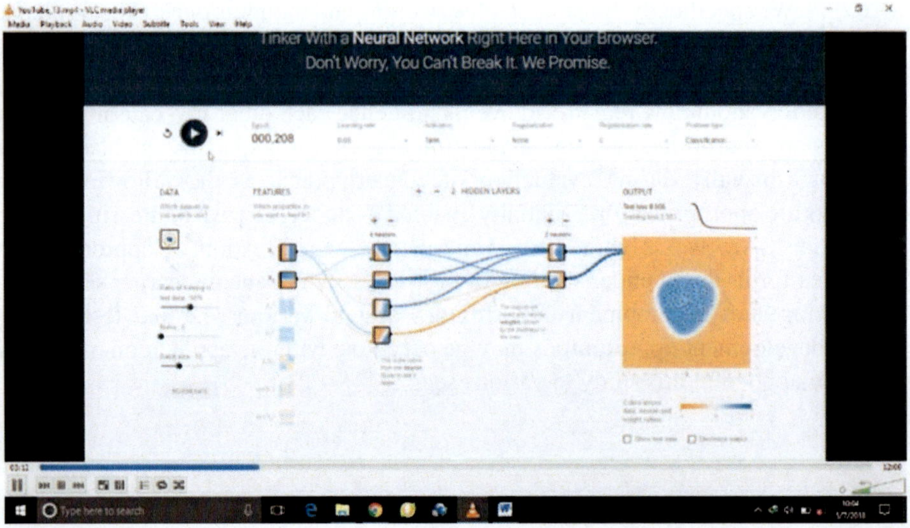


**Fig. 2** Output demo

Preceding to the demonstration Fig. 1, illustrates necessary settings such as Learning rate which determines the learning speed the player ought to select appropriately.

In the demonstration, see Fig. 2, it was made clear that every neuron is connected with other neurons through a connection link. Each connection link is associated

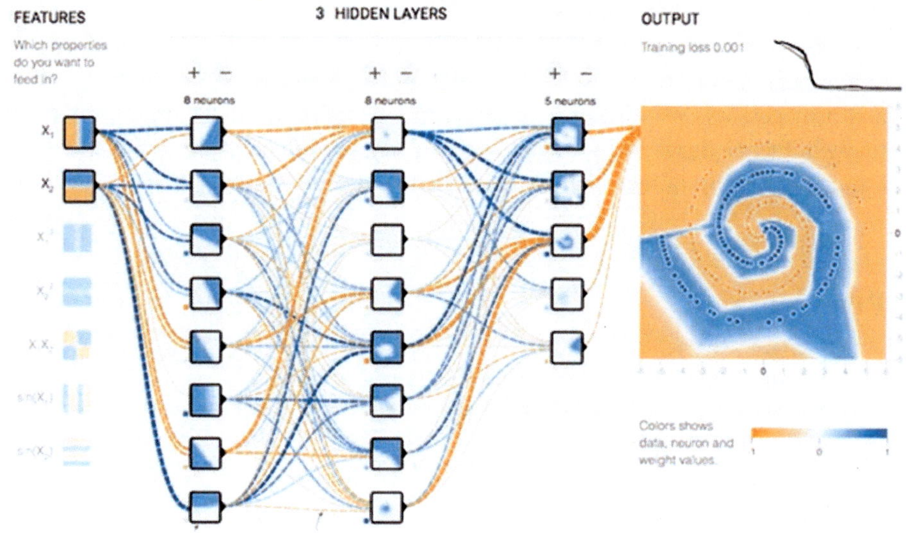


**Fig. 3** Solution

with a weight that has information about the input signal, whereby thickness represents the weights with higher priority.

Whereas the solution, (see Fig. 3), indicates how difficult it could be for the programmer to generate a classification code for a given dataset. The neurons in the first hidden layers are doing the same simple classifications, whereas the neurons in the second and third layers are composing complex features out of the simple features, eventually coming up with the double spiral pattern. The learners highlighted that it could require dozens of IF statements with many conditions, each checking which small area a given data point is in. Hence, the learners were in a position to identify the relevance of machine learning as a performance measure compared to human programmer.

Learners argued that for both games, the learning experiences are more intense in a home compared to a school context due to the strict school schedules. Thus, context or environment should be considered as a significant determinant of learning experience. However, just like any other simulation games or tools, we anticipate excessive gaming may lead to players to suffer from various shortcomings such as isolation, anti-social habits, psychological disadvantages, anxiety and ultimately affect learning/academic outcomes.

## 4 Conclusion

The major contribution has been in syllabi, courses and activities enhancements so that they are seen as an indispensable part of the curriculum entitlement of all learners. Future research must turn its attention much more towards empirical studies that not only analyze a single course but also scrutinize on a large scale how learning games are understood and practiced in different contexts and/or countries. We recommend application of virtual realities and augmented realities were applicable and/or necessary. We also commend careful selection of games that do not negatively influence our learners' mental process, cause violence, learners need not to be trapped in the virtual world away from the real world.

## References


1. E. A. Özbek, International open and distance learning conference proceedings book, in *Digital Transformation, MOOCS, Micro-Credentials and MOOC-Based Degrees: Implications for Higher Education*, (2019), pp 37–50
2. R. Puentedura, Transformation, technology, and education, in *Strength. Your Dist. Through Technol.*, (2006), p. 8
3. R. Puentedura, *A Brief Introduction to TPCK and SAMR Metaphors* (Ruben R. Puentedura's Weblog, 2011)
4. D. Romrell, L.C. Kidder, E. Wood, The SAMR model as a framework for evaluating mLearning. J. Asynchronous Learn. Netw. **18**(2) (2014). https://doi.org/10.24059/olj.v18i2.435
5. A. Nordby, K. Øygardslia, U. Sverdrup, H. Sverdrup, The art of gamification; teaching sustainability and system thinking by pervasive game development institute for digital media and computer science, Norges Tekniske og Naturvitenskaplige. Electron. J. e-Learning **14**(3), 152–168 (2016)
6. P. Sajjadi, J. Vlieghe, O. De Troyer, Exploring the relation between the theory of multiple intelligences and games for the purpose of player-centred game design. Electron. J. e-Learning **15**(4), 320–334 (2017)
7. A. Gaskell, How can we maximise the potential of new technologies to meet the Sustainable Development Goals (SDGs)? Int. J. Learn. Dev. **6**(2), 91–94 (2015)
8. T. Pløhn, Pervasive learning – Using games to tear down the classroom walls. Electron. J. e-Learning **12**(3), 299–311 (2014)
9. J. Mayanja, A. Tibaingana, P.M. Birevu, Promoting student support in open and distance learning using information and communication technologies. **6**(2), 177–186 (2016)
10. A. Nordby et al., Learning via game design: From digital to card games and back again multimedia and IT, Lillebaelt Academy for Professional Higher Education. Electron. J. e-Learning **14**(2), 301–311 (2012). https://doi.org/10.1212/01.wnl.0000168910.48309.4a
11. E. Marchetti, A. Valente, Learning via game design: From digital to card games and back again multimedia and IT, Lillebaelt Academy for Professional Higher Education. Electron. J. e-Learning **13**(3), 167–180 (2015)
12. C. Fabricatore, X. López, Sustainability learning through gaming: An exploratory study. Electron. J. e-Learning **10**(2), 209–222 (2012)
13. C. Auer, A. Srugies, *Public Diplomacy in Germany*, vol. 5, (2013)
14. A. Srugies, Competition or cooperation? A comparative analysis of the public diplomacy of the European Union and selected member states, (2016)
15. S. Jedrinovi, L. Anja, F.S. Vesna, Online course as an encouraging tool for didactic use of ICT in higher education, in *International Scientic Conference Innovative Approaches to the Application of Digital Technologies in Education and Research*, (2019, May), pp. 20–23
16. A. Oulasvirta, K. Hornbæk, HCI research a s problem – Solving, (2016)
17. T. Almpanis, Using a mixed methods research design in a study investigating – learning. **14**(5), 301–311, (2016), https://doi.org/10.1212/01.wnl.0000168910.48309.4a

18. B. Ahlstrand, H. Mintzberg, J. Lampel, V. Vaishnavi, W. Kuechler, B. Kuechler, Design science research in information systems. Last updated: November 15, 2015. Des. Sci. Res. Meth. Patt. Innov. Inf. Commun. Technol. **406**, (2004), doi: 0684847434
19. V. Vaishnavi, W. Kuechler, B. Kuechler, Design science research in information systems. Last updated November 15, 2015. 2004
20. K. Vijay, *101 Design Methods: A Structured Approach for Driving Innovation in your Organization* (Wiley, Hoboken, NJ. Published simultaneously in Canada, 2013)

# A User Experience Design Toolkit for Citizen Designers

**Apala Lahiri Chavan and Girish Prabhu**

**Abstract** In a world where runaway consumerism is leading to the real possibility of destruction of planet Earth, User Experience (UX) designers and the design processes they follow contribute to the constant creation of new and desirable experiences and fear of obsolescence. While there is new thinking by some sections of the international design community to create responsible/conscious design processes, these voices need to be reinforced if one is to see significant shifts in current design practices. This chapter discusses the creation of citizen designers as a possible approach to lend bottom-up support to hasten the dissemination of top-down design community initiated changes to make design more responsible and inclusive. Drawing from Participatory Design and Capability Approach, a design toolkit in the form of a game is proposed, to engage and motivate citizens to become citizen designers.

**Keywords** Citizen designer · Capability centred design · Participatory design

## 1 Introduction

At the very outset, we would like to define User Experience (UX) as used in this chapter. According to Interaction Design Foundation's definition, the process used to provide meaningful and relevant experiences through products and services is called user experience design. This process includes elements such as design, usability, function and branding across the entire journey from when users first buy the product through various stages of usage [1].

*This chapter presents ongoing doctoral research work by the first author that is part of an unpublished thesis.

A. L. Chavan (✉)
Design and Innovation, Human Factors International, Mumbai, India

Ph.D. Candidate, Manipal Academy of Higher Education, Manipal, India
e-mail: apala@humanfactors.com

G. Prabhu
Design and Innovation, Human Factors International, Mumbai, India

 
E. Eilu et al. (eds.), *Digital Literacy and Socio-Cultural Acceptance of ICT in Developing Countries*, https://doi.org/10.1007/978-3-030-61089-0_2

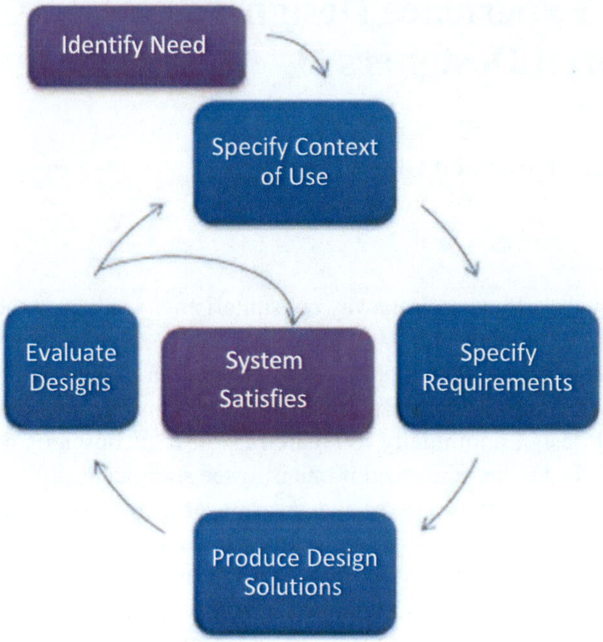


**Fig. 1** User experience design process
Image Source: https://www.usability.gov/what-and-why/user-centered-design.html

**Fig. 2** Factors that influence UX – The UX Honeycomb
Image Source: https://www.usability.gov/what-and-why/user-experience.html

To this description, we would like to add that in this chapter we refer specifically to the User Experience of digital products (whether off the shelf or customised) and not to the User Experience of non-digital products (Figs. 1 and 2).

In the last few years, several critiques have emerged about the mainstream frameworks being used for User Experience (UX) Design – the process involved and the influencing factors within the process. These critiques have applied a variety of filters to revisit current UX design frameworks, resulting in diverse perspectives/alternative frameworks such as Post-colonial Design [2], Feminist Design [3], Transition

Design [4], Design Ethics [5], Design Justice [6], Sustainable Design [7] to name some of the prominent ones.

What is common across these new/alternative frameworks is their concern about excluding users and points of view that do not act as enablers of return of investment in the short term for organisations who fund UX design.

In fact, "users" for whom experiences are designed are either, at best, included in the design process in short duration sessions to understand how the funding organisation's objectives can be successfully "sold" to users through the design or at worst, they are not included at all in the design process.

The mainstream design process, while attempting to include users in the process of designing the experience, merely looks at pre-decided functionings (goals/objectives) and not at capabilities that provide freedom to users to decide their own functionings. Hence the outcome is about designing features/functions that facilitate the functionings that have been decided upon by the organisation/brand providing (funding) the experience and has no concern about the user's capabilities and freedom to decide and in many ways, through persuasive techniques these experiences are pushed out to users as much "needed" experiences.

Take for example, an entertainment app that streams shows and films. When the app is being designed, a conceptual design of the app is already made, after discussion with business stakeholders, with the features that represent the business goals and objectives (these goals and objectives is what we term as "functionings", borrowing from the capability theory vocabulary [8]). It is at this point that users are involved in research sessions with designers where varying degrees of participatory processes are used to understand what the users' reactions are to the concept of the entertainment app and its features, which aspects need refinement, what is not resonating and hence needs persuasion techniques to be used to ensure adoption and usage. There is very little room to deeply understand and accommodate users' own goals and objectives (their preferred functionings). The focus of the sessions with users is usually to understand how their goals and objectives can be best adapted to the goals and objectives already decided by the business, even if that means the use of manipulative persuasion design techniques [9].

This inward looking design process, based largely on what businesses need and how creatively designers can convert those needs into successful and persuasive experiences for people without any serious dialogue with users and understanding of their ecosystem, leads to a myopic view of the world. "The contemporary is the one whose eyes are struck by the beam of darkness that comes from his own time", said philosopher Giorgio Agamben in his essay on contemporaneousness [10].The field of UX design and its current processes continue to be struck by that "beam of darkness" and ignore the fallout of continuously designing new and shiny experiences.

The question that becomes critical to answer is whether the move to bring about change to the current UX design process by including ways to balance the needs of business with responsible design decisions be strengthened by making citizens aware of the role of design in their lives and the impact on the life of the planet. Will this enable citizens who have been viewed just as users so far, to contribute to a more responsible design process as well as make them more aware of their own role in making sustainable purchase and usage decisions in their everyday lives?

This chapter aims to present a possible way, in the form of a game, to enable citizens to become citizen designers by helping them to acquire important capabilities that they would need, to actively contribute to the design process like citizen journalists [11] contribute to journalism.

## 2 Designing for a Better World: New Considerations for User Experience Design

UX Design frameworks being used today take no cognisance of a set of globally relevant and comprehensive set of human values, needs or capabilities when designing product and service experiences. The heartening news is that several companies are at this point in time beginning to relook at their current corporate vision and mission to incorporate tenets of conscious capitalism [12], ethical service [13]. The financial crisis that began in the USA in 2007 and the impact of which is still being felt, the ongoing climate change debate, the forecasts of dire food and water shortage in the not-distant future have all led to some shifts in corporate thinking. In addition, many corporates are realizing that millennial consumers are more likely to engage with a brand if they see it as ethical and conscious of the needs of the planet. A win-win scenario can therefore emerge if corporates who sponsor UX design can see that adopting a more responsible design approach with involvement of citizen designers can increase profit along with purpose [12].

Hence this is perhaps a better time than ever before for UX Design to also make a turn towards more responsible design using democratic processes and that makes the existence of citizen designers an absolute necessity. A new UX design framework that looks at the following, as important factors (other than what the business stakeholders/sponsors consider as important) to arrive at concepts for product/service that users would buy and experience (Figs. 3 and 4).

Three Factors to Consider:

1. A set of values that all stakeholders in a design process must select from. An example would be the Schwartz Universal Framework. There has been a renewed interest in the research of a universal values construct due to the theoretical and methodological work done in the last few decades [14, 15]. The recent theory identifies a set of ten distinct types of values and then details the dynamic relationship between them. The construct of values is structured similarly across diverse cultures but the relative importance assigned to each value may differ in different cultures.
2. A set of needs that all stakeholders are free to prioritise and select for themselves. Max Neef's Human Scale Development Model [16] could be one needs framework that could be used as a way of categorisation of a superset of human needs.

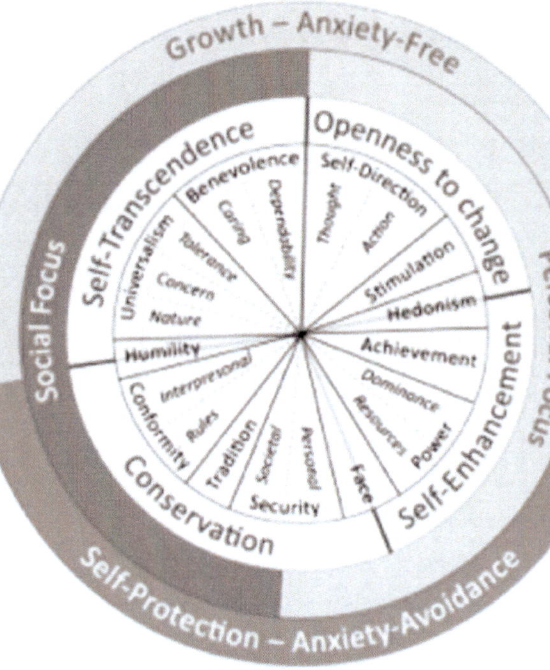


**Fig. 3** Universal Values Framework
ImageSource: http://citeseerx.ist.psu.edu/viewdoc/download?doi=10.1.1.832.177&rep=rep1&type=pdf

Considering human needs within a model of human development, Max-Neef has questioned earlier theories, such as Maslow's [17], proposing instead a matrix of intertwining needs (axiological and existential needs, in which the first are representations of the latter), whose priority changes in time and context. By placing axiological and existential needs in a matrix, intersections are created between each, which can be filled with what the author calls satisfiers. The model is certainly an excellent way to get a big picture view of universal user needs from an ecosystem level rather than dividing into hierarchies and narrow categories.

Max-Neef comes from the perspective of understanding human needs so that any development process can bring about the most improvement in people's quality of life. He further notes that the interrelated nature of human needs makes it necessary to view human needs as a system.

3. Martha Nussbaum's list of 10 capabilities [18] namely life, bodily health, bodily integrity, senses, imagination and thought, emotions, practical reason, affiliation, other species, play and control over one's environment (both political and material). She specified these capabilities as "must have" for living a life with human dignity and having the ability to exercise one's agency could form the third set from which all stakeholders decide which capabilities (of users) are to be primarily enhanced by the product/service and its design.

| | Need | Being (qualities) | Having (things) | Doing (actions) | Interacting (settings) |
|---|---|---|---|---|---|
| 1 | subsistence | physical and mental health | food, shelter, work | feed, clothe, rest, work | living environment, social setting |
| 2 | protection | care, adaptability, autonomy | social security, health systems, work | co-operate, plan, take care of, help | social environment, dwelling |
| 3 | affection | respect, sense of humour, generosity, sensuality | friendships, family, relationships with nature | share, take care of, make love, express emotions | privacy, intimate spaces of togetherness |
| 4 | understanding | critical capacity, curiosity, intuition | literature, teachers, policies, educational | analyse, study, meditate, investigate, | schools, families, universities, communities, |
| 5 | participation | receptiveness, dedication, sense of humour | responsibilities, duties, work, rights | cooperate, dissent, express opinions | associations, parties, churches, neighbourhoods |
| 6 | leisure | imagination, tranquility, spontaneity | games, parties, peace of mind | day-dream, remember, relax, have fun | landscapes, intimate spaces, places to be alone |
| 7 | creation | imagination, boldness, inventiveness, curiosity | abilities, skills, work, techniques | invent, build, design, work, compose, interpret | spaces for expression, workshops, audiences |
| 8 | identity | sense of belonging, self-esteem, consistency | language, religions, work, customs, values, norms | get to know oneself, grow, commit oneself | places one belongs to, everyday settings |
| 9 | Freedom | autonomy, passion, self-esteem, open-mindedness | equal rights | dissent, choose, run risks, develop awareness | anywhere |

**Fig. 4** Max Neef's Human Scale Development Model
Image Source: http://www.buddhaxbondi.com/new-page-1/

In order to shift the narrow and limited involvement of users in the design process as is prevalent today to a truly democratic and empowering co-creation exercise will need a thorough revamp of current processes and this revamp can happen only if citizens are aware of their right to participate in design of products/services that will impact their lives and are capable of confidently collaborating with designers and other stakeholders. This is the most important step needed to democratize design.

## 3 How Can Citizens Be Included in the Design Process?

As John Rheinfrank [19] pointed out almost two decades ago, "the designer is moving from being the detached expert to be a collaborator". This will imply that project goals will be defined by also taking into consideration the views of users. The value of co-creation with users will thus gain momentum. "In the past we designed for users, today we design with users, tomorrow we will have design by users" [19].

This shift towards integration of participatory design methods in the current UX design process, to ensure that users views are taken into consideration, has indeed gained some momentum. However, the empowerment of users in the process remains rather unevenly distributed in participatory design and in recent times there have been concerted attempts to "revitalise" and shift focus in participatory design to making impact on big issues by envisaging new ways in which partnerships and roles of researchers can unfold with more emphasis on research activism [20]. Hence, we feel that the approach presented in this chapter, of working bottom up and raising the design awareness of all citizens and not just those who are invited to participatory design sessions, is important. We have done this through a scalable option of an easily accessible game as the first step in creating a toolkit for citizen designers (as opposed to the numerous UX toolkits that exist today for student and/or professional designers). We hope that this would contribute to hasten the transformation of the designer that Rheinfrank had envisioned two decades ago, from being a detached expert to a collaborator who designs with informed users and facilitates design by users, where the users have now become citizen designers [19]. The existence of citizens who are aware and informed about design's role and impact in their lives will also help to revise participatory design that contributes to "big issues" of sustainable and responsible choices and question participatory design when used as a way to merely focus on banal issues around usability [21].

Two frameworks that have been applied to design and have provided the inspiration to look for answers to the question of including citizens in the design process are Participatory Design and Capability Centred Design. Even though the objectives of these two frameworks are very different, with one emphasising full participation of the end user in the design process and the other focusing on designs that expand the capabilities of the end user, both provide useful ideas about how to make people from all walks of life have the capability to understand what role design plays, what they can expect and demand and hence are able to be citizen designers who can critique existing ideas and help generate new ones.

## 3.1 Participatory Design

The idea of citizen designers is inspired by the original ideology of participatory design, as practiced in Scandinavian countries. As MIT Technical Review [22] observed about the pioneering participatory design project UTOPIA and how this project showed that it is possible to design industrial technology/products with users of the technology/product.

Scandinavian participatory design practices originate from a very strong political commitment by designers and researchers in the 1980s [23], to see design and democracy work hand in hand by building a collaborative relationship between designers and the communities of users who would be the final consumers of the technology/product (and now service) being designed. The strong belief in the ability of a democratic design process yielding a democratic result that would lead to a collective evolution of the lives of both, designers and users enabled Scandinavian advocates of participatory design to move this new design philosophy (at that time) ahead.

"Design is about changing: changing artefacts as well as changing people, organisations, communities" [24]. This emphasis on change and evolution of the entire ecosystem consisting of people, their practices as well as organisations and technology is another differentiator of the Scandinavian participatory design approach.

It is this dialogic exchange and emphasis on mutual learning that also makes action research and participatory action research sources of methodological inspiration for the participatory design movement. There is also an epistemological similarity between the two [25–27].

In addition, as Lucy Suchman articulates, recognizing that designing of technology has involvement and impact in the political movements of the ecosystem that technology is situated in could lead to an alternative assessment of the critical role that design and development of technology systems can play in furthering of democracy [28].

This political stance that formed the basis of participatory design along with the importance laid on explicitly articulating the values being embedded in design of systems is very much in line with what we consider as missing in mainstream UX design frameworks. Participatory design, especially in the context of India and other emerging countries could lead to a complete rethinking of the role of designers and that is very much needed to change the current perspective of the designer as the disinterested and detached expert. This current power dynamic between the designer and the user makes it difficult to empower those for whom the design solutions are aimed at, to participate and co-create along with designers.

There needs to be a rethinking of the current UX Design frameworks such that citizens become more than mere users and are able to have a voice in design decisions that have an impact on them rather than be passive consumers of decisions taken by designers and business owners.

Participatory design processes are also expected to help users who participate in the process, to experience what it is like to "overcome the habit of submission, a frame of mind that curtails people from fully and critically engaging with the world and participating in civic life" [28].

### *3.2 Capability Centred Design*

The concept and measurement of human welfare has undergone a massive change with Amartya Sen's capabilities approach. Several measures such as the Human Development Index, Better Life Index, Happiness Index, etc. are now a part of social welfare measures, in addition to the traditionally preferred measure of national income.

The Humanist Revolution in economics was ushered in by Amartya Sen and Martha Nussbaum [29], with their capabilities approach. What people can *do* instead of what they simply *have* is what Sen and Nussbaum focused their attention on in the capability approach. In his search for the most suitable measures for people's well-being, Sen found that income or possessions or fulfilment of desires were not adequate indicators of people's well-being. His argument was that people's actual capability of being and doing is what determines their well-being or the lack of it [30].

The defining elements of the capability approach are functionings and capabilities. One way to look at functionings is that these are all the things that someone aspires, desires and chooses to be and do in life while capabilities are the superset of these "things" called functionings that someone possesses and can decide on which functionings to actually pursue [31, 32].

In contrast to capabilities and functionings, goods are merely things that one possesses. It is one's capabilities that allow one to decide how to use goods in fulfilling ways. Both Sen and Nussbaum also emphasised that people have the freedom to decide the types of functionings that they value and not be forced to follow someone else's pre-decided criteria for what functionings are important for them [15, 16, 18].

Nussbaum [18] also, unlike Sen, believes that a "decent social minimum" can and must be ensured for all citizens, in terms of capabilities. A list of 10 categories was identified by Nussbaum, of those capabilities that are a must to accord every human being a life of dignity and freedom [18].

Nussbaum's List of Capabilities include [18]:

1. Life,
2. Bodily health,
3. Bodily integrity,
4. Senses, imagination and thought,
5. Emotions,
6. Practical reason,
7. Affiliation,
8. Other species,
9. Play and.
10. Control over one's environment – both political and material.

This list of 10 capabilities provides a tangible minimum set of capabilities that can form part of the core purpose of any UX design activity and hence a critical component of the citizen designer's toolkit.

## *3.3 How Does the Capability Approach Impact Design?*

Since technology in its various product/service avatars is closely connected with the enriching of human capabilities, this approach provides a construct to make the design process expand its objectives and outcome to positively impact human capabilities and agency and not just business outcomes. Nussbaum's list of central human capabilities that she felt must be protected by public policy could provide the seeds of what could become the core purpose that helps transform UX design and consequently citizen design. The list of capabilities would provide a starting point in the design process, to reflect on which of the 10 should form part of the outcome, that is, the product/service user experience being designed. On the other hand, building awareness of this list of capabilities and their role in user experience design from the user's perspective would be an important component of the game being conceptualised as part of the citizen designers' toolkit.

There is also an interesting connection between capability theory and participatory design because capability theorists view "agency" not just as the expansion of people's capabilities but also people's active participation in the process that determines that expansion. Just as in participatory design, the capability approach doesn't look at people as helpless and needing experts / outsiders to make decisions about their capability expansion. Hence, the process needs to be participatory and democratic so that the process itself becomes a way for people to exercise and strengthen their agency [31].

An example of how the capability approach can be applied to design has been described by some authors while exploring the design of "care robots" that are meant to be part of the ecosystem of caregiving for elderly. Coeckelbergh [33, 34], a philosopher of technology, wrote about evaluating products/technologies like the care robot by measuring their impact on their users' capabilities, using Nussbaum's capability list [18]. If this evaluation criterion was to be used, then these human capabilities need to be considered early in the design process. According to Coeckelbergh [34], the main advantage of the capabilities approach is the manner in which it can influence design to meet core human needs, as in the case of robot caregivers. While the reason to introduce robot caregivers is usually to do with cost savings that are obtained by automation, the capabilities approach could be used to ensure that the human element in the interaction between the robot caregiver and the person receiving care is also given utmost importance.

As Oosterlaken has discussed, design can indeed enhance capabilities of individuals thereby enabling them to realise their desired life goals. Design can also collectively change attitudes and help shift behaviour towards a more sustainable and inclusive world [35–38].

## 4 Making of Citizen Designers: The Citizen Designer's Toolkit

But how do we ensure that citizens have the required skills to be able to actively participate in the design process? This participation should be such that it provides citizens the opportunity to overcome the lack of confidence when interacting with designers. It is critical that citizen designers are able to not only critique concepts / ideas presented by designers but also present alternative ideas/concepts by themselves. Citizen designers should, eventually, be able to build their own capacity to act for themselves and their communities.

In order for citizens to actively participate in the design process, there have to be ways in which they can be made capable of understanding at least the basics of design. As Dong has pointed out there are a set of important capabilities that are needed to make citizen designers. He categorised these capabilities as information, knowledge, abstraction, evaluation, participation and authority [39].

Connecting participatory design and the capability approach may lead to ways in which citizens can become citizen designers and hence change the design process to create a shared means of expression and reflection between designers and citizen designers. This would perhaps make it possible to have a truly collaborative way for designers and citizen designers to codesign their lives and the world they both aspire to make better.

In order to shift the narrow and limited involvement of users in the design process as is prevalent today to a truly democratic and empowering co-creation exercise will need a thorough revamp of current processes and this revamp can happen only if citizens are aware of their right to participate in design of products/services that will impact their lives and are capable of confidently collaborating with designers and other stakeholders. This is the most important step needed to democratize design.

The concept of enabling all citizens (and not just designers) to have access to the language and process of design and provide them explicit ways to question and reconfigure current designed experiences led to conceptualizing a "game" that can be played at three different levels depending on what the objectives of the players are. A game is the first step in creating an entire toolkit for citizen designers. Using a toolkit such as this would also provide citizens with additional skills and may even lead some citizens to become professional UX designers, especially at a time when the field of UX Design is in its growth phase and demand for UX designers is rising every day.

## *4.1 Games Reviewed*

In the area of learning and skill development, serious games have become an important mechanism that is transforming learning. Serious games have as their main goal the imparting of education. However, being a game, the format enables the learning to happen in an engaging, fun way. Serious games can range from subjects like history, science, math to teaching soft skills like communication, empathy, etc. Younger learners in particular find serious games a motivating way to learn.

As Val Shute, PhD, and professor in the educational psychology and learning has noted, the popularity of games cutting across demographics and socioeconomic class is what makes it a very potent medium [40].

One of the earliest serious games, Microsoft's Flight Simulator, launched in 1982 still remains very popular. Today, there are hundreds of serious games but some that regularly make it to the top 10 list [41] include:

- Microsoft Flight Simulator.
- Tiltfactor Laboratory.
- A Force More Powerful.
- Darfur is Dying.
- PeaceMaker.
- World Without Oil.
- FoldIt.
- IBM City One.
- Superbetter.

Serious games also have social awareness as a goal along with educating learners about larger issues related to social impact, Katsaliaki and Mustafee have reviewed 35 serious games that deal with the topic of sustainable development. It is pertinent to note that of the 35 games on the theme of "SD" that they reviewed, there were 5 games where the player had to be play the role of a citizen and learn various ways of creating a more sustainable ecosystem [42]. Panzl, Classe, Araujo, and Vossen discuss serious game design as an approach to bring about change in citizens' behaviour towards and engagement with public service improvement. They propose the design of a sample serious game [43].

A review of serious games such as the following (see excerpts from reviews and write ups below) reinforced the value of using a game [44], to enable citizens to become aware of the impact of design in their lives and to also learn how to apply elements of design to bring about change in their own ecosystems.

- Fernando Quiroga, D. & Lucía Agudelo, M. "I got the power".

- *This is a game that helps to understand the way in which a social leader emerges within a community in Colombia. The idea is to allow anyone to interact with each proposed scenario and make decisions that help the community. The main motivation is to teach about the importance of social leaders in Colombia. Additionally, the game teaches about some of the risks that social leaders can suffer. And this is a subject that should not be taken lightly and more in these post-conflict moments in which Colombia lives* [44].
- González, F., Barrera, S., Albornoz, J. López, H. "La Guerra del Centavo".
- *The project "La Guerra del Centavo" seeks to create a game that puts the user in the place of a public service bus driver of the old transportation system in Bogotá. In this system, drivers receive money according to the number of passengers they collect, which causes problems for everyone involved and causes drivers to make decisions about how to drive in order to pick up a significant number of passengers while they meet the traffic rules. On occasion, service users blame drivers for system problems, but in reality they have limited options to work with* [44].
- A single Ramírez, S., Alejandro Silva Ariza, J., Zambrano Votto, C., Bautista Cabrera, D. "Una de Dos".

- *Player video game that simulates the decisions that a Colombian rural family must face in their daily lives. Every decision will affect the economic resources of your family and their well-being* [44].
- Eberhardt, R., Feeley, C., Leung, S., & Seidman, M. "Untitled Policy Game".
- *A game about the local decisions that can be made in Puerto Rico, in which players are tasked to make decisions that are best for the island but are ultimately restricted by the resources and agency that they have. Each player has a role that gives them access to two piles of resources: a teacher might have access to social capital and community cards, a housing justice advocate might access government and social capital, a zoning board officer might get government and economy. In the center are random policies that have requirements – if all of the requirements are met based on the cards played in a round, the policy is passed and whatever consequences that policy enacts change the state of the board* [44].
- "Fair Play", *enables the player to adopt the persona of an African American student and in that role playing mode experientially learn about biases people have and how they manifest* [44].

## 4.2 A Game for Citizen Designers – The Concept

The current concept is that of a puzzle with multiple solutions that contains all the elements/pieces of the framework. Based on some underlying rules, the "player" would need to use the "design pieces" of the framework to explore how different combinations would lead to different ways of reconfiguring their experience.

- Level 1 would be for those who would like to understand how design can lead to different configurations of everyday life, ranging from status quo to fundamental change. Playing the game would make players aware of the role of design, the alternative solutions that can be designed to create a more equitable and sustainable future and enhance personal capabilities.
- Level 2 would be for those who would like to also, in addition to what Level 1 offers, understand how business decisions impact design outcomes.
- Level 3 would be for players who have mastered levels 1 and 2 and hence have the capability to function as citizen designers and now would like to wear a designer's hat and understand the role a designer plays.

All three levels, when experienced would provide potential citizen designers with a holistic view of all the elements and stakeholders who impact the design ecosystem and hence enable a citizen designer to navigate through the various points of view and express his/her perspective when involved in participatory design processes.

**Level 1** The game, in the form of an online puzzle to be played on phones or computers would have "question cards" stating a design problem on each card, to be solved by the player and "design pieces" that the player could combine to get ideas/

concepts as possible solutions to the design problem that is stated on the selected "question card". Different combinations of the "design pieces" would generate different ideas/concepts for the same design problem.

The "design pieces" would be components from the "Universal Values" set, the "Human Needs" set and the "Capability" set mentioned earlier. The player would pick one of the readymade question cards and to generate suggested ideas/concepts he/she would be allowed to combine any number of "design pieces" from each of the three sets. The selected combination of "design pieces" would reveal a specific set of concepts/ideas (Fig. 5).

Example of a Question Card that the player could select – "What ideas can I generate for designing a cinematic entertainment product/service for myself and my friends?" On selecting the above card, the game would then present the various "design pieces" that belong to the three different sets and let the player decide on what "design pieces" she decides to combine. Let us imagine that the player decides to select the following (Fig. 6 and 7):

- Values – self-direction and self-transcendence.
- Needs – participation, leisure, identity and freedom.
- Capabilities – senses, imagination and thought, affiliation and play.

As soon as this combination is locked down, the game would provide possible ideas/concepts that meet this specific combination of values, needs and capabilities. For example:

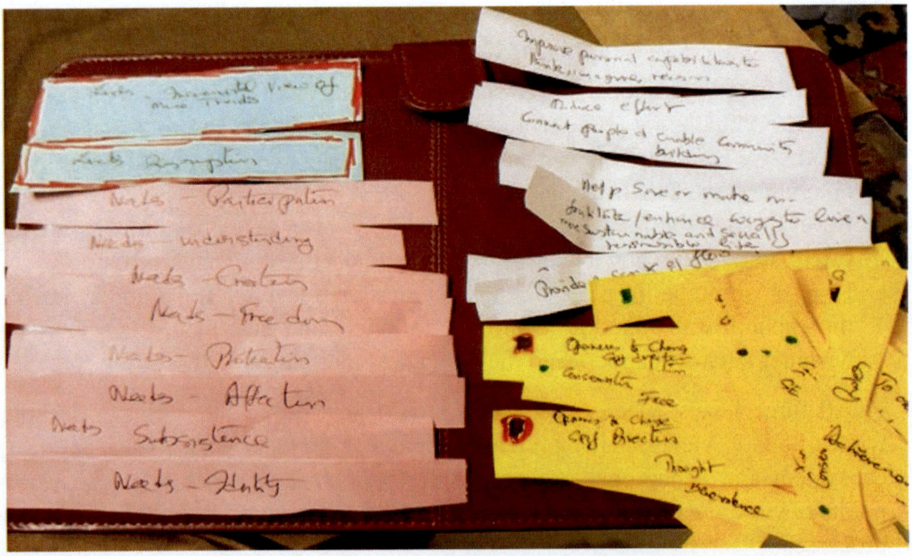

**Fig. 5** Prototyping the concept of the game with different coloured strips representing pieces from different sets (Values, Needs, Capabilities, etc)

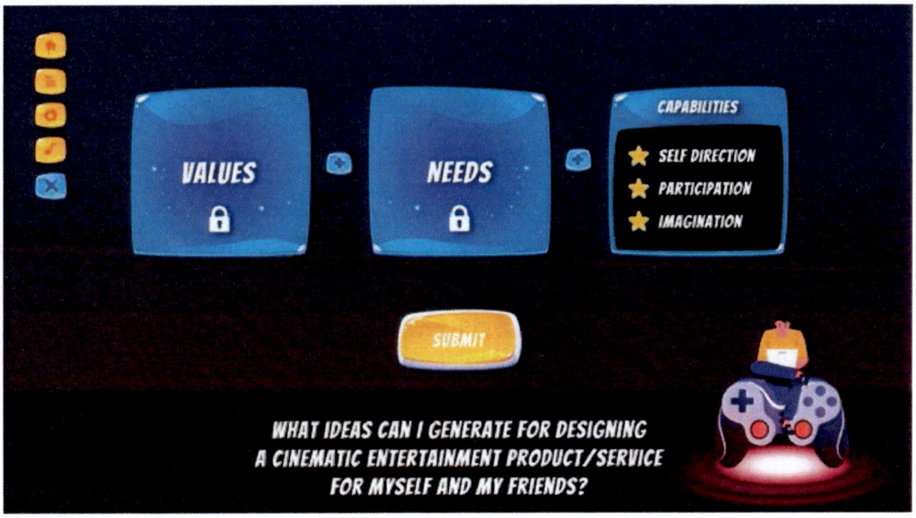


**Fig. 6** Prototyping the concept of the game with different coloured strips representing pieces from different sets (Values, Needs, Capabilities, etc)

**Fig. 7** Prototyping the concept of the game with different coloured strips representing pieces from different sets (Values, Needs, Capabilities, etc)

- A channel with interactive docudramas that can have different plot progression depending on the interaction by the viewer.
- A channel with crowd-sourced ideas for each show that is commissioned, at both the level of the overall storyline and the details of dialogue, locations, props, character development, music, etc.
- A massive multi-creator online futuristic 24/7 show.

On the other hand, if the player now combines the "design pieces" in a different manner, the resulting ideas would be different. Let us suppose that the player now selects the following combination to be the foundation of ideas:

- Values – stimulation and hedonism.
- Needs –affection, leisure and freedom.
- Capabilities – emotions and play.

With the above combination of the puzzle "pieces", a different set of ideas would be generated for the same question – "What ideas can I generate for designing a cinematic entertainment product/service for myself and my friends?"

Possible ideas generated by the game:

Netflix-like streaming service

- with similar shows (emotional and dramatic) but with a way to watch as a group even if everyone is not in the same place.
- with shows and films categorized according to viewers' moods,
- with shows and films organised according to private categories that cannot be seen by anyone else other than the viewer and public/ family categories that can be seen by others.

As one can see, one combination of elements could provide ideas that are very different from what is current and is experienced by the player today while another combination provides ideas that are mainstream. This experiential awareness that design solutions depend on what elements are considered as fundamental building blocks should lead citizens to awaken to the myriad possibilities that can be generated. Playing the game will also build the citizen designers' capability to identify and demand to know why a certain combination of elements have been privileged over other combinations in generating the design solutions.

In addition, the game would also ask the players for their own ideas and give players the options to generate their ideas and add to the idea pool that the game's algorithms draw upon. Ideas sourced from players could be rewarded/ recognised. The game would also provide deeper levels of exploration and learning for each of the 3 sets from which players can select "design pieces". Hence, players would be able to learn about the role of Universal Values, Human Needs and Capabilities, the origin of these concepts and their impact on design (and life). They would be shown successful designs that have taken into account "design pieces" from these 3 sets and hence can act as exemplars from around the world and demonstrate how specific combinations of design pieces can make the world a better place or not.

**Level 2** Players would be exposed to additional categories of "sets" such as business modelling and marketing from which they need to select "design pieces" in addition to the 3 sets they drew upon in Level 1. At this second level, citizens when playing this game would learn to use components such as conscious capitalism, Cleantech, Greentech, Carbon Disclosure Rating, Social Audit, Green Levy, Social Impact Bond, revenue planning, cost structures, USP, target market, competition, distribution channels, etc. and see how the ideas generated change (or not) based on the combinations the player selects. Hence at Level 2, the player would start to understand the business/financial ecosystem of decision-making that affects the design of products / services and their usage and experience.

**Level 3** Players who have been exposed to what constitutes design and what can be expected from design can now change places from being a citizen designer to get an idea of what a professional designer needs to deal with. Hence, at this advanced level, those citizen designers who are considering a professional level of involvement in the design process as a designer would benefit from role playing as a designer in the fictional design world created by the game.

## 5 Conclusion

A game such as this would help raise awareness of all citizens about how "designed" their lives actually are and increase their capability to "demand" that design enables desired reconfigurations to happen. Perhaps even work on the reconfigurations by themselves instead of depending on designers. Playing the game would also enable building citizens' capability to discern the impact of their decisions (as they select different combinations of values, needs and capabilities) on the resulting concepts that emerge and how those concepts help the world become a more sustainable and inclusive planet.

Playing this game would also prepare citizens to contribute meaningfully to participatory design processes and thereby help move citizen participation in design and democratisation of design to become reality. Design that has involvement of citizen designers has the possibility, as Oosterlaken said, "to create products that improve people's real opportunities or capabilities" [36].

## References

1. Interaction Design Foundation, What is user experience design, https://www.interaction-design.org/literature/topics/ux-design, Last Accessed 31 May 2020
2. L. Irani, J. Vertesi, P. Dourish, K. Philip, R. Grinter, *Postcolonial Computing: A Lens on Design and Development, CHI 2010* (ACM, 2010)
3. S. Bardzell, *Feminist HCI: Taking Stock and Outlining an Agenda for Design, CHI 2010* (ACM, 2010)
4. T. Irwin, Transition Design: A Proposal for a New Area of Design Practice, Study, and Research, Design and Culture, (2015)
5. J. Chan, Design ethics: Reflecting on the ethical dimensions of technology, sustainability, and responsibility in the Anthropocene. Design Stud. **54** (2018)
6. S. Constanza-Chock, *Design Justice, Community-Led Practices to Build the Worlds We Need* (MIT Press, 2020)
7. TUDelft, Sustainable design engineering, https://www.tudelft.nl/en/ide/about-ide/departments/sustainable-design-engineering, Last Accessed 28 May 2020

8. I. Robeyns, *The Capability Approach* (The Stanford Encyclopedia of Philosophy, 2011)
9. Amoli, The ethical boundaries of persuasive design, https://uxdesign.cc/the-ethical-boundaries-of-persuasive-design-c0b040906386, Last Accessed 31 May 2020
10. G. Agamben, What is the contemporary? in *What Is an Apparatus and Other Essays*, (Stanford University Press, 2009)
11. K. Bulkley, The rise of citizen journalism, https://www.theguardian.com/media/2012/jun/11/rise-of-citizen-journalism, Last Accessed 10 Dec 2019
12. G. Fox, The rise of conscious capitalism, https://www.forbes.com/sites/gretchen-fox/2019/03/26/the-rise-of-conscious-capitalism/#7c3280f3139d, Last Accessed 27 May 2020
13. F. Callejas-Albiñana, I. Martínez-Rodríguez, A. Callejas-Albiñana, I. de Vidales-Carrasco, Assessing the growth of ethical banking: Some evidence from Spanish customers. Front Psychol **8** (2017)
14. S.H. Schwartz, W. Bilsky, Toward a universal psychological structure of human values. J. Pers. Soc. Psychol. (1987)
15. S.H. Schwartz, An Overview of the Schwartz Theory of Basic Values, Online Readings in Psychology and Culture, (2012)
16. M. Max-Neef, *Human Scale Development, Conception, Application and Further Reflections* (Apex Press, New York/London, 1991)
17. S. McLeod, Maslow's hierarchy of needs, https://www.simplypsychology.org/maslow.html, Last Accessed 1 June 2020
18. M. Nussbaum, *Women and Human Development: The Capabilities Approach* (Cambridge University Press, 2001)
19. J. Rheinfrank, The Philosophy of (User) Experience, Presentation at CHI2002/AIGA Experience Design Forum, (2002)
20. S. Bødker, M. Kyng, Participatory design that matters – facing the big issues. ACM Trans. Comput.-Hum. Interact. **25**(1), 1–30 (2018)
21. L. Bannon, J. Bardzell, S. Bødker, Introduction: Reimagining participatory design—Emerging voices. ACM Trans. Comput.-Hum. Interact. **25**(1), Article 1 (2018)
22. R. Howard, Utopia – Where workers craft new technology. Technol. Rev. **88**(3) Massachusetts Institute of Technology (1985)
23. S. Bodker, P. Ehn, D. Sjogren, Y. Sundblad, Co-operative design - Perspectives on 20 years with 'The Scandinavian IT Design Model', Proceedings NordiCHI, (2000)
24. T. Bratteteig, J. Gregory, Understanding design, paper for IRIS 24, the Information Systems Research Seminar in Scandinavia, Ulvik, (2001)
25. M. Elden, R. Chisholm, Emerging varieties of action research: Introduction to the special issue. Hum. Relat. (1993)
26. J. Greenwood, M. Levin, *Introduction to Action Research: Social Research for Social Change* (Sage, Thousand Oaks, CA/London, 1998)
27. W. F. Whyte (ed.), *Participatory Action Research* (Sage, Newbury Park, 1991)
28. L. Suchman, Strengthening our collective resources: a comment on Morten Kyng's 'A Contextual Approach to the Design of Computer Artifacts'. Scandinavian J. Info. Syst. (1998)
29. J. Alexander, *Capabilities and Social Justice: The Political Philosophy of Amartya Sen and Martha Nussbaum* (Ashgate Publishing, 2008)
30. A. Sen, *Development as Freedom* (Oxford University Press, 1999)
31. D. Crocker, *Functioning and Capability: The Foundations of Sen's and Nussbaum's Development Ethic* (Sage Journals, 1992)
32. S. Alkire, *Measuring the Freedom Aspects of Capabilities, Global Equity Initiative* (Harvard University, Cambridge, MA, 2005)
33. M. Coeckelbergh, *How I Learned to Love the Robot: Capabilities, Information Technologies, and Elderly Care Article* (University of Vienna, 2012)
34. M. Coeckelbergh, Personal robots, appearance, and human good: A methodological reflection on roboethics. Int. J. Soc. Rob. (2009)

35. J. Borenstein, Y. Pearson, *Robot Caregivers: Harbingers of Expanded Freedom for All?* (Springer, 2010)
36. I. Oosterlaken, Design for Development: A Capability Approach, Design Issues, (2009)
37. I. Oosterlaken, The capability approach and innovation/technology/design, Maitreyee, No. 20, (2012)
38. I. Oosterlaken, Taking a capability approach to technology and its design – A philosophical exploration, Doctoral Dissertation, (2013)
39. A. Dong, C. Nichols, Re-conceptualizing design through the capability approach, (2012)
40. S. Pappas, Games with impact: Psychology-savvy video games address health, education and social-emotional skills, https://www.apa.org/monitor/2019/09/games-impact, Last Accessed 29 May 2020
41. Growth Engineering, 10 serious games that changed the world, https://www.growthengineering.co.uk/10-serious-games-that-changed-the-world/, Last Accessed 25 May 2020
42. K. Katsaliaki, N. Mustafee, A Survey of Serious Games on Sustainable Development, Proceedings of the 2012 winter Simulation Conference, (2012)
43. N. Pflanzl, T. Classe, R. Araujo, G. Vossen, Designing Serious Games for Citizen Engagement in Public Service Processes, International Workshop Series on Business Process Management and Social Welfare, (2016)
44. M. Flanagan, M. Jakobsson, *Games and Colonialism*, To be published by MIT Press

# Contextual Inquiry Related to Cab Booking: Using Uber and Ola Cabs in India

**Akash Rode and Ganesh Bhutkar**

**Abstract** Ola Cabs and Uber are leading Cab Booking Apps in India that offer services like cab booking, ridesharing, and food delivery. To cater to the increasing demand from customers and to expand their horizons, these tech giants are involved in constant competition with each other to gain new users and retain existing ones. This research paper is a result of an extensive study of the notion of cab booking in India. The objective is to understand the nature and the perception of Indian users who avail these cab services via mobile apps, especially users of Uber and Ola Cabs. In doing so, the motive is to empathize with Indian cab users, both drivers and travelers, by understanding their problems and frustrations. This is achieved by using the ethnographic approach of Contextual Inquiry and work models are developed based on the insights found therein.

**Keywords** Contextual inquiry · Work models · Ola Cabs · Uber · Cab booking · Traveler · Mobile apps · India

## 1 Introduction

A cab (also known as taxi) is a vehicle for hire with a driver, which may be used by a single passenger or a group of passengers. The vehicle hired as a cab can be a public hire, a private hire, or local buses. For the scope of this paper, the term "Traveler" is been used to identify passengers or users commuting via cab.

A. Rode (✉) · G. Bhutkar
Centre of Excellence in Human-Computer Interaction, Vishwakarma Institute of Technology (VIT), Pune, India
e-mail: ganesh.bhutkar@vit.edu



E. Eilu et al. (eds.), *Digital Literacy and Socio-Cultural Acceptance of ICT in Developing Countries*, https://doi.org/10.1007/978-3-030-61089-0_3

## *1.1 History of Cab*

The first-ever documented form of cab service, a public horse carriage coach service for hire started in London in 1605 [1]. After 30 years, in 1635, the British Parliament passed the Hackney Carriage Act to legalize horse-drawn carriages for hire by local innkeepers to merchants and visitors. Further in 1834, the hansom cab was designed and patented by an architect which was two-wheeled, fast and light enough to be pulled by a single horse making the journey cheaper. By the nineteenth century, the modern taximeter was invented and perfected by a trio of German inventors and used in the world's first gasoline-powered taximeter cab. Thereafter, cabs proliferated around the world in the early twentieth century [2].

## *1.2 Cabs in India*

India is the second most populated country in the world residing people from diverse cultural backgrounds. India has more than 2000 ethnic groups with major religions including Hindus, Muslims, Christians, Sikhs, Buddhists, Jains, and others [3].

For fellow Indians, the word "taxi" is reminiscent of the classy and iconic Premier *Padmini* cabs. These cabs were colored in dual shades of black and yellow and were manufactured by Premier Automobiles Limited under a license from Fiat [4]. Those were the real glory days for cab drivers. Travelers faced problems due to factors like monopolistic attitude, poor services, refusal to provide services, and high rates. For the first time, in the early 2000s, well-organized cab services with high traveler satisfaction and cheaper rates came into existence in the form of Radio Cabs. Initially, cab operations were started on small scales by companies like Mega Cabs and Fast Track Taxi. By 2006, other companies joined the bunch. These were Meru Cabs, Easy Cabs, and Savaari. Each of these companies operated in different ways. Some companies made the driver place a deposit with the company. The company would have ownership of the car. Drivers would be trained in soft skills, and in return, the company would charge a fixed monthly amount from them. In such cases, the driver had to bear the fuel cost and the company would bear car maintenance charges. Other companies like Easy Cabs let the driver own the car after 5 years of nonstop service with the company [5].

## *1.3 Rise of Ola Cabs and Uber in India*

A historic shift happened in the Indian cab business when initially Ola Cabs entered the Indian market with its fleet aggregation model followed by Uber. Ola Cabs is a company developed by ANI Technologies Pvt. Ltd. based in Bengaluru, India.

While Uber is an American multinational company based in San Francisco. Ola Cabs was founded in 2010, while Uber, founded in 2009, entered the Indian cab business 3 years after Ola Cabs. As of FY 2018, Ola Cabs generated a revenue equivalent to US $310 million, while Uber generated about US $11.27 billion (as of 2018) [6]. In January 2018, Ola entered the overseas market through Australia followed by New Zealand and also began operations of auto-rickshaws in the UK in March 2019. Uber is a giant with operations in over 785 metropolitan areas worldwide. As of 2019, Uber has 110 million users worldwide while Ola has over 150 million users [7].

Ola Cabs and Uber both categorize cars based on their power and capacity. This helps the user to efficiently decide a car-type for their travel requirements. While Uber in the United States of America may have some interesting car-types like UberBLACK (Black Luxury Car), UberWAV (Wheelchair Accessible Vehicle), and UberJUMP (Electric Bike), Uber in India has common car-types categorized as auto-rickshaws, micros, hatchbacks, sedans, and SUVs [8, 9]. In January 2019, Uber pilot launched Uber Boat in the city of Mumbai which includes a ride along the city's popular coastal routes using two variants of speed boats, a 6–8-seater and other 10+ seater [10]. Similar car-type categorizations exist for Ola Cabs vehicles as well.

### *1.4 Women's Safety*

As Indians began to shift from using traditional cab services in the early 2000s to today's Ola-Uber services, several personal and cultural issues came into play, which directly affected their perception of using cab services in their daily lives. The most crucial issue was women's safety in traveling by cabs.

According to one of the articles referred [11], an Uber driver in Delhi molested a woman returning from work at night and threatened to kill her if she reported the incident to the police. In August of 2016, a 12-year-old girl was allegedly gang-raped and strangled by two men in an Ola Cab in Kolkata city. The article stated that both Ola and Uber repeatedly condemn such crimes and help the police find culprits. Even so, nothing seems to work as the crimes recur with increasing frequency. One of the police officers commented saying that the route of the vehicle was changed mid-way during the ride and he was surprised that the company could not detect this abnormal behavior. He suggested that these companies should do random checks on their drivers about following the right route notice any uncertainty in their travel behavior. This seems to be an issue that both Ola and Uber are determined in solving, but haven't yet resolved it completely. One article [12] also talks about incidents where women reported such nasty crimes to Ola-Uber helplines, but such events continue to repeat with ever-increasing frequency.

### 1.5 Exploring New Horizons – Uber Eats and Food Panda

Soon after launching their cab services, Ola Cabs and Uber both launched their food delivery franchises. Ola Cabs acquired Food Panda in December 2017, while Uber founded UberEats, an American online food ordering and delivery platform in 2014.

Ola seems to fail at food ordering business. Initially, it had launched Ola Café in 2015, but failed and was shut down a year later. Food Panda was Ola's second attempt in the food delivery business. Since, Indian food delivery market already has giants like Swiggy, Zomato, and UberEats, Food Panda struggled to survive and started scaling down its marketplace. After laying off several employees, Food Panda now focuses on running the cloud kitchen brands instead of food delivery business. UberEats on the other hand ranks second with 4.2 stars on the Google Play Store after Zomato. It has good brand recognition as compared to Food Panda. UberEats being a newcomer in the Indian market, it offers good discounts for consumers and low commissions for merchants in an attempt to break into the market. Even so, it seems that, with Swiggy and Zomato dominating the Indian food delivery market, it would be tough for these two newcomers to create their own place in the market [13, 14].

## 2 Related Work

Initially, several research papers related to customer perception and satisfaction levels of Ola Cabs and Uber were studied. Several blog articles were also read which talked about issues that individuals using Ola and Uber face on a daily basis.

One research paper [15] talks about consumer perception of Ola and Uber cab services in the city of Mumbai. This study was focused on identifying differences in between Ola and Uber customers where data was collected via a structured questionnaire. This paper presents an interesting approach to consumer perception using specifically designed hypotheses. This study identified the difference in between services of Ola and Uber in India with an intention to help the cab service industry design their marketing strategies and customer relationship plans. The study concluded that the female working population in Mumbai preferred Uber over Ola. The conclusion helped in developing a perspective to design questionnaire for women. This further helped in deriving relevant insights related to the needs of women travelers which was crucial to this study.

Another study [16] dives into the details of understanding the dynamics of the cab market in Pune city of India, where it focuses on studying various technical factors like pricing, market share, revenue models, and app convenience. It collects data using a structured interview questionnaire. Interview results convey valuable user feedbacks on parameters such as waiting time, satisfaction levels, safety levels, and payment options. This paper represents data derived in visual format by ample use of graphs and charts. This makes it easy for the nontechnical reader to derive insights and make sense of the information. It also represents the complaints and suggestions for Ola and Uber in crisp and clear distinguishing tables. It concludes that Ola has more market share as compared to Uber in Pune city of India. In addition, it says that consumers are generally satisfied with these apps and 68% respondents did not have any complaints whatsoever. Also, it adds the fact that maximum respondents prefer safety to be the most important factor while choosing a cab service. Interestingly, only 23% respondent "strongly agreed" that Ola and Uber are safe and secure for women. The conclusion derived by this research study helped understand various factors that an Indian traveler looks for in a cab service. Several suggestions provided by travelers as a part of this study helped in understanding certain behavior traits of Indian cab travelers.

One interesting research paper [17] studies how cab aggregators such as Ola and Uber have impacted society through various methods like offering coupons, quality service, mobile applications, 24×7 user support, and much more. This paper inculcates within it a "suggestions" section where it simply enlists actionable suggestions retrieved from the questionnaire. These suggestions which are titled as "*To Aggregators*," "*To Cab Owners*," and "*To Drivers*" make it easy for the reader to understand insights derived from the research work. Commendable work has been done in identifying specific benefits enjoyed by online cab booking apps and features expected in near future. Some important features mentioned that are expected in future are "CCTV to monitor the driver" and "female drivers for female customers" Additionally, this study derives that half of the travelers would prefer cabs over public transport in the city of Coimbatore in India. Even so, it points out the fact that cab industry revenues are gradually declining in the Indian society due to increasing expectations of the user with respect to add-on benefits. This paper helped highlight certain inherent expectations of drivers and travelers before booking a cab.

Another study titled OLA vs. Uber: The Battle of Dominance [18] was referred, which tries to understand the dynamics of the Indian cab market by studying factors like pricing, market share, and revenue models. This study puts forth interesting insights by enlisting a SWOT (Strengths, Weaknesses, Opportunities, Threats) Analysis of both apps. Thereafter, several research papers [19–21] related to Ola and Uber services were referred. In addition, several research papers [22–25] related to Traveler Classifications and Analysis were studied and analyzed. Also, research papers related to Contextual Inquiry were referred [26–28]. All of the above papers studied provide valuable observations related to Indian travelers and drivers. Several research papers [22–25] and online blogs [29–31] were referred to understand traveler categorizations.

## 3 Research Methodology

The methodology of research work for studying Indian Ola-Uber travelers was done in two major steps; first of which includes conducting a Contextual Inquiry of travelers and, second, developing work models that effectively convey insights related to their behavior and underlying issues.

### 3.1 Contextual Inquiry

Contextual Inquiry is a research technique that emphasizes studying users in their natural environment to understand key insights that may help identify core requirements. Insights thus derived can also put light on certain usability problems that may exist in the environment.

Fieldwork is essential for observing user interaction and understanding user problems. The objective of this research paper was to gather qualitative data. Since several other studies have already derived quantitative statistics. So, instead of deploying a questionnaire, random users (travelers) were asked for their experience of commuting through Ola and Uber cab services. These travelers represented in Table 1 below comprised of 25 users who consisted of close relatives, friends, work colleagues, and strangers. Out of these 25 users, 11 were female and the remaining 14 were male. These users were selected by the method of convenient sampling [26, 27].

Most insights were derived from interaction with drivers. These included 14 car and auto-rickshaw drivers. Three drivers were intentionally included who weren't a part of the Ola-Uber family. The reason for this was to understand their views and opinions related to Ola and Uber as companies, as these companies are slowly invading the market which initially belonged solely to them. For each of them, a rapport was built with initially while traveling by their cab. In such an environment, they were asked about their experiences and frustrations related to these apps. The insights derived were noted as the conversation progressed.

### 3.2 Development of Work Models

Work Models sort information in different manners producing flow diagrams that help build a truly holistic view of the system. They explore various types of users, their roles, major and minor cultural influences, and primary/secondary actions.

**Flow Model** Flow Model focuses on identifying the roles and responsibilities of different types of users. This is useful in determining workflow hierarchies and understanding interaction links between these roles. These links can be of direct or indirect nature.

**Table 1** Traveler demographics in user study

| Age range (in years) | Number of female travelers (11) | Number of male travelers (14) |
|---|---|---|
| 18–25 | 5 | 4 |
| 26–35 | 2 | 3 |
| 36–45 | 3 | 4 |
| 46–55 | 1 | 1 |
| 56–65 | – | 2 |

**Cultural Model** A person's beliefs, values, and surroundings play a major role in shaping how they perceive the world around them. This further determines their actions. Cultural Model is crucial when it comes to understanding the user because it helps to understand not only their beliefs and values but also their frustrations and pressures. Gaining this knowledge about the user is important because it helps us empathize with the user, and only then, we can design effective experiences for them.

**Sequence Model** The Sequence Model represents tasks and related action sequences involved in task completion. This task completion is from the point of view of an Indian user. It focuses on identifying the intent of the task and, in the process, tries to uncover any problems that may be faced by the user while executing the specific action sequence.

**Sensory Model** The Sensory Model includes a listing of all materialistic and non-materialistic aspects that contribute to the experience of using the system.

**Artifact Model** Artifact Model describes the use of artifacts that can potentially make a difference towards effective working of the system and its user experience.

**Physical Model** Physical Model involves analysis of workplace dynamics by considering materialistic elements like buttons, tablets, and printers.

Out of these, four work models, namely flow model, cultural model, sequence model, and sensory model, have been developed to understand Indian Ola and Uber cab travelers.

## 4 Work Models

Below are the four work models that have been developed to understand the Indian Ola and Uber cab travelers and their usability issues.

### *4.1 Flow Model*

When thinking of users of Ola and Uber cab services, we have defined them as "travelers," as they use cabs as a means of traveling from one location to another. Based on the task they are trying to accomplish, Ola-Uber cab travelers are classified into nine distinct categories. They are as follows:

1. **The Professional** – The Professional travelers choose to use cab services strictly for business or work purposes. Such travelers generally have their e-wallet configured with Ola/Uber mobile apps. This helps them effortlessly claim their travel expenses from their respective companies.
2. **The Frugal** – The Frugal travelers are extra-careful while spending money on cab services. A major chunk of middle-class Indian users falls into this category.

It was observed that such travelers compare prices on both apps (Ola and Uber) before they book a ride. These travelers are also extra-vigilant about offers and discount codes that are displayed to user via Push Notifications. With a motive to save money, these users make sure that they grab such offers. They frequently travel via carpool services such as Uber Pool and Ola Share.

3. **The Socialite** – The Socialite travelers travel solely for recreational or entertainment purposes. They may be going for an outing or a movie with their friends. These people generally travel in groups.
4. **The Shopper** – The Shoppers use cab services only when they go for a day-long shopping. When asked the reason for the same, they say, after a long day of shopping they find driving in traffic tiring, hence they prefer to travel first class by cab services.
5. **The Confidential** – The Confidential travelers travel only to achieve a specific pre-decided personal chore. They may have to accomplish a single task or a list of tasks before they head back home.
6. **The Newcomer** – The Newcomers are visiting a certain place for the first time. Since they may not be well-versed with the local language or may not know anyone local, they prefer searching for their desired destination online and hiring a cab to get there.
7. **The Unfortunate** – The Unfortunate are individuals who face a sudden breakdown of their primary vehicle in heavy traffic and are hence forced to book a cab. They may also have an old bike or car which faces frequent breakdowns when commuting. At such times, they acquire the assistance of a cab.
8. **The Challenged** – The Challenged are individuals who face a certain medical emergency with their family members or close friends. In such a scenario, they are forced to book a cab to get to a nearby emergency room.
9. **The Transitionary** – The transitionary are travelers who use cabs as a means to switch from other modes of travel. For instance, someone might hire a cab to travel from the train station to the airport or walk to the auto-rickshaw stand after getting off a cab.

In Fig. 1, five most frequently observed user types out of nine namely – the professional, the frugal, the socialite, the shopper and the unfortunate are shown enclosed in ellipses. All possible activities associated with user–cab interaction are represented using rectangles. If the user is involved in the activity, a link has been created between that activity and the user. Each user type has been assigned a color. Every activity has certain subtasks that are written inside the rectangle. Adjacent to the subtasks are the color-coded rectangles that represent user association with that subtask. For instance, to book a cab, *the professional* user type would not consider searching for offers and discounts in the app as the travel cost is taken care of by her/his company.

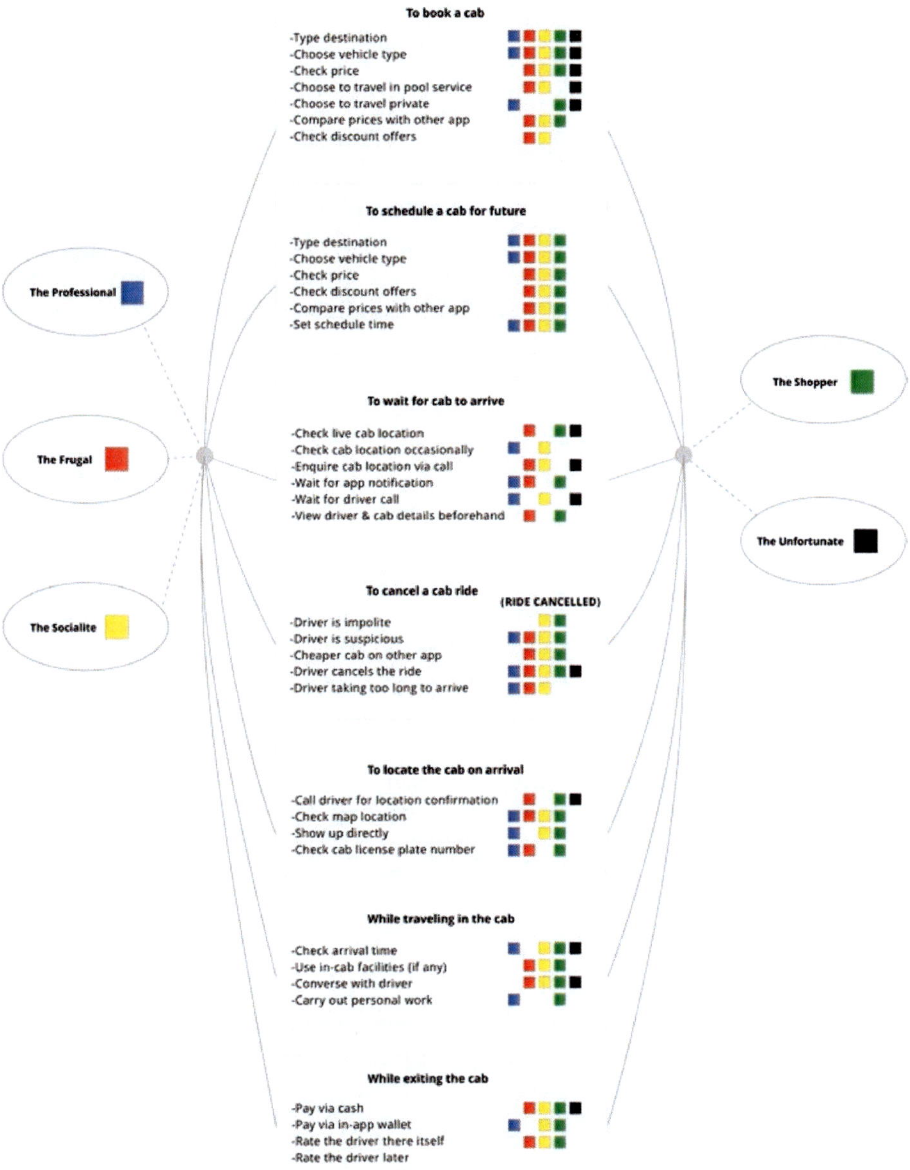


**Fig. 1** Flow Model for Indian Ola and Uber cab users

## 4.2 *Cultural Model*

When it comes to Indian users, defining a cultural model was a must because it put light on several factors that come into play when using a cab service like Ola or Uber. The scope of culture in this paper is restricted to the traditional culture of India and the daily lives of Indian citizens. These individuals are several types of

Ola-Uber users that have been previously categorized in the Flow model. The findings of the ethnographic study are reflected and presented as follows:

### 4.2.1 Women's Safety

This is by far the most serious issue that Indian women face today. There have been numerous cases of sexual assault on women by cab drivers. These issues range from verbal abusing to molestation, to abduction, to gang rape. To address this issue, several Ola and Uber cabs have easily accessible Panic/Emergency buttons fitted near the doors of the car.

### 4.2.2 The Confirmation Call Syndrome

A certain kind of behavior that has been observed frequently among both drivers and travelers is the habit of confirming a ride by calling the other person. Even after booking a ride, traveler calls the driver for confirmation and asks about the driver's location and Expected Time of Arrival (ETA). Often times, if the traveler doesn't call the driver, the driver takes the initiative of calling the traveler to "*seal the deal*" and enquire about landmarks near the pickup location. It seems that Ola and Uber have taken note of this cultural trait and have inculcated the feature of free internet calls in their apps. In addition to calling, they have also embedded an in-app chat feature where traveler and driver can text each other.

### 4.2.3 The Indian Cab Driver – Traits and Frustrations

There are certain behavioral traits of Indian drivers that have been observed by several customers which they find undesirable. In addition, while interviewing auto-rickshaw drivers, certain frustrations that these drivers face were also revealed.

1. Drivers Refuse Long-Distance Rides

If you ask Indian travelers about this behavior, they would agree with the fact that local auto-rickshaw drivers deny certain rides just because they are long distance. They worry that they won't get a customer on a ride back from the drop location. The same behavior has been observed with Ola and Uber cab drivers. Customers complain that several times drivers cancel the ride if the drop location is far from the driver's zone of operation and convenience. Many times, drivers request the passenger to use the same cab while coming back saying that they would wait for them till they finish off their chore. Drivers may or may not charge external additional charges for waiting.

2. Drivers Start a Trip Without the Traveler

Initially, as customers complain, there have been several cases where the driver started the trip without the traveler. There was no way of knowing this until Ola introduced the OTP verification in their system. Uber also claims that their algorithm checks for the location of the driver and the traveler before starting the ride.

3. Drivers Confirm the Next Ride in Advance

It has been experienced by several travelers that drivers accept the request for a new ride while they are currently dropping off another traveler. They convince the new traveler by saying that they have a current drop in the same location and will reach their pick-up location in less than 10–15 min. This behavior is sure to fetch the driver more rides which helps them reach their daily incentive quota but, travelers despise such behavior. They may choose to cancel the ride if the driver takes too long to reach for the pickup.

4. "Can You Please Give Me a 5-Star Rating?"

Driver ratings are extremely valuable to drivers as compared to customer ratings for customers. It was observed on several occasions that at the end of a ride, drivers request the customer upfront to give them a 5-star rating. This behavior is rarely observed amongst drivers outside of India.

5. Indian Drivers Love Chatting

No one can disagree with the fact that driving all day long can get monotonous and boring. Indian drivers love to strike up an interesting conversation, whatever may be the topic. Travelers might sometimes find this behavior irritating. It has also been observed that drivers give a better rating to customers who strike up a conversation with them as compared to those who do not. The vice versa is the case for travelers, which means that a traveler will rate a driver better if he does not chat monotonously. It generally depends on several factors related to individual personalities of travelers and drivers.

6. Is Ola Better or Uber? Who Cares?

Although these tech giants (Ola and Uber) might be involved in tough competition to gain more users and profits, drivers don't seem to care about it. A majority of drivers (and travelers) use both apps simultaneously. Such drivers have two smartphones with them, each running either of the apps. Whichever service gives them a

**Fig. 2** An auto-rickshaw driver using two smartphones fixed the handle

quicker ride, they choose to go with it. Figure 2 shows an auto-rickshaw driver using two smartphones to run both apps at the same time to get quicker customers.

7. Do We Call It the Extra Change or a Tip?

While interviewing travelers, most agreed to the fact that, after payment, Indian cab drivers are reluctant to return the change. They respectfully ask the traveler if they can keep the change. Such behavior is also observed among general drivers in India (drivers not specific to Uber and Ola) On the contrary, it was observed that international cab drivers return the exact change every time. But, international culture also follows the concept of the tip where it is up to the user to tip the driver according to their convenience.

8. Delayed Payments Affect Drivers

In an interview with an auto-rickshaw driver, an insight regarding the transfer of money was revealed. According to the driver, revenues earned by Ola Money (which is Ola Cabs e-wallet) rides get transferred the next day itself. While revenues earned by Uber e-wallet take around a week to get transferred to the driver's account. The driver confessed saying that such late payments make it tough to manage home expenses. At such times, he prefers switching his online service and starts opening his service to local passengers.

9. Night Charges

Drivers complained that they do not get paid extra for night trips. Even though these ride-booking apps charge extra for night trips, drivers do not receive their additional share for such trips. An auto-rickshaw driver commented saying "Night times are risky for us also. We should be paid the incentives."

10. Local Auto-Rickshaw People Despise Tech-Giants Like Ola and Uber

This insight was revealed by booking an auto-rickshaw directly on road. Observing no smartphones on the dashboard, a conversation about Ola and Uber was initiated. The driver responded saying "Due to these tech companies invading local markets, we local *rickshaw-wallas* have lost our business. Even though these apps are convenient for the user, they have affected us local people whose livelihood depends on driving *rickshaws*. So, we do not use these apps."

11. No Additional Charges for Extra Distance Traveled for Pick-Up

Drivers complain that often the actual location of the pickup is a bit far from the location pinned on the map. This is mostly due to the complexity of streets in India. In such cases, drivers have to travel a little farther to pick-up the customer. These individually insignificant extra distances traveled when added up together lead to an increase in petrol expenses for the driver.

### 4.2.4 The Indian Traveler

The Indian Ola-Uber app user is unique because it consists of people from diverse religions and different mindsets. Indian people have found an interesting way to voice their issues to the public. They do this through the medium of stand-up comedy. These videos reach out to millions through social media and are thus, sure to reach the intended people.

1. Hatred Towards Certain Specific Cars

Indian travelers despise certain cars and hate it when they show up for their Ola or Uber ride. The main reason for this is the bad comfort level of the car. Users complain saying that if they are traveling by car, they expect it to provide comfort. Generally, cab business owners purchase old cars at cheap rates and hire drivers to earn better profits. This tendency ruins the cab experience for the end-user.

2. Ola Share and Uber Pool

Indian users find services like Ola Share and Uber Pool cheap, which makes them opt for it. At the same time, they complain about the excessive time it takes to reach their destination. They say that this mainly occurs due to impromptu out-of-the-way pickups and drops as a result of poor algorithm.

3. Using Ola and Uber Parallelly

All users had both mobile apps (Ola Cabs and Uber) installed on their smartphones. Most users confirm the fact that they check for cab rates on both apps before confirming a ride. Even so, there exist a few cases where users confirm the ride on both Ola and Uber. Whichever ride has the shortest time, they opt for it and cancel the other ride. Figure 3 shows a user who has booked a ride parallelly on both Uber and Ola apps.

4. In-Car Music Matters

Travelers claim that the kind of music that is being played in the cab affects the experience of the journey. If the driver is playing cheap songs in regional languages, there are high chances of him getting a poor rating.

5. Drivers with Bad Habits

Several users conveyed a key issue that ruins the cab experience for them which is bad habits of the driver like chewing tobacco, smoking, or drinking. A user comments having commuted through numerous cabs which are full of tobacco smell. In such a scenario, she had no choice but to tell the driver to open all the windows. "Such drivers definitely end up with a 1-star rating," she says.

6. Drivers Asking for Directions

Most of the drivers are not well educated or trained, which makes it hard for them to understand English pronunciations by the Global Positioning System (GPS) voice assistant. As a result, they may miss a turn and get lost. In such cases, they have no option but to ask local people for directions. This adds up to the frustration of the traveler who may have to reach her/his destination early and cannot afford to lose extra time in transit.

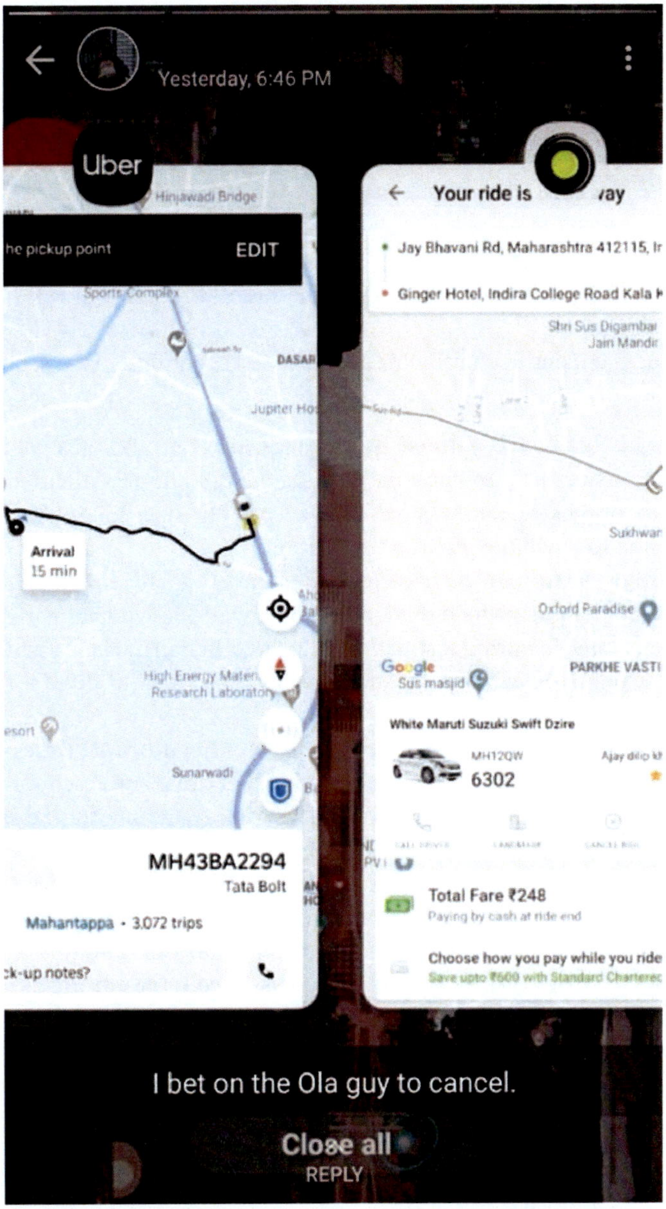


**Fig. 3** WhatsApp status of a Pune user with cabs booked simultaneously on both apps – An insight into how Indian users book cabs

7. Retrieving Lost Belongings

If a traveler is in hurry, it may happen that she/he forgets belongings or important documents in the cab. Travelers complain that it becomes hard to get back the lost belongings as the driver may already be on his next ride. They also suggest that there should be a helpline number through which they can convey necessary details to proper authorities instead of depending solely on the driver.

### *4.3 Sequence Model*

The most crucial and frequently performed task is booking a cab. Actions sequence diagram has been created in Fig. 4 for the tasks. Problems at relevant steps have been identified.

Let's look at identified problems at every step according to the sequence diagram in Fig. 4:

1. **Incorrect location is pinned by the traveler** – It was observed that, when a traveler was in hurry to book the cab, she/he did not pay attention to the pinned location for pick-up and was pinned inaccurately. As a result, the driver required more time to reach the pick-up destination.
2. **Exact drop loction of the traveler is inaccurate** – The drop location is searched using Google Maps which recognizes most well-known landmarks. But, on several occasions, the drop destination is located in narrow lanes which are hard to pinpoint on Google Maps. In such a case, driver has to drive a few additional meters to drop the traveler.
3. **Undesirable increase in ride fare after checking multiple times** – Most Indian travelers compare ride prices on Ola and Uber apps before confirming the ride. They complain about increase in ride fare after searching for the same ride multiple times.
4. **Cabs are not available** – Due to high demand in popular areas or low cab frequency in less popular areas, a traveller may not find cab. In such a case, she/he may choose to opt for local cab services like auto-rickshaws.
5. **Traveler having to walk a significant distance to reach the cab due to inaccurately pinned location** – Several times, a driver is not able to reach the exact pinned location due to narrow lanes. A driver may also disagree to reach the exact pinned location due to certain inconveniences. In such a case, the traveler has no choice but to walk to the cab's current location.
6. **No Internet connection to pay via E-wallet** – There are certain category of travelers who depend on E-wallets to pay for their cab rides. So, in absence of good internet connection, they may not be able to pay for their cab rides using E-wallets.

### *4.4 Sensory Model*

In the case of Ola and Uber, both apps convey completion of specific action sequences via push notifications and a specific sound that accompanies it. In an interview, the traveler mentioned the peculiar alert sound of Ola Mobile app. When

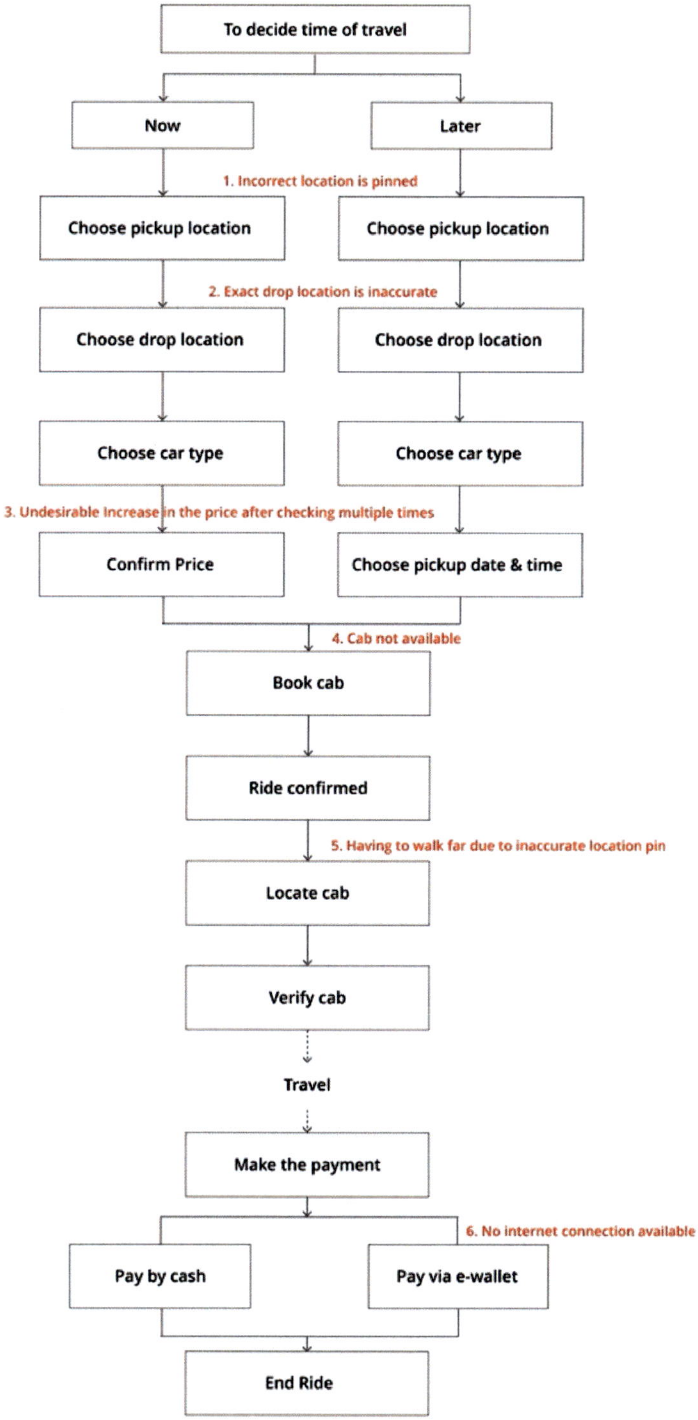


**Fig. 4** Sequence diagram for the task of booking a cab

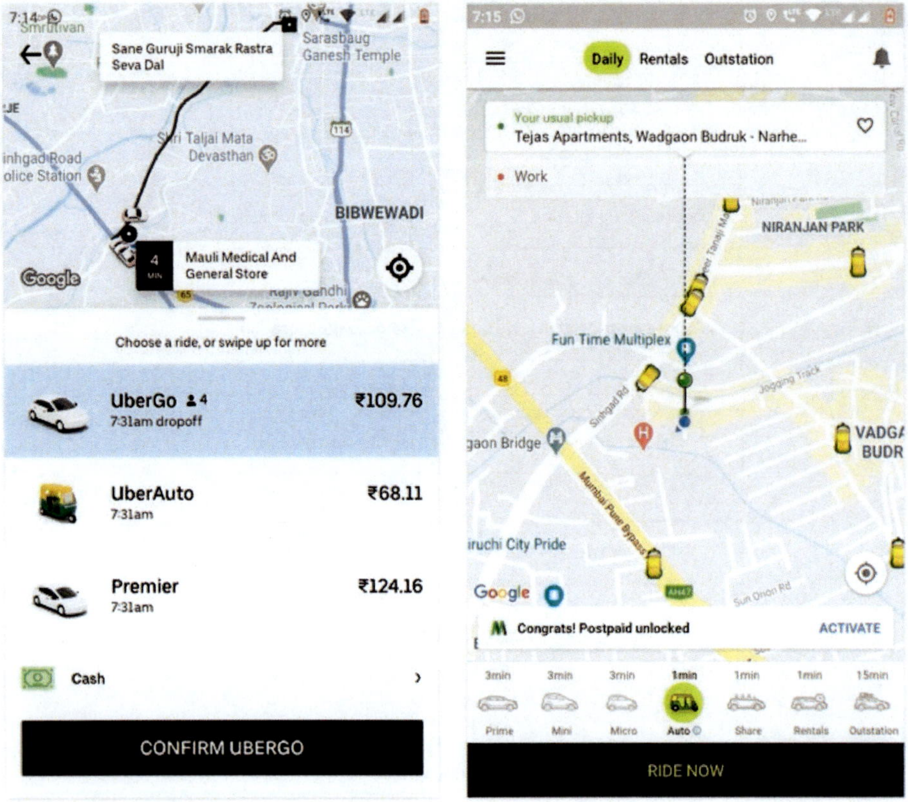


**Fig. 5** Uber and Ola Android App screenshots

the cab arrives at the desired pickup location, a specific ringtone is played that resembles the sound of shutting the car door. This sound is accompanied in unison with the horn of a car. As this sound is not that popular in phone ringtones, it does an effective job of attracting the user's attention. Figure 5 shows screenshots of Uber and Ola Android app user interfaces.

Uber and Ola both provide an effective user experience in their mobile apps. The User Interface of Uber is of a minimalist nature incorporating ample amounts of white space in their design which makes it visually appealing for certain users. Users were also attracted by the 3D icons of cars used in the Uber app. Some users were also attracted to the bright yellow color used throughout the Ola Cabs mobile app.

## 5 Aspects Derived from CI and Related Work Models

The following aspects were derived from the Contextual Inquiry and related work models for the notion of cab booking in India.

## 5.1 *Aspects Derived Related to Indian Travelers*

Below are the aspects that define the unique Indian traveler.

- Five most frequently observed user types for Indian cab travelers were derived from Flow Model as seen in related Fig. 1. These **traveler types—professionals, frugals, socialites, shoppers, and unfortunates**—help us understand pain-points of the traveler.
- **A women's safety is of utmost concern for Indian travelers** as discussed in Sect. 4.2.1. There exist several issues that women face even today while commuting by cabs, which have led to numerous cases of sexual assault on women by cab drivers.
- As seen in Fig. 3 and discussed in Sect. 4.2.4.3, Indian travelers have both apps (Uber and Ola) installed in their smartphones and generally compare fare prices before booking a cab. **Through fare comparison, they ensure that they cash-in the cheapest cab ride.**
- As discussed in Sect. 4.2.4.2, Indian travelers find Cab Pool services cheaper, but are irritated about excessive time it takes to reach the destination due to impromptu pickups. **The excessive time due to pickups is a major cause of annoyance among Indian travelers.**
- Indian travelers call the cab driver for confirmation after a ride is booked. As discussed in Sect. 4.2.2, this **call to cab driver is their way of confirming the driver's location and Expected Time of Arrival (ETA).**
- As discussed in Sect. 4.2.4.5, **Indian travelers despise traveling with drivers having bad habits like chewing tobacco, drinking, or bad hygiene** in general. Such drivers end up with a poor rating from travelers.
- **Indian travelers count in-car music as a significant factor in determining their ride experience** as discussed in Sect. 4.2.4.4.
- **Indian travelers are annoyed when drivers repeatedly ask them verbally for a 5-star rating** as discussed in Sect. 4.2.4.6.
- **Indian travelers find it hard to retrieve lost belongings**. As discussed in Sect. 4.2.4.7, they also suggest authorities to setup a dedicated helpline number which would assist them retrieve their belongings.

## 5.2 *Aspects Derived Related to Indian Cab Drivers*

- Indian drivers use this technique to maximize their profits. As seen in Fig. 2, **they have both apps (Uber and Ola) installed on their smartphones and switch their services in between the two apps**, based on which service gives them a quicker customer.
- As discussed in Sect. 4.2.3.1, **Indian drivers are always on the look-out for short-distance rides and try to avoid long-distance rides through cancellations or delays.** They use this strategy frequently to get out of inconveniences like long-distance pickups or long-distance drops.

- **Indian drivers complain about having to drive a few extra meters from the pinned location to pick-up the traveler**, as discussed in Sect. 4.2.3.11. They complain that, this adds to their fuel costs and results in lower profits.
- Indian drivers confirm the next ride while they are dropping off a current traveler, as discussed in Sect. 4.2.3.3. **They may request the next traveler to wait for a period up-to 15 min.**
- As discussed in Sect. 4.2.3.4, Ratings are of utmost importance to drivers. **The Indian drivers do not shy away from asking travelers upfront for a 5-star rating.**
- As discussed in Sect. 4.2.3.7, **Indian drivers respectfully ask the traveler if they can keep the extra change in the form of a tip after the ride.** This behavior is common among drivers outside of services such as Uber and Ola as well.

## 6 Conclusion and Future Work

This research paper aims to define and categorize the Indian cab traveler. It makes use of Contextual Inquiry and Work Models to study and understand Indian travelers and cab drivers. It makes an effort to understand specific cultural aspects that separate Indian cab travelers from others around the globe. It also tries to understand some problems faced by Indian cab travelers. The study of cultural separation will prove a game-changer while designing experiences for them in the future.

In future, this study will be referred to improve Indian cab booking system as a while by suggesting several tangible modifications in design of related apps and services.

## References

1. Drife, Taxis in India – The History, Geography and Economics of it. Medium, @Drife Official, May 6, 2019
2. Taxi, Wikipedia, https://simple.wikipedia.org/wiki/Taxi, Accessed on Jan 07, 2020
3. Sachitanand, R. The challenges before cab aggregators in India. The Economic Times, Industry, Transportation, Shipping, October 28, 2018
4. Utkarsh. The Evolution of the Indian Taxi Market - Comparing the Biggies. Inc42, Features, Datalabs, 25 Nov, 2015
5. Demographics of India. Wikipedia, https://en.wikipedia.org/wiki/Demographics_of_India. Accessed on Jan 12, 2020
6. Uber. Wikipedia, https://en.wikipedia.org/wiki/Uber. Accessed on 04 Jan 2020
7. Helling, B. Uber cars: Your guide to the various types of Uber cars. Ridester, Brett Helling, May 20, 2020
8. What do different types of Uber cars mean. Quora, Uber, August 22, 2018
9. IndSA, Comms. Uber Launches Uber Boat in Mumbai. Uber Newsroom, India, January 30, 2019
10. Arun. Know 20 different types of travelers you can come across. Treebo, Blog, Travel Hacks, February 20, 2018

11. Reddy, R. Uber and Ola are still not safe for women. Livewire, Personal category, June 27, 2019
12. Sharma, I. Panic buttons won't fix Ola and Uber's sexual-assault problem. Quartz India, Double-edged Sword, June 8, 2018
13. N. Borkar, Competitive analysis of the Indian food delivery industry – Zomato, Swiggy, UberEats and food panda. Medium, @Nikhil Borkar, October **19** (2018)
14. S. Bansal, *Ola Drops Food Panda Delivery, Lays off Several Employees* (Livemint, Companies, Startups, April 2020)
15. R. Rajesh, S. Chincholkar, A study on consumer perception of Ola and Uber Taxi services. Indian J Comp Sci **3** (2018)
16. Khade, A., Patil, V.: A Study of customer satisfaction level of Ola and Uber paid taxi services with special reference to Pune city. Int J Manag Technol Eng, 8, VIII, 1596–1603. (2018)
17. S. Yamuna, R. Vijayalakshmi, K. Jeeva, D. Boopathi, R. Kumar, A progressive study on users perception and satisfaction towards online cab service with reference to Coimbatore. Int J Res Appl Sci Eng Technol **7, IV**, 156–157 (2019)
18. R. Shukla, A. Chandra, H. Jain, OLA VS UBER: The battle of dominance. IOSR J Bus Manag **1, XI**, 73–78 (2017)
19. N. Khatoon, N. Kumar, P. Singh, S. Kumari, S. Lakhimsetti, *The Platform Economy: A Case Study on Ola and Uber from the Drivers' Partners Perspective* (Tata Institute of Social Sciences, 2019)
20. G. Manjunath, Brand awareness and customers satisfaction towards OLA cabs in Bengaluru North and South region. RJSSM **5** (2015)
21. D. Uthira, A diagnostic study of cab aggregation industry using Serviqual gap model. Int J Market Technol **8**(6), 59–84 (2018)
22. Y. Cui, Q. He, A. Khani, Travel behavior classification: An approach with social network and deep learning. Transportation Research Record: J Transport Res Board **2672**(47), 68–80 (2018)
23. G. Ayyappan, A. Kumaravel, A novel classification approach of travel review dataset based on health. Int J Comp Sci Eng **11**(6), 30–33 (2019)
24. Joseph, S, Chen., Yu-Chih, Huang., Jen-Son, Cheng.: Vacation lifestyle and travel behaviors, J. Travel Tour. Mark., 26, 494–506. (2009)
25. J. Golob, F. Golob, Classification of approaches to travel-behavior analysis. TRB Special Rep **201**, 83–107 (2015)
26. P. Kale, G. Bhutkar, V. Pawar, N. Jathar, Contextual design of intelligent food carrier in refrigerator: An Indian perspective, in *Human Work Interaction Design. Work Analysis and Interaction Design Methods for Pervasive and Smart Workplaces*, Advances in Information and Communication Technology, 468, ed. by J. Abdelnour Nocera, B. Barricelli, A. Lopes, P. Campos, T. Clemmensen, (Springer, Cham, 2015)
27. V. Shinde, G. Bhutkar, V. Pawar, *Overview of Garbage Monitoring System, Humanizing Work and Environment, 2017* (AMU, India, 2017)
28. Uber Eats. Wikipedia,. https://en.wikipedia.org/wiki/Uber_Eats, Accessed on 14 Jan 2020
29. Pal, B. Among these 8 categories, what type of traveler are you?. Travind, Blog, Types Of Travelers, December 30, 2017
30. King, S. What kind of traveler are you?. Conde Nast traveller, blog, Steve King
31. N. Wani, G. Bhutkar, S. Ekal, Conducting contextual inquiry of twitter for work engagement: An Indian perspective. Int J Comp Appl **168**(9), 975–8887 (2017)
32. U. Pandya, R. Rungta, G. Iyer, Impact of use of mobile apps of Ola cabs and taxi for sure on yellow and black cabs. Pac Bus Rev Int **9**(9), 91–105 (2017)
33. Bhattacharya, A. Ola Vs Uber the latest score in the great Indian taxi app game. Quartz India, Be Fast or Be Last, February 12, (2019)
34. Bellis, M. Hailing: History of the taxi. Thoughtco, Humanities, History & Culture, February 5, 2020
35. Taxicab, Wikipedia, https://en.wikipedia.org/wiki/Taxicab, Accessed on 07 Jan 2020
36. Bharadwaj, S. Taxi industry in India: A story of bootleggers and Baptists. Swarajya Magazine, Politics, November 20, 2016

37. Lobo, J. In the age of Uber and Ola, Mumbai's iconic taxis are becoming works of art. Quartz India, The One And Only, April 20, 2018
38. Deo, V. Everything you wanted to know about the big 5 cab services in India. Techstory, Business, May 5, 2015
39. Uber Driver & Grandmother By Urooj Ashfaq. YouTube, https://youtu.be/EzJhTb564iA. Accessed on Jan 12, 2020
40. Uber Pool & Wagon R Standup Comedy by Ashutosh Gupta. YouTube, https://youtu.be/omy0JIUNuAw. Accessed on 12 Jan 2020
41. It's My Birthday Standup Comedy by Sumit Anand. YouTube, https://youtu.be/6CdhE3hh2CQ, accessed on 12 Jan 2020
42. Five Star Rating Vinay Sharma Standup Comedy. YouTube, https://youtu.be/TxmGrsT0v3c, accessed on Jan 12, 2020
43. Ola Share ka Suffer Standup Comedy by Rajat Chauhan. YouTube, https://youtu.be/bj-uhD9jjWg, accessed on Jan 12, 2020
44. Ola Cabs. Wikipedia, https://en.wikipedia.org/wiki/Ola_Cabs. Accessed on 04 Jan 2020

# Data Set Reduction to Improve Computing Efficiency and Energy Consumption in Healthcare Domain

**Priyadarshan Dhabe, Param Mirani, Rahul Chugwani, and Sadanand Gandewar**

**Abstract** Many data sets used in the healthcare domain are huge and thus, need a considerable amount of computations, memory, electrical energy, and processing time for training and testing of Machine Learning (ML) algorithms. In this chapter, we propose a new data set reduction technique to reduce the size of data sets without affecting much of classification and recognition accuracy. This can be done by identifying and removing the redundant data samples from the training data set, which leads to the creation of reduced data sets, which are termed as "*Bonsai data sets*." As per our experimentation with the Skin segmentation data set, we reduced 39,993 samples out of 1,83,793 samples, which is around 21% samples, without sacrificing much on percentage classification and recognition. We used Logistic Regression (LR) K-nearest Neighbor (KNN) and Support Vector Machine (SVM) for testing classification and recognition accuracy using original as well as *Bonsai data sets*.

**Keywords** Data set · Data points · Bonsai · Recognition · Distance factor

## 1 Introduction

Nowadays, computer-assisted medical diagnosis is pervasive due to better accuracy and lesser efforts. Various ML algorithms are used to train models that can be used for reasoning about the diagnosis. We need a data set to develop a model using ML algorithm. The more the samples in the data set, the better will be the developed model and hence the diagnosis. But a very large number of unnecessary samples in a given data set may lead to the following problems.

P. Dhabe (✉) · P. Mirani · R. Chugwani · S. Gandewar
Department of Information Technology, Vishwakarma Institute of Technology, Pune, Maharashtra, India
e-mail: priyadarshan.dhabe@vit.edu; param.mirani17@vit.edu; rahul.chugwani17@vit.edu; sadanand.gandewar17@vit.edu



E. Eilu et al. (eds.), *Digital Literacy and Socio-Cultural Acceptance of ICT in Developing Countries*, https://doi.org/10.1007/978-3-030-61089-0_4

1. Increased computing time: - Many ML algorithms need multiple passes through the data set for model learning.
2. Increased number of computations.
3. Increased electrical power consumption.
4. Increased memory space.

Data set plays an important role in developing an ML model. It is a collection of n-dimensional data points. Each such data point is defined with $n - D$ vector containing $n$-values, one for each attribute. These data points are grouped into classes based on the similarity of their features. An input to the ML algorithm is such a data set and output is the developed model and is used for prediction. These data sets are created and maintained by organizations and/or repositories without taking much care of redundant samples within them. It is not much of their fault since data collection is their prime goal. Thus, we are proposing a new approach to identify and remove redundant data samples for preparation of "*Bonsai data sets*" and their obvious advantages like energy efficiency, reduction in computations, improved execution time and memory space, without affecting much on the accuracy.

While reducing the data set one must have to consider that, the reduction should not degrade substantially recognition and classification accuracies of the trained model. To accomplish this, we identify and remove redundant data points. A sample is called as redundant if it does not contribute to training since their peculiarities are already represented by some another pattern/sample. Thus, their removal will definitely reduce the size of data set but not have an adverse effect on the recognition and classification accuracy.

Our proposed approach is very simple, we remove all the data samples of a pattern class that are falling close to the mean of that sample population with a user-defined distance parameter $d$, which needs to be learned (tunable) or decided for each data set. Since mean consists of all the properties of those removed data points it will contribute in the same way to the learning as compared to the removed redundant data points. There are many classifiers available for testing/verification of the classification and recognition accuracy. But, due to their popularity/wide applicability, we used LR [1, 2], KNN [3, 4] and SVM [5, 6] for verification of classification and recognition accuracy, before and after removal of redundant patterns from the training set. Independence of the proposed technique from the classifier is also another feature and provides the freedom to the user to choose any classifier.

### *1.1 Related Work*

Reducing the data set has two-fold meaning. One interpretation is to reduce the dimensionality of the data set and the other is to reduce the number of data samples. The former approach is called as *dimensionality reduction* [6] and the latter is known as *sampling* [7] [8]. Reducing data sets for optimization is a common practice while building the ML model. Each of the approaches has its specific method and process of reducing the data set. There are seven methods reported in [6] for dimensionality reduction. Our proposed method, the *Bonsai data set*, aims to remove data points that do not contribute to the learning.

Data columns with little changes in the data carry little information. Missing Value Ratio [6] is a technique in which the data columns having too many missing values are removed. Low Variance Filter [6] is another technique in which all data columns with variance lower than a given threshold are removed. Since variance depends on range, we have to normalize the data set before the application of this technique. High Correlation filter [6] is another method in which we eliminate based on the correlation of two attributes. This is because if two attributes have a high correlation, only one of them will contribute to learning. Hence one of the two columns is removed.

Decision Tree Ensembles [10], also referred to as random forests, are useful for feature selection in addition to being effective classifiers. One approach to dimensionality reduction is to generate a large and carefully constructed set of trees against a target attribute and then use each attribute's usage statistics to find the most informative subset of features.

Principal Component Analysis (PCA) [1] is also a statistical procedure that orthogonally transforms the original n coordinates of a data set into a new set of n coordinates called principal components and then eliminates unnecessary attributes. Backward Feature Elimination [6] is a method where attributes are removed, one at a time, and then the model is rebuilt. Then the error rate of the new model and previous model is compared. This process continues until the error rate of a new model is more than the older model. The older model then becomes the best performing model. Forward Feature Construction [6] is the inverse process to the Backward Feature Elimination. We start with one feature only, progressively adding one feature at a time, i.e. the feature that produces the highest increase in performance.

Patterns recognition algorithms such as LR [1, 2], KNN [3, 4], and SVM [5, 6] commonly suggest that their computations increase with the increase in the number of datapoints in the data set. These methods are also used for experimentation purposes for computing classification and recognition accuracy of *the Bonsai data set*.

There are some sampling algorithms given in [7] [8] [9] for the selection of a subset of samples using various criteria from bigger data sets. Our proposed approach of *Bonsai data set preparation* is somehow more similar to *sampling* than *dimensionality reduction (feature selection)* [6], but different than sampling too. In sampling, one sample at a time is randomly/systematically chosen and a decision is made to add it to the reduced set or not. We can consider the proposed approach of *Bonsai data set preparation* as a *batch sampling* (we are coining this term and refers to), where a batch of patterns i.e. subset of patterns within a given class are declared redundant and removed from that pattern class.

While developing these algorithms, everyone proposed/improved these algorithms without considering any limitations on electrical energy consumption. Due to this, a lot of electrical energy is consumed in building and using these algorithms. Based on [11], almost 28–35% of the total energy is spent in moving data across memory hierarchy (e.g. register to RAM, RAM to Disk). Algorithms such as SVM consume more power [12] than other learning algorithms [13]. However, this consumption of energy is not even constant over its time of training. The following papers [12, 13] were also helpful in understanding the consumption of power for training and using ML algorithms.

The main research objective of this paper is to propose a method that detects and removes redundant data samples from a data set, which do not (very little) affect the reasoning accuracy, for optimizing computations, time, space, and electrical energy also. Cross-checking the accuracy of using redundant and reduced Bonsai data set using classifiers is another research question that is discussed in this paper.

## 2 Proposed Method of Preparation of Bonsai Data Set

In this section, we are presenting the proposed method of preparation of *the Bonsai data set*, which is the smallest subset of nonredundant data samples from the training set, such that, it must provide comparable classification and recognition accuracy. The proposed method is explained as follows.

Let $S$ be the $K$ sets of training pairs $(r_h, d_h)$, where $r_h = (r_{h1}, r_{h2}, \ldots, r_{hn})$ be an n-dimensional input pattern for $n > 0$ and $d_h$ is the class label. Each $r_{hi}$ is one of the numeric features of the pattern $r_h$ $i = 1, 2, \ldots, n$. An input pattern $r_h$ has a class label $d_h \in \{d_{h1}, d_{h2}, \ldots, d_{hC}\}$ i.e. set $S$ has training patterns of $C$ classes$C > 0$.

While deciding the redundant patterns we consider all the patterns belonging to a given class from training set, say $m \leq C$, and computed the mean pattern $R_m = \left\{r_{m1}, r_{m2,\ldots,} r_{mn}\right\}$ as an arithmetic mean of all the patterns of $m^{th}$ class. We define a maximum distance $d_{\max}$ as the maximum distance of any pattern of $m^{th}$ class from its mean $R_m$. We use a user-defined, tunable, fractional parameter $0 < \alpha \leq 1$ for deciding the redundant patterns. A pattern of a $m^{th}$ class is redundant if its distance $d$ from mean $R_m$ is satisfying condition $d \leq (\alpha * d_{\max})$. The parameters $d_{\max}$, $(\alpha * d_{\max})$ and $R_m$ used for computation of *the Bonsai data set* are pictorially illustrated in Fig. 1.

In our experimentation we start with a smaller value of $\alpha$, identify and remove the redundant patterns from each class. Then, we check the classification and recognition accuracy. By gradually increasing $\alpha$, we decide the final value of $\alpha$, where we are getting maximum classification and recognition accuracy nearly the same as that of exhibited by the original training data set. We settle at this value of $\alpha$ and measure the total number of redundant patterns from the training data set. We recommend all the researchers not to use these patterns for their work to save computations, time, memory space, and electrical energy. Thus, the proposed approach provides benefits in all possible ways and hence strongly recommended.

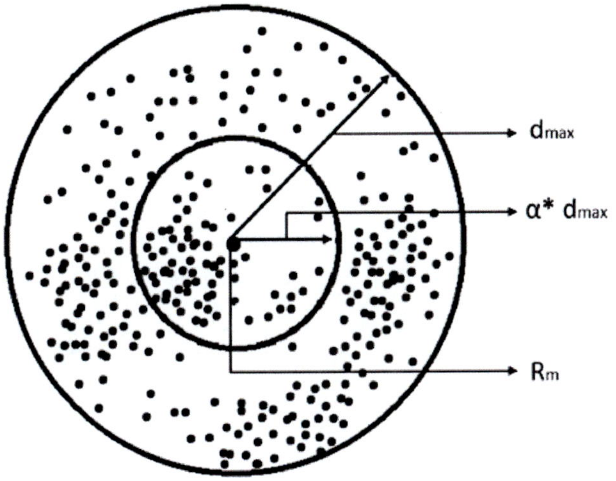


**Fig. 1** Pictorial description of parameters used for computation of the Bonsai data set

# 3 Experimental Results

We experiment with the Fisher Iris data set [14] for generating *proof-of-concept* (*POC*) of our approach. Then, we worked with the Skin Segmentation data set [15]. For both the data sets, we used 75% (roughly) randomly selected patterns as a training set and the remaining 25% as a testing set. The training set is used for classification accuracy and testing set for knowing recognition performance.

As per our study, there is no consensus on how much must be the split of training and testing set. A lot of debate is found on StackOverflow [16], between ration 75:25%, 80:20% and even on 50:50%. We learned that this ratio needs to be decided based on the size of the data set. If the data set is very large then even 50% samples in the training set and the remaining 50% samples in the test set will work fine. Due to the large data set, 50% of samples are also enough to fit the training data. But, if the data set is small to medium 80% to 20% and 75% to 25% is common practice. The 80% training and 20% testing set is also supported by the "*Pareto principle*" [17]. But we used [18, 19] as a logical base of 75% samples in the training set and 25% in the testing set, this is also a default setting in *sklearn* library in Python [19].

## *3.1 Fisher Iris Data Set*

In this sub-section, we compare the recognition and classification accuracy of the Fisher Iris data set, as described in Table 1, using both the Bonsai set and originally provided data set.

**Table 1** Description of Iris Data Set [14]

| | |
|---|---|
| Number of datapoints | 150 |
| Number of attributes | 4 |
| Number of classes | 3 |
| Associated task | Classification |

**Table 2** Experimental results of the Fisher Iris data set

| Distance factor ($\alpha$) | Number of samples used in training | %Rd | %C | % R |
|---|---|---|---|---|
| 0 | 115 | 0 | 96.52 | 97.37 |
| 0.1 | 112 | 2.61 | 96.43 | 97.37 |
| 0.15 | 108 | 6.09 | 96.30 | 97.37 |
| 0.2 | 97 | 15.65 | 95.88 | 97.37 |
| 0.25 | 86 | 25.22 | 95.35 | 97.37 |
| 0.3 | 75 | 34.78 | 94.67 | 94.74 |
| 0.35 | 60 | 47.83 | 93.33 | 94.74 |
| 0.4 | 48 | 58.26 | 93.75 | 97.37 |
| 0.45 | 42 | 58.26 | 92.86 | 92.11 |

We start experimentation with a value of $\alpha$ from 0 to 0.5, and the percentage reduction of patterns, obtained percentage classification, and recognition are shown in Table 2. In Table 2, %Rd. represents percentage reduction, %C indicates percentage classification, and %R represents recognition.

The value $\alpha$ is used to identify and eliminate redundant data samples from the training set. For $\alpha = 0$, i.e. without any reduction (using all 115 samples from the training set), we obtained 96.52% classification and 97.37% recognition accuracy, respectively. This result is the first highlighted row of Table 2. Then by increasing $\alpha$ up to 0.25, we obtained the same % recognition accuracy of 97.37 as far $\alpha = 0$ and there is a small drop in % classification from 96.52% to 95.35%. This small drop in classification accuracy should not be treated as a serious handicap of the proposed method since most of the time we need recognition than the classification. Thus, by reduction of 25.22% of the training set (86 samples out of 115), we can provide exactly the same recognition accuracy and comparable classification accuracy. This is shown in the fifth highlighted row of Table 2.

Such a *Bonsai data set* is recommended for training various ML systems rather than using the original data set.

If we increase $\alpha$ further from 0.25, we observe a decrease in both classification and recognition accuracy and thus it is not recommended to use these values. The plot of $\alpha$ versus classification and recognition accuracy is as shown in Fig. 2. The plot of percentage reduction in samples and accuracy is explained in Fig. 3. We suggest choosing the maximum value of $\alpha$ where we obtained classification and recognition accuracy close to the original training set.

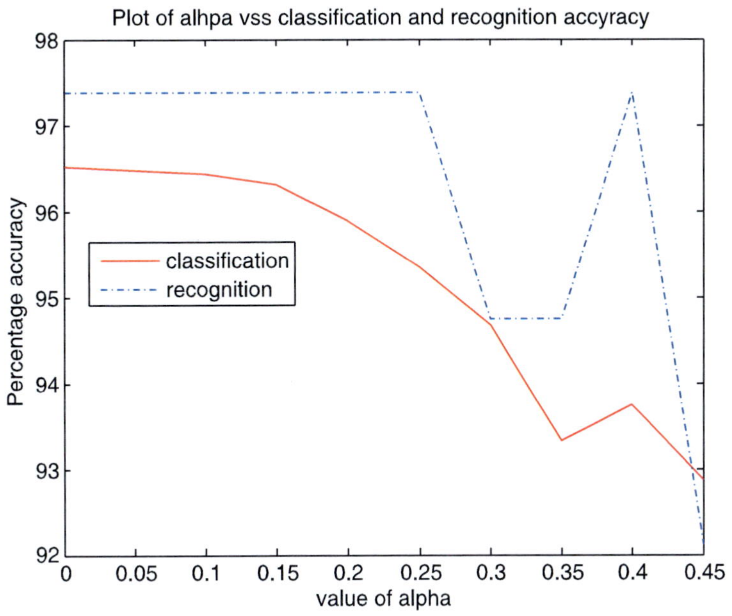


**Fig. 2** Plot of alpha versus classification and recognition accuracy

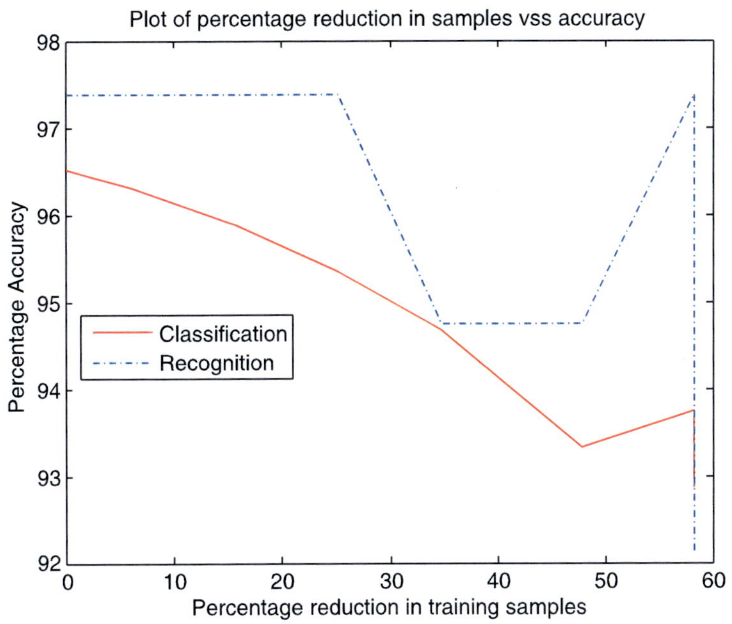


**Fig. 3** Plot of percentage reduction in samples versus classification and recognition accuracy

**Table 3** Accuracy of Bonsai data sets using different algorithms

| Algorithm | Distance factor ($\alpha$) | | | | | | | |
|---|---|---|---|---|---|---|---|---|
| | 0.1 | | 0.25 | | 0.4 | | 0.5 | |
| | C% | R% | C% | R% | C% | R% | C% | R% |
| KNN(3) | 96.43 | 97.37 | 95.35 | 97.37 | 93.75 | 97.37 | 97.14 | 86.84 |
| KNN(5) | 96.43 | 97.37 | 95.35 | 97.37 | 93.75 | 97.37 | 97.14 | 86.84 |
| LR | 98.21 | 97.37 | 97.67 | 97.37 | 95.83 | 94.74 | 97.14 | 86.84 |
| SVM | 98.21 | 97.37 | 97.67 | 97.37 | 97.92 | 94.74 | 94.29 | 86.84 |

Let's summarize the different results which were found after training the Bonsai data sets for different distance factors $\alpha$. At the initial state distance factor is $\alpha = 0$ which means that 0% reduction has been done i.e. we are using the complete training data set and the result obtained is shown as the first row of Table 2. As we go down in the table you will observe that distance factor $\alpha$ is increasing thereby, increasing the percentage reduction. Till the distance factor $\alpha = 0.25$ results are quite good and surprisingly close to the initial results, which is a great thing because the percentage reduction is nearly 25.22% which means only 75% of the original training set is sufficient to provide around the same classification and recognition accuracy.

After the distance factor $\alpha = 0.25$ 5 we observe the drop in recognition percentage so for this data set we suggest that you should not go beyond a 25.22% reduction to get the best possible results. The 25% reduction means that out of 115 in Iris data set, we are training from only 86 patterns i.e. we can remove 29 patterns without compromising much on the accuracy.

We also measured the performance of the *Bonsai data set* using various ML algorithms and the performance is as shown in Table 3.

As shown in the above Table 3, we applied different training algorithms on Bonsai data sets having different distance factor. As you can see all the algorithms give a really good recognition percentage till the distance factor is 0.25 when you go beyond that percentage recognition drops for every algorithm.

The motive behind applying different algorithms on Bonsai data sets was to show that independent of any specific classification algorithm *Bonsai data sets* performs well. From the above experimentation, one can conclude that without losing much on the performance we can reduce the data set and it works with any ML algorithm.

## *3.2 Skin Segmentation Data Set [15]*

Features of this data set collected from the UCI repository [15] are mentioned below in Table 4. It is one of the famous data set from the healthcare domain.

To summarize the results mentioned in Table 5, highlighted first row indicates 0% reduction for $\alpha = 0$, we obtained 99.97% classification and 99.96% recognition accuracy. When we increase the distance factor $\alpha$, percentage reduction will also

**Table 4** Features of the Skin Segmentation data set

| | |
|---|---|
| Number of datapoints | 245057 |
| Number of attributes | 4 |
| Number of classes | 2 |
| Associated task | Classification |

**Table 5** Results of the Skin Segmentation data set

| Distance factor ($\alpha$) | Number of samples used in training | %Rd | %C | % R |
|---|---|---|---|---|
| **0** | **183,793** | **0** | **99.97** | **99.96** |
| 0.1 | 176,769 | 3.82 | 99.97 | 99.96 |
| 0.15 | 161,330 | 12.22 | 99.96 | 99.82 |
| 0.175 | 156,249 | 14.99 | 99.96 | 99.42 |
| **0.2** | **143,793** | **21.76** | **99.97** | **99.96** |
| 0.25 | 133,794 | 27.20 | 99.97 | 96.08 |
| 0.3 | 118,091 | 35.75 | 99.97 | 92.25 |
| 0.5 | 57,261 | 68.84 | 99.98 | 78.06 |

increase and percentage recognition goes down slowly. The drop observed in performance is in decimal points $\alpha = 0.2$. For this particular data set we suggest ideal reduction percentage would be 21% for $\alpha = 0.2$, the reason behind this is that after this point there is more drop in recognition accuracy. This means that only 79% of patterns of the existing training set is sufficient for comparable classification and recognition accuracy. Thus, the remaining 21.76% patterns i.e. around 39,993 patterns out of 183,793 are declared redundant and can be removed. Removal of this 21%, huge number, of redundant patterns leads to a reduction in 21% number of computations, 21% computational time, and 21% amount of memory space and 21% less energy consumption. Figure 4 shows the plot of $\alpha$ versus accuracy for *Skin segmentation data set*.

From the experimentation using both the data sets, we conclude that there can be considerable redundant samples in the training data sets used for ML applications. We proposed a new sampling [7] [8] [9] type method of preparation of *the Bonsai data set* to detect and eliminate the redundant samples that do not contribute much in ML training. In the proposed approach we need to decide the best value of the parameter $\alpha$ experimentally, such that we can get around comparable classification and recognition accuracy using minimum samples from the training set. Such reduction will be helpful for a reduction in the number of computations, time, memory space, and consumption of electrical energy for training ML algorithms.

Reduced data set (Bonsai data set) will cut down the training time and recall time of the system used for medical diagnostic purposes, and thus keeps their users happy without compromising the quality of output. It also supports green computing [20] by reducing the number of computations to be done, since more computations need more consumption of electrical energy. More energy consumption can have a cascading effect of heat generation in the system and may need cooling.

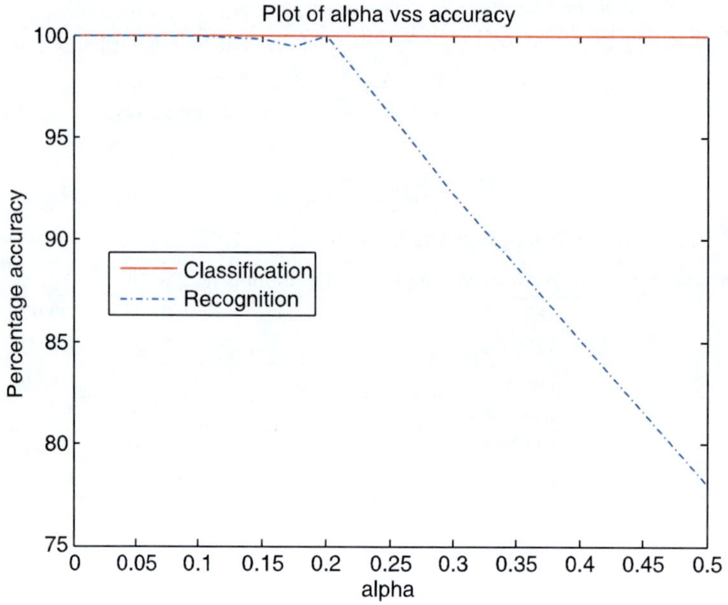


**Fig. 4** Plot of alpha versus percentage accuracy

# 4 Discussion of the Results

As per our experimentation with both Iris data [14] and Skin Segmentation data [15], the proposed method of Bonsai data set, suggests the redundant patterns within each pattern class can be removed so that ML algorithms can be trained quickly, with less memory space and thus need less consumption of electrical energy. A tunable parameter $0 < \alpha \leq 1$ needs to be decided for each data set for determining the redundant training samples. The more the value of $\alpha$, the more will be the redundant training samples, like *sampling technique* [7–9]. By experimenting with $\alpha$, we need to decide the value of $\alpha$, where we get comparable recognition and classification accuracy as compared to the original training data set. Results show that the proposed approach allows us to reduce 21% computations, 21% computational time, 21% of memory space, and a similar amount of electrical energy (since 21% fewer computations are performed), for the skin segmentation data set [15]. Similar results can be expected from other ML data sets too, and this proves the importance of the proposed approach of the *Bonsai data set*.

We suggest to all the websites like *UCI machine learning repository* [21] and kaggle [22] hosting huge data sets for ML purpose to also host reduced Bonsai data sets, such that many researchers can use the Bonsai data set and can work quickly with it by saving computations, time, memory space, and electrical energy.

## 5 Conclusion

It is concluded from the proposed sampling method of *the Bonsai data set* that we can remove the redundant patterns from ML data sets and can prepare the reduced Bonsai data sets. We obtained Bonsai data set with a 21% reduction in training samples for Skin Segmentation data set [15] and hence the same percentage reduction in computations, time, memory space, and electrical energy. Then, we can train the ML models on these Bonsai data sets without compromising much on classification and recognition accuracy. It is observed that using Bonsai data sets for model training, we can achieve reduced computations, memory space, and reduced electrical power consumption as compared to data sets containing redundant patterns. Because of these advantages *Bonsai sets* help users to train ML models efficiently. Thus, they are strongly recommended as opposed to original data sets, in the health care domain.

## References

1. H. Abdi, L.J. Williams, Principal component analysis. Wiley Interdisciplinary Reviews: Computational Statistics **2**(4), 433–4598 (2010)
2. J.T. Austin, R.A. Yaffee, D.E. Hinkle, Logistic regression for research in higher education. Higher Education: Handbook of Theory and Research **8**, 379–410 (1992)
3. T.M. Cover, P.E. Hart, Nearest neighbor pattern classification. IEEE Trans. Inf. Theory **13**(1), 21–27 (1967)
4. B.E. Boser, I. Guyon, V. Vapnik, A training algorithm for optimal margin classifiers, in *Proceedings of the Fifth Annual Workshop on Computational Learning Theory*, (ACM Press, 1992), pp. 144–152
5. Srivastava D. K., Bhambhu L. : Data classification using support vector machines. JATIT J. (2009). Available online http://www.jatit.org/volumes/research-papers/Vol12No1/1Vol12No1.pdf
6. Ray S.: Seven Techniques for Data Dimensionality Reduction, available online https://www.analyticsvidhya.com/blog/2015/07/dimension-reduction-methods/
7. A. Yıldırım, C. Ozdogan, D. Watson, *Parallel Data Reduction Techniques for Big Datasets* (2014). https://doi.org/10.4018/978-1-4666-4699-5.ch004
8. P. Tan, M. Steinbach, A. Karpatne, V. Kumar, *Introduction to Data Mining, Second Edn* (Pearson, 2019)
9. H.: Sampling: An approach to solve the bird counting problem, (2019), available [online] https://towardsdatascience.com/sampling-79075e9176cb
10. L. Breiman, J.H. Friedman, R.A. Olshen, C.J. Stone, *Classification and Regression Trees* (Wadsworth International Group, Belmont, California, 1984)
11. G. Kestor, R. Gioiosa, D.J. Kerbyson, Hoisie (2013). Quantifying the energy cost of data movement in scientific applications. IEEE International Symposium on Workload Characterization (IISWC)
12. E.G. Martin, C.F. Rodrigues, G. Riley, H. Graham, Estimation of energy consumption in machine learning. J. Parallel Distrib. Comp. **134**, 75–88 (2019)

13. E. García-Martín, N. Lavesson, H. Grahn, E. Casalicchio, V. Boeva, How to measure energy consumption in machine learning algorithms, in *ECML PKDD 2018 Workshops. ECML PKDD 2018. Lecture Notes in Computer Science*, ed. by C. Alzate et al., vol. 11329, (Springer, Cham, 2019)
14. Fisher Iris Data Set – Available online on. https://archive.ics.uci.edu/ml/datasets/iris
15. Skin Segmentation Data Set – Available online on. https://archive.ics.uci.edu/ml/datasets/skin+segmentation
16. Train-test data split- Available online on. https://stackoverflow.com/questions/13610074/is-there-a-rule-of-thumb-for-how-to-divide-a-dataset-into-training-and-validatio
17. Pareto principle- Available online https://en.wikipedia.org/wiki/Pareto_principle
18. Guyon I., A scaling law for the validation-set training-set size ratio, AT & T Bell Laboratories, Berkeley, Calif, USA, 1997
19. sklearn library- Available online https://scikit-learn.org/stable/
20. Kurp, P. :Green Computing. Commun. ACM. 51. 11–13. 2008, 10.1145/1400181.1400186
21. UCI Machine Learning Repository. https://archive.ics.uci.edu/ml/index.php
22. The Kaggle data sets. https://www.kaggle.com/datasets

# Design Strategies for Improving Anticipated User Experience in a Developing Country Setting: Case of Uganda

**Emmanuel Eilu**

**Abstract** Whereas the use of mobile phones could improve the low voter turnout in many developing countries, different studies show that there is a lot of negative perception and responses towards the use of mobile phones for voting. This study assesses the extent to which six design strategies namely, reciprocity, tailoring, social and political validation, demos, ease of use, and perceived usefulness, would influence positive perceptions and responses towards the use of mobile phones for voting in a developing country like Uganda. Expert evaluation and user studies were used to assess the extent the six strategies influence positive perception and response towards the use of mobile phones for voting prior to their implementation. There was an average of 78% agreement by experts and 72% agreement from user studies that the design requirements could influence positive perceptions and responses towards the use of mobile phones for voting. This paper, therefore, recommends design strategies that can be used by practitioners to design products or services for the market where the product or service has not been well received or has been rejected before, especially in a developing country setting like Uganda.

**Keywords** Anticipated user experience · Mobile phone voting · Expert reviews · User studies

## 1 Introduction

In electoral democracy, the electorate hold within their authority, the supremacy to accord political legality, or withhold it from any leader, institution, and regime [1]. In order to accord political legality or withhold it, people have to fully participate in the electoral process. It is through voter participation that the will of the people can be truly demonstrated. Therefore, the number of people who turn up to express their

E. Eilu (✉)
Uganda Christian University, Mbale University College, Mukono, Uganda



E. Eilu et al. (eds.), *Digital Literacy and Socio-Cultural Acceptance of ICT in Developing Countries*, https://doi.org/10.1007/978-3-030-61089-0_5

preferences in a voting process is of great importance to the very core definition of electoral democracy [2]. Lijphart [3] observes that, if there is low voter turnout, it would then often translate into "unequal participation," and therefore, the vital principle of democracy is not demonstrated.

However, the number of citizenry interested in participating in a democratic electoral process more especially in African is low. Recent statistics show that Africa registers the lowest voter turnout in the world with an average of (65%) [4]. The voter turnout in some African countries like Uganda is even lower. For example, in 2005, the government of Uganda held a referendum which registered a 47% voter turnout [5]. The 2011 and 2016 presidential elections in Uganda registered less than 60% voter turnout [6, 7]. A number of African countries with the same trend include Nigeria with 43% in the 2015 elections and Tanzania with 43% in the 2010 elections [8]. Violence, long voter queues at the polling stations, and intimidation have been registered as some of the factors hindering high voter turnout [4, 6, 7, 9]. Several efforts are underway to improve voter turnout in Africa and one of those efforts is the use of Information and Communication Technologies to increase voter turnout. One of the most available and affordable technology that can be deployed in an electoral process to improve voter turnout is the mobile phone [10]. Uganda, for example, has over 23 million mobile phone users and a network penetration of 73%. Therefore, the use of mobile phone as a voting tool would increase voter turnout in the following ways. Anyone who owns a mobile phone can vote from anywhere and at any time than to make a long trip to the polling station and queuing for hours [11]. The use of mobile phones for voting would also limit violence and intimidations that normally take place at the voting. The sick and the disabled who could not reach the polling station would be able to vote through their mobile phones.

Whereas the use of mobile phones could improve the low voter turnout in many African countries, different studies show that there is a lot of negative perception and responses towards the use of mobile phones for voting [12–15]. This is largely influenced by social, political, and technological factors, though political factors are highly considered a major hindrance. These perceptions and responses (be it positive or negative) towards the use of mobile phones for voting are generally called User Experience (UX), one of the fields in Human-Computer Integration.

## 2 User Experience and Anticipated User Experience

ISO DIS 9241-210 defines UX as *a person's perceptions and responses that result from the use and/or anticipated use of a product, system or service*. It, therefore, means that UX is closely related with a person's perceptions (what a person thinks about a product) and responses (feedback) as the client is using a particular product or service. UX is also closely related with perceptions and responses that arise as a result of anticipating the use of a product or service (before use). The second part of the ISO definition (perceptions and responses that result from anticipated use) is the type of UX of interest for this paper (anticipated use of mobile phones for voting).

This kind of user experience is referred to as Anticipated User Experience (AUX). Yogasara et al. [16] define AUX as the *needs, wishes and feelings that the user expects to occur when imagining an encounter with an interactive product or system*. AUX is a nonphysical interaction with a product or service, it is when a client fantasizes with the product or service and imagines possible interactions [17]. For Pakanen [18], AUX are needs and wishes that result from expected interaction with the concept of a product before the actual product exists. Law et al. [19] call it "imagined experience". In this paper, AUX is portrayed as a positive imagination that someone has towards a product or service. AUX is a form of fantasy or excitement that a user feels about a given product or service before using it. When fully achieved, AUX influences positive perceptions and responses towards a given product/service, i.e. it creates a desire for the product or service even before using it. The next section discusses six design requirements for influencing positive perceptions and responses towards a given product/service.

## 3 AUX Design Requirements

In his paper entitled the "science of persuasion" published in 2001, Robert B. Cialdini [20] proposed a series of strategies in marketing that can have a positive influence on a potential customer. In 2002, J. B Fogg [21] adapted some of Cialdini's strategies to develop persuasive technology model which can be used to shape, reinforce, or change behaviors, feelings, or thoughts about an issue, object, or action." Other scholars such as Davis (1989) also forwarded strategies that influence positive perceptions prior to the product use as discussed below.

- Reciprocity- is when someone is rewarded after or before performing a certain action. Reciprocity employs rewards to influence a user to use the product [20, 21].
- Tailoring- means designing a product according to the user's preferences, needs, and profiles [21]. For example, designing according to a user's favorable colors and user's language.
- Social and Political Validation- where individuals follow or conform to the actions of others within a group [20, 22]. It uses peer pressure to influence positive perceptions towards a certain product or service.
- Demos- means step-by-step guidance on how to use a product [23]. It reduces "uncertainty" that the user had towards a product, therefore influencing positive perceptions towards the use of the product or service.
- Ease of Use- the degree the user believes that the product or service is easy to use. It focuses on a simple-to-learn and easy-to-use product or service to influence positive perceptions towards a product or service [23, 24].
- Perceived Usefulness- means the degree to which the user believes that the product or service is useful. It highlights the usefulness of using a product or service to influence positive perceptions [24, 25].

However, the success of these design strategies in a developing country like Uganda may not be guaranteed given the fact that these six design strategies were derived from the context of a developed country. There was a need to assess the extent the six design strategies would influence positive perceptions towards the use of mobile phone for voting in a developing country like Uganda before the strategies are finally used for design a mobile phone voting system. There are a number of methods that can be used to measure the extent the six strategies influence positive perception and response towards the use of mobile phones for voting prior to their implementation. However, due to time and financial limitation, this research carried out a pilot study and used two approaches to evaluate anticipated user experience. These approaches include expert reviews and user studies.

## 4 This Research

The purpose of this research was to assess the extent to which each of the six strategies would influence positive perceptions and responses towards the use of mobile phones for voting in a developing country like Uganda, where the use of mobile phones for voting was rejected.

This study is of great importance, for it provides vital ideas on how a product or a service can be designed for acceptance in a developing country setting. The design strategies can be used by practitioners or entrepreneurs to redesign a product or service for acceptance particularly where the product or service has not been well received or has been rejected before, especially in a developing country setting like Uganda.

## 5 Methodology

In the pilot study, two evaluation methods were used as discussed below.

### *5.1 Expert Reviews*

Expert review involves a usability or user experience specialist who, using her/his expert knowledge on a product, service or procedure, performs the review from the angle of a typical user so that they can highlight any user experience issues that might spoil a genuine user's experience. In this case, UX experts use their expertise of users and UX theories to perform an evaluation. According to Baguma [26], most techniques used for validation do not involve end users but rather the designer or human factor experts assessing the impact of the design on a typical user. For best results, Microsoft [27] and Tory & Moller [28] recommend a minimum of five experts to review the system or framework. However, Nielsen [27] argued that expert reviews should be performed by at least three experts. Their argument was based on an extensive study that found out that three experts could identify about 60% of the problems. The study found out that as the number of experts increased in the evaluation process, there were very few additional problems found. For example, it took 10 experts to find 85% of the problems and 15 experts to find 90% of the problems.

In this study, the researcher used six user experience experts. These user experience experts had published extensively in the field of user experience. Since these user experience professional resided both in and outside the country (Uganda), it was difficult and expensive to bring them together in a workshop. The easier way was to e-mail evaluation questionnaires to them, and their opinions were expressed in writing. Questionnaires allowed the user experience experts to work independently and to form and express their own opinions freely.

The first task during the evaluation was to brief the experts on the preamble and rational that led to the development of these design strategies. These design strategies were therefore presented to user experience experts to determine their validity, the extent to which each of each design strategy, if used for designing a mobile phone voting systems, they would influence positive perception and make a voter who has a negative perception towards the use of mobile phones for voting, to willingly accept to use his/her mobile phone for voting.

On a scale of 1–10, the user experience experts were kindly requested to assess the extent each of the requirements identified can influence positive perceptions towards the use of mobile phones for voting, and the major limitations and flaws of each of the requirements. A total of 21 evaluation questionnaires were sent by e-mail to 21 user experience experts. Only six experts responded to the questionnaire. There was no response from the other 15 experts.

## 5.2 *User Studies*

This involves the application of the design strategies on a limited basis in a pilot setting while results are closely being monitored [29]. User studies are more accurate than survey and expert reviews, as they reflect real users attempting common tasks on a product or system [30]. Pakanen [18] observes that there are only a handful of user evaluation methods that can be used to evaluate Anticipated User Experiences in the early development phase. However, the anticipated eXperience Evaluation by Gegner & Runonen [31] is one of best suited for the evaluation. This enables the participant to visualize and imagine a use situation and give feedback on how he or she feels about the experience. Furthermore, AXE provides deep insights into the participant's real-life context. AXE is an early study that reveals the user's perception about a product before the product is made available to the user for consumption. It involves using users in an interview setting and can be used both as an evaluative method and a method for gathering ideas for improvement.

The next section describes how the areas of study and users/participants for the evaluation exercise were chosen.

### 5.2.1 Selection of the Area of Study and Participants for User Studies

#### 5.2.1.1 Criteria for Selecting Area of Study and Participants

During the user study, there was a need to select participants from both the rural setting and the urban setting. The rural and urban areas were identified based on the Uganda National Bureau of Statistics report of 2014. Sonde Village, in Misindye Parish, Goma Sub-county in Mukono District was identified as a typical rural setting in Uganda. While Kansanga located in Kansanga-Muyenga Parish, Makindye Division Sub County, in Kampala district was identified as an urban area. In Kansanga, user studies were conducted in a university setting, that is, at International University of East Africa (IUEA), one of the private universities in Uganda founded in 2010. The main reasons for the choice of IUEA in Kansanga was the ease of recruitment of participants (teaching staff and students) given that the primary researcher worked there as a lecturer in the Faculty of Science and Technology. Participants in the evaluation study, both in rural and urban settings, were purposively selected based on the following criteria;

- Ugandan nationals of 18 years and above (voting age).
- Own a mobile phone.
- The participants should have negative perceptions towards the use of mobile phones for voting. It was necessary to use these kinds of participants so as to determine the extent to which the design strategies would influence the participant's perception towards using mobile phones for voting.

#### 5.2.1.2 Sample Size of the Participants

Nielsen [22] recommends a minimum of 5 participants for a usability/user experience evaluation of a product or service. Using 5 users is expected to find 85% of the usability or user experience problems in a test iteration [22]. Macefield [32] states that for problem discovery, anywhere between 3–20 participants is valid, with 5–10 participants being a workable baseline range. Faulkner [33] used yet another benchmark task and found that although tests with 5 users revealed an average of 85% of usability problems, the total percentages for each set of 5 participants ranged from nearly 100% down to only 55%. Groups of 10 participants did much better, finding 95% of the problems with a lower bound of 82% as shown in Table 1.

Based on the evidence from existing research on sample size for user experience evaluation studies and the criteria for the kind of participants needed, a total of 15 participants were selected for the evaluation. All the 15 participants were not in favor of using mobile phones for voting. Five students were purposively chosen

**Table 1** Number of problems found in a project relative to the number of participants (Faulkner 2003)

| Number of participants | Minimum % found | Mean % found |
|---|---|---|
| 5 | 55 | 85.55 |
| 10 | 82 | 94.686 |
| 15 | 90 | 97.050 |
| 20 | 95 | 98.4 |
| 30 | 97 | 99.0 |
| 40 | 98 | 99.6 |
| 50 | 98 | 100 |

from IUEA, 3 students were male and 2 were female. Five teaching staff were chosen for the evaluation. All the staffs were Master's degree holders; 3 female and 2 male.

On the side of the rural participants, five participants were randomly selected from Sonde village. The researcher got participants at the village trading center. Three had completed Ordinary Level (O-Level) certificates (General Certificate of Secondary Education-GCSE) and 2 had Advanced Level (A-Level) certificates. Of the 5, only 1 female met the criteria during the sampling process carried out in the area. The 5 participants then took part in the user evaluation studies.

### 5.2.2 User Evaluation Process

The Anticipated User Experience evaluation approach is divided into three major steps: concept briefing, concept evaluation, and data analysis as briefly described below.

#### 5.2.2.1 Concept Briefing

Concept briefing is done at the beginning of the evaluation session. The whole idea is shared with participants. The purpose of a concept briefing is to allow participants to deepen their understanding of the concept. Typically, the concept is conveyed through description and use scenarios. Scenarios are one of the ways a researcher explains defined tasks in the test situation to the participants or users [34]. When using scenarios, the narratives can be accompanied with other material than text such as illustrations or prototypes. In addition, mock-ups and visual expressions or illustrations can be used to clarify and better illustrate the concept.

The mobile phone voting prototype was coded in PHP scripting, hosted on MySQL and the USSD functionality was simulated using the Africa stalking (http://africas-talking.com/sandbox) developer platform (sandbox). A short scenario of using a mobile phone for voting in the next 2021 presidential elections in Uganda was performed. Each feature of the mobile phone voting prototype was clearly described namely, reciprocity, customization, ease of use, perceived usefulness, and endorsement. This short scenario was presented using simple common terms to make it easy for all participants to understand. The scenario session was followed by a demonstration of the mobile phone voting prototype.

After demonstrating how the mobile phone voting prototype works, participants were informed that their political, particular party leaders like presidential candidates themselves, had tested and endorsed the system. Participants were then given an ID number and PIN. As the participants used the prototype, the researcher keenly observed and listened while taking notes on how each of the participants reacted to each feature, their perceptions, fears, and what they would have liked in regards to the way the prototype functions and its features. Data obtained from this process was used to assess the extent to which each feature influenced the participant's perception towards the use of mobile phones for voting.

#### 5.2.2.2 Concept Evaluation

After the user has used the prototype, the researcher goes through all the visual displays (Prototype) again with the participants. Then in an interview setting, the researcher begins to conduct an in-depth investigation of participant's views, experiences, attitudes, opinions, and beliefs towards a given product concept and visual displays. During the concept evaluation process, the researcher had an interview with each participant to get their views about the utility of the system for national elections, ease of use, its limitations, its influence on their attitude towards voting using mobile phones in national elections and general perceptions. The participants were asked to indicate the features in the prototype that positively influenced their perceptions towards the use of mobile phones for voting, and why. The answers to this question provided a platform for the researcher to start a conversation around the concept. As the participant explained his or her choice, the facilitator went ahead to probe deeper the participant's view regarding each feature of the prototype.

### 5.2.3 Data Analysis

Content analysis was the main method used to analyze data obtained from both user studies and feedback from user experience experts. Context analysis is a research procedure used to make replicable and acceptable inferences by interpreting and coding textual material. This is normally done by systematically evaluating and classifying texts such as documents, oral communication, and graphics. Using context analysis, the researcher reviewed data collected from user studies and experts. The feedback from the respondents was carefully analysed, common themes were identified as well as areas of disagreements. Results of data analysis are presented below.

### 5.2.4 Presentation Results of the Evaluation

The results of expert reviews and user studies are summarized in Table 2.

From Table 1, it is worth noting that expert review evaluation on the extent reciprocity, demos, customization, endorsements, perceived usefulness, and perceived ease of use influence positive perceptions and responses towards the use of mobile phones for voting showed higher percentages than user studies. Reciprocity received 80% positive influence from the experts, and 60% from user studies. Although some of the concerns identified include the risk that it can stimulate the use of voting software/app, without voters being serious (just to get the reward). For example, voting blanks may be higher. Reciprocity is also less effective with older people.

User studies gave the tailoring strategy a slightly higher influence (73.3%), while the experts gave it a 70%. However, some of the limitation identified included; it is hard to think of any other politically neutral way of tailoring. For example, working with default votes or preferred choices at the top would be not neutral and not acceptable.

Experts, on the other hand, gave higher percentages to endorsement (80%) and demos (80%), while user studies showed that endorsement and demos had a 53% and 73% positive influence, respectively. However, participants pointed out that endorsements could work very well, but the endorsers should be people that really have authority to the specific voter, someone whose opinion they really value (not necessarily a political leader, it could be a musical star, or elderly in a village, etc.). Demos on the other hand should be made optional: a user can check/watch

**Table 2** The extent the design strategies influence positive perceptions towards the use of mobile phones for voting

| Design strategies | Expert reviews | User studies | Total average |
|---|---|---|---|
| **Reciprocity**: A voter given a reward after voting say free SMS, Mobile data, free airtime | 80 | 60 | 70 |
| **Tailoring:** The voter's preferences, needs, and profiles are identified and the voting service customized based on those preferences. E.g. voting application in the local language | 70 | 73.3 | 72 |
| **Endorsement by influential social and political leaders:** Use of influential peers, political and technical acquaintances, like the head of political parties, influential persons to encourage voters to use their mobile phones for voting. | 80 | 53 | 67 |
| **Demos:** Step–by-step guidance through the mobile phone voting process using various means, say on TV, radio, or the use of mobile phone voting test prototypes. | 80 | 73 | 77 |
| **Ease of use:** The proceedure should be short, easy to learn and use. | 80 | 84 | 82 |
| **Perceived usefulness:** Highlighting the usefulness of using mobile phones for voting is crucial. Benefits such as time-saving, privacy, and flexibility to vote from any place. | 80 | 89 | 85 |

guidance/demos when she/he wants or not check it at all. Another limitation for demo is that it does not work with older people.

User studies indicated higher positive influence percentages in perceived ease of use (84%) and perceived usefulness (89), while expert reviews indicated 80% influence for perceived ease of use and 80% positive influence on perceived usefulness. However, participants pointed out that positive anticipated user experience will only result from perceived ease of use if the people already know or can easily see that indeed it is easy to use the tool or system.

Therefore, the average positive influence towards the use of mobile phones for voting for all the design strategies was 72% for survey studies and 78% for expert reviews, while the overall average percentage for the two methods was 75%.

## 5.3 Discussion and Recommendations

The design strategies presented in this paper can be used to design products and services for acceptance. These strategies lead to acceptance by influence positive perceptions towards a given product or service. The design strategies can be used by practitioners or entrepreneurs to redesign a product or service for the market where the product or service has not been well received or has been rejected before, especially in a developing country setting like Uganda. This paper, therefore, draws its conclusion from the results presented in the earlier section, and presents recommendations for practitioner and entrepreneurs.

Going by the average percentages from all the two pilot studies (expert reviews and user studies). The most influential design strategy was perceived usefulness. The participants also found the prototype useful, it would save time and it would be very convenient, it could also allow them to cast their votes effectively. The study found out that an average of 85% of the responses in the expert reviews and user studies said that the use of mobile phones for voting would save time, very convenient, and would enable them to cast their votes effectively. There would no need to travel to the polling station and follow a long queue so as to vote. From the study, perceived usefulness was closely linked to proximity and convenience that the mobile phone provides to the voter. Battarbee & Mattelmäki [35] recommend a need for a portable object for product acceptance to be achieved. A study by Google shows that there is 67% chance for a product to be bought if it can be accessed on a mobile device [36]. Therefore, proximity and convenience are key factors to be considered by practitioners who want to offer a product or service. From the study, we see that highlighting the benefits of using mobile phones for voting, like it takes a shorter time to cast a vote, facilitates voting from anywhere at any time within the voting time, and that the voter will not have to go to the polling station where sometimes violence and intimidation occurs- influences positive perceptions towards the use of mobile phones for voting.

The second most influential design strategy was ease of use (easy to learn and operate). This had an average of 82% for the two studies. The study found out that ease of use, easy to learn, and familiar interfaces were of such a great significance in influencing positive perceptions towards the use of mobile phones for voting in Uganda. That means that simpler steps need to be used, for example, Amazon.com's "one-click shopping" uses reduction technique: it retains user details obtained during registration, which is later used to reduce the number of steps necessary to buy a product [23]. Therefore, when designing for a product or service, practitioners should note that, in a developing country setting, keeping task processes easy and short is of great importance to the product success.

The third most influential strategy was the Demo. A demonstration on how the system worked made participants feel confident about the system. The study revealed that an average of 77% of the participants in the two studies said that the demo was useful in building their confidence and created a positive perception about the future systems. Therefore, practitioners should note that prospective customers or clients should always be provided with an opportunity to try out the product before they buy it. This resonates with the famous Chinese verb " *Tell me and I'll forget; show me and I may remember; involve me and I'll understand"* A survey conducted in the United States found out that about 58% of the participants would buy an item if they tried it out [37].

The fourth most influential strategy was customization, with an average of 72%. Customizing the product or service to the customer's needs and preferences is highly recommended. For example, the name and language, has a potential to positively influence the customer to purchase a product. Cialdini [20] agrees and observes that having a customer's name on the product increases the chance of buying the product. A study carried out in the United States found out that there was a higher chance (20% more) that a customer would pay more for a customized product [38].

The fifth influential design strategy was reciprocity, with an average of 70%. Rewards do influence positive responses. A study in the United States found out that a waiter's tips increased by 23% when waiters gave guests two mints as they waited for their order [20]. Therefore, there is need to consider rewarding customers for using a product or service. These rewards can be inform of bonuses, annual recognition or any other favours the business can give its clients.

## 6 Conclusions

As stated earlier, the major purpose of this study was to assess the extent each of the six strategies would influence positive perceptions and responses towards the use of mobile phones for voting in a developing country like Uganda. There was very little difference between the expert review and user studies. During the study, the experts were provided with questionnaires, no real mobile phone voting system was given to the respondents to try out. The experts were asked to imagine a mobile phone voting system, and give feedback on the extent each of the design strategies would influence positive perception and responses toward the use of mobile phones for voting, and limitations of each requirement. There was an average of 78% agreement by experts that the design strategies would influence positive perceptions and responses towards the use of mobile phones for voting. On the other hand, when the same strategies were used to develop a mobile phone voting prototype and piloted/ tested with users in user studies, there was a 72% agreement that the requirements could influence positive perceptions and responses towards the use of mobile phones for voting. Practitioners in other fields can borrow a leaf from this study and embrace the various design strategies discussed in this paper, and use them to design for acceptance by positively influencing perceptions and responses for their various products and services.

## References

1. M. Bratton, Voting and democratic citizenship in Africa (2013) ISBN: 978-1-58826-894-5 hc
2. M. Kuenzi, G.M.S. Lambright, Who votes in Africa? An examination of electoral turnout in 10 African countries, Working Paper No. 51 (2005)
3. A. Lijphart, Unequal participation: Democracy's unresolved dilemma. Am. Polit. Sci. Rev. **91**, 1–14 (1997)
4. C. Vergne, Turnout in developing countries: The effect of mass media on national voter participation, (2009)
5. C.E. Petersen, Uganda: Presidential, Parliamentary and Local Council Elections, (2006)
6. EU–European Union Election Observation Mission, Uganda, general elections, preliminary statement Uganda 2011 elections: Improvements marred by avoidable failures, 20 February 2011, Kampala
7. O. Oola, Insight on conflict: Uganda's 2011 election: Any lessons for local peace builder (2011), http://www.insightonconflict.org/2011/03/uganda-2011-election-lesson-for-local-peacebuilders/
8. IIDEA-International Institute for Democracy and Electoral Assistance, Strömsborg, SE-103 34 Stockholm, Sweden (2014)/15
9. P. Collier, C.P. Vicente, Violence, bribery, and fraud: The political economy of elections in Sub-Saharan Africa draft, October (2010)
10. C.E. Lallana, mGovernment and eDemocracy in conjunction with mGovernment event (2007). www.worldbank.org/edevelopment/mgov2. Accessed August 2016
11. A.R. Zaharov Can e-voting problems be solved by SMS text messaging? (2017)
12. C.K. Ayo, U.O. Ekong, N.A. Ikhu-Omoregbe, V.A. Ekong, M-voting implementation: The issues and trends (2011).. https://www.academia.edu/610732/m-Voting_Implementation_The_Issues_and_Trends. Accessed January 2016
13. E. Eilu, R. Baguma, Designing a reality fit M-voting. *Presented at the 7th International Conference on Theory and Practice of Electronic Governance*, 22–25 October, (2013). Seoul, Korea. ACM, 978-1-4503-2456-4/
14. R. Kailasam, M-governance … leveraging mobile technology to extend the reach of e-governance. *Proceedings of the TRAI conference on Mobile Applications for Inclusive Growth and Sustainable Development*
15. N.K. Sin, A. Sulaiman, A.H.S. Zolait, Analysis of mobile users' perception towards SMS voting. Comm. IBIMA **1**(1), 182–190 (2008)
16. T. Yogasara, V. Popovic, B. Kraal, M. Chamorro-Koc, Anticipating user experience with a desired product, (2012), The AUX framework
17. P.M.A. Desmet, P. Hekkert, Framework of product experience. Int. J. Des. **1**(1), 57–66 (2007)

18. M. Pakanen, Visual design examples in the evaluation of anticipated user experience at the early phases of research and development, (2015)
19. E.L.C. Law, V. Roto, M. Hassenzahl, A.P.O.S. Vermeeren, J. Kort, Understanding, scoping and defining user xperience: A survey approach; CHI (2009) User experience
20. R.B. Cialdini, Influence: The psychology of persuasion. HarperCollins Publishers Inc. (2007) New York 10022
21. B.J. Fogg, Persuasive technology: Using computers to change what we think and do. Morgan Kaufmann (2002)
22. J. DeMers, The importance of social validation in online marketing (2015), *Forbes*, http://www.forbes.com/sites/jaysondemers/2015/02/19/the-importance-of-social-validation-in-online-marketing/
23. R. Khaled, Culturally-relevant persuasive technology (2008) PhD Thesis
24. A.L.D. Slatten, An application and extension of the technology acceptance model to nonprofit certification. J. Nonprofit Manag. **14** (2010)
25. J.C. Riedel, V.C. Fransoo, S. Wiers, Modelling Dynamics in Decision Support Systems (2006)
26. R. Baguma, Accessible web design through web accessibility guidelines, (2009) PhD Thesis
27. Microsoft Cognitive Walk-through Process. (2012). http://msdn.microsoft.com/en-us/library/windows/desktop/bb189003.aspx. Accessed June 2016
28. M. Tory, T. Moller, Evaluating visualizations: Do expert reviews work? IEEE Comp. Grap. Appl. **25**(5), 8–11 (2005)
29. A. Inglis, Quality assurance in education: Approaches to the validation of quality frameworks for e-learning. **16**(4), 347–362 (2008)
30. T. Lang, M. Zrobok, Usability testing versus expert views: a comparison of usability evaluation methods, (2016), Retrieved April 2018 from https://www.peakxd.com.au/articles/usability-testing-versus-expert-views
31. L. Gegner, M. Runonen, For what it is worth; Anticipated experience evaluation. Proceedings of 8th International Design and Emotion Conference London (2012) Central Saint Martins College of Arts & Design, 11–14 September
32. R. Macefield, How to specify the participant group size for usability studies: A practitioner's guide. J. Usab. Stud. **5**(1), 34–45 (2009)
33. L. Faulkner, Beyond the five-user assumption: Benefits of increased sample sizes in usability testing. Behav. Res. Methods Instrum. Comput. **35**, 3 (2003)
34. L. Arhippainen, Capturing user experience for product design. IRIS26, the 26th Information Systems Research Seminar in Scandinavia. (2003) Porvoo, Finland 9–12.8.2003
35. K. Battarbee, T. Mattelmäki, Meaningful product relationships. Proceedings of design and emotion conference 1–3 July (2002) Loughborough, England
36. P. Dholakiya, Want conversions? Start with user-friendly, useful landing pages (2013). https://blog.kissmetrics.com/want-conversions/ Accessed April 2016
37. S. Lovell, 'Try before you buy' – why brands should use product demonstration? (2015), http://gottabemarketing.co.uk/try-before-you-buy-product-deminstration/. Accessed June 2016
38. E. Spaulding, C. Perry, Making it personal: Rules for success in product customization, (2013), Retrieved March 2018 from http://www.bain.com/publications/articles/making-it-personal-rules-for-success-in-product-customization.aspx

# Does Technological Context Support Academic Entrepreneurship Activities in Algeria?

**Soraya Sedkaoui, Mounia Khelfaoui, and Nadjat Kadi**

**Abstract** This paper aims to examine the effect of technological context on academic entrepreneurial activities in Algeria. For this, we adopted a structural perspective taking into consideration a systemic framework. It provides also an opportunity to examine the nature of the entrepreneurial activities in Algerian universities and identify the technical elements which may influence academic entrepreneurs. To this end, a survey methodology was used to collect data. The research results demonstrated that academic entrepreneurs in Algeria have reacted positively to patents production and consulting activities, more than on the entrepreneurial activities related to enterprise creation. The results also show that R&D activities can be counted as one of the missing gaps in fulfilling academic entrepreneurial activities. The results show also that Algerian universities must develop their internal ICT infrastructures and gain appropriate systems and technological skills to increase the level of academic entrepreneurship. Besides, at any level of the university, academics must be more creative and entrepreneurial.

**Keywords** Academic entrepreneurship · Technological context · Algerian University · Academic entrepreneur

S. Sedkaoui (✉)
Department of Economics, University of Khemis Miliana, Khemis Miliana, Algeria

Industry, Organizational Development, and Innovation Laboratory, Khemis Miliana, Algeria

SRY Consulting, Montpellier, France
e-mail: s.sedkaoui@univ-dbkm.dz

M. Khelfaoui
Department of Economics, University of Khemis Miliana, Khemis Miliana, Algeria

Industry, Organizational Development, and Innovation Laboratory, Khemis Miliana, Algeria

N. Kadi
Department of Economics, University of Khemis Miliana, Khemis Miliana, Algeria

Algerian Digital Economy Laboratory, Khemis Miliana, Algeria



E. Eilu et al. (eds.), *Digital Literacy and Socio-Cultural Acceptance of ICT in Developing Countries*, https://doi.org/10.1007/978-3-030-61089-0_6

## 1 Introduction

The current context indicates that the traditional role of the university, namely teaching and research, is being challenged and now it is extended to a socioeconomic development project. Universities have changed their position in economic growth in time and have gone beyond their traditional teaching and research activities. It is a recent change in the university's traditional model, where knowledge transfer and collaboration between industrials and universities plays a key role in the tasks and responsibilities of the university.

Several activities carried out by universities across the world demonstrate that they are working to become more entrepreneurial. This is to say that the universities are going through a major transformation that attests to a transition towards what is called an '*entrepreneurial university*' in the sense of Clark [1, 2] and Etzkowitz [3]. This new definition of the university has a positive influence, not only in the academic world but also in the society as a whole.

Universities are the primary source of new knowledge and play a fundamental role in creating and growing an economy geared towards entrepreneurship. They are expected to develop needed solutions to the industry and the society, by leveraging researchers' knowledge and expertise. Therefore, the role of the university in a technological and regional development environment is to be a *knowledge hub*. Such a *hub* university creates intermediaries that span the boundary between the academic and business worlds.

Moreover, universities are also exposed to the universe of digitalization. With the advent of IT tools, it is therefore difficult to ignore the role of digitalization in maintaining an organization's competitive advantage. It will provide innovators and entrepreneurs with exciting opportunities [4–7].

The effect of digital technologies on various academic activities, described as academic entrepreneurship [8–11], has been intense and encompassing activities such as industry research collaborations, patent applications, creative ideas commercialization, transfer of technology, and more.

Algerian universities are not excepted by this changing context [7]. Their roles lie in fostering the rapid diffusion of technology, increasing the interactions between the academic and business world, and enabling the dissemination of tacit knowledge that is essential in the innovation process.

This entrepreneurship concept needs to explore the strategic technological context that supports entrepreneurial activities and determine its influence on the academic entrepreneurs' intentions. Especially, with the rising number of entrepreneurship activities which reach more than 1.2 million entrepreneurs in 2019 according to the Algerian Statistical Information Report [12]. Therefore, there's the need to actively improve the entrepreneurial capacity within universities.

For this, this paper aims to analyze the state of the art of academic entrepreneurship in Algeria. Its main purpose is to identify the role that digitalization can play in establishing a positive academic entrepreneurship context. In other words, this paper seeks to explore the effects of the technological context on academic entrepreneurship.

This study will be also an opportunity to analyze the implications and challenges of digital context on academic entrepreneurship in Algeria and provide a guideline for enhancing a variety of available opportunities. To this end, the main research objective can be formulated as follows:

> Which practices of academic entrepreneurship are more affected by the Algerian technological context?
>
> What forms of research collaboration between the academic and business world are best served by digital technologies?

The remainder of this chapter is structured as follows. Section 2 reviews the relevant literature on academic entrepreneurship. Section 3 highlights the opportunities and challenges of academic entrepreneurship for Algerian universities and presents the different factors that can support its various activities. Section 4 describes the conceptual model and the adopted methodology to support the research hypothesis. Section 5 outlines and displays the results and discussion. Section 6 ends with conclusions, implications, limitations, and further research directions.

## 2 Academic Entrepreneurship: A Literature Review

Entrepreneurship is one of the key driver elements for a healthy productive economy. Since the seminal work of Schumpeter [13, 14], entrepreneurship has been regarded as a positive driving force for regional economic growth and development [15–18], and [19].

Also, it is one of the major channels for performing the universities' third missions: socioeconomic support functions [20]. Etzkowitz [21] claimed that entrepreneurship within universities is not an unusual trend.

Entrepreneurial ventures have increased since the 1970s when biomedical industry researchers started to explore ways to exploit their work by forming or involving private companies. This characterized, according to Stuart and Ding [22], a new entrepreneurial division called academic entrepreneurship.

Based on such initiatives, many new technology-based companies slowly gain a position for academic entrepreneurship on the growth agenda of policymakers.

Academic entrepreneurship is defined as [23]:

> The transformation of advanced scientific research into commercial products, with a focus on high-technology ventures

As several new technology-based firms have grown from such initiatives, academic entrepreneurship is gradually gaining a place on policymakers' development agenda. This entrepreneurial branch helps academic entrepreneurs to build commercial value out of scientific and technological knowledge [24].

Universities' evolving roles in economic growth have received great attention from academics and have been studied for decades in developed countries in several disciplines. Many researchers tried to define the essence of academic entrepreneurship.

The analysis of empirical studies shows that the entrepreneurial university model has its origins in the US, with the launching of the *Bayh-Dole Act* or "*University and Small Business Patent Procedures Act*" (12/12/1980). This act allowed universities to patent inventions made by their researchers, even if the research has been fully funded by the government.

The study carried out by Shane [25], in the American university context, revealed that after the *Bayh-Dole Act*, the patenting of university research increased in general in the efficient scientific fields and carrying business opportunities such as the basic sciences sector and, especially, the biomedical inventions sector.

It should be noted that from the 1970s, several US universities, such as Harvard, Stanford, or MIT, began to adopt internal policies and strategies to facilitate and support the commercialization of scientific research outputs.

From 1970 to nowadays, the entrepreneurial culture initiated in American universities seems to have spread in many countries or geographic areas, and several researchers have carried out pioneering studies on academic entrepreneurship.

The first works on the university's third mission date from the late twentieth century, and only in 1995 and 2000 the term was included in the implementation of the Europe of Knowledge 2020 [26, 27], and [28].

Globally, the literature under the heading "Academic Entrepreneurship" looks at activities, people, programs, and organizations that serve the economic exploitation of scientific knowledge.

By analyzing the number of the scientific publications indexed in Web of Science (all database) during the last decade, and containing the concepts 'Academic Entrepreneurship', we can notice that more than 2500 publications were made as shown in the following figure.

Figure 1 reflects the interest of Academic Entrepreneurship among the scientific community and shows that the number of scientific contributions has been increasing. Also, the results of this analysis show that the publications were maximal in the field of Business Economics (46.40%), educational research (20.76%), Business, and Engineering (14.55%). It is worth noting that entrepreneurial university has attracted less attention in other research areas such as Social Sciences, Environmental Sciences, and Computer Science.

The next sections will provide some keys to understand the reasons for adopting such an attitude of openness towards the industry and to define the technological context in which it operates.

## 3 Academic Entrepreneurship for Algerian Universities: Opportunities and Challenges

Universities play an important role in the growth and development of an entrepreneurial ecosystem [7], as they constitute a vital source of knowledge and continually regenerate the stock of students and scientists.

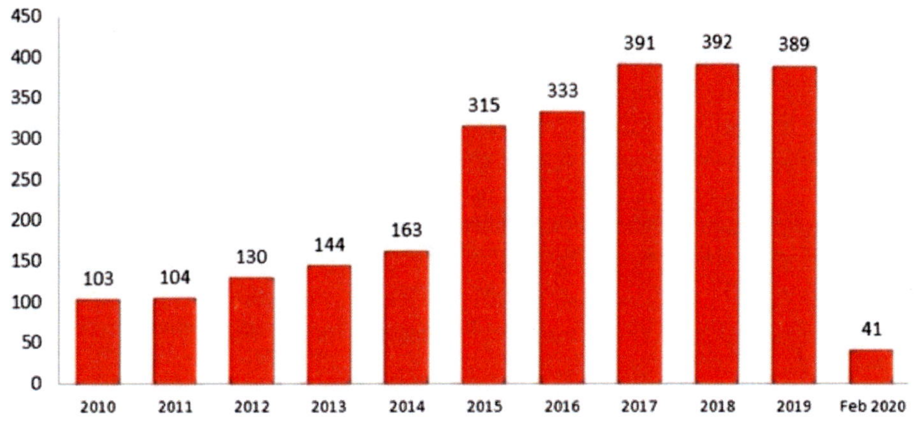


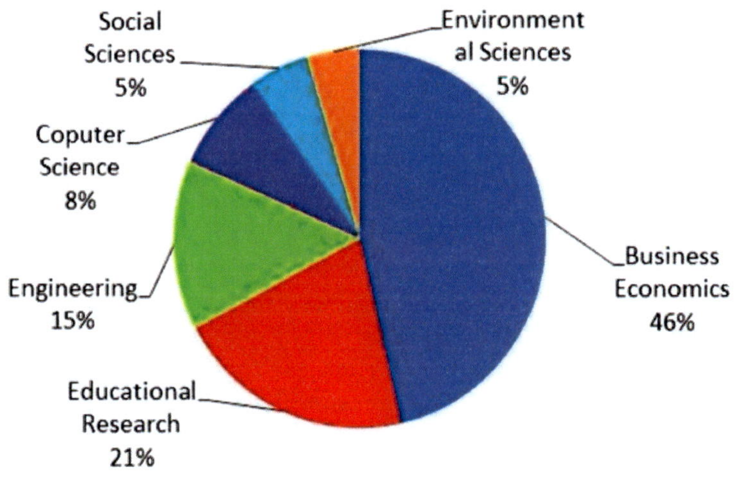


**Fig. 1** Number of publications and % of academic entrepreneurship literature by research areas (2010–2020)

The idea of academic entrepreneurship is regarded to be an important technological and regional development factor for the national innovation system with multiple stakeholders such as the government, local authorities, and the business sector.

Academic entrepreneurship affects economic development because it creates product innovation ideas, stimulates investment in new technology, provides opportunities for knowledge workers, and enables local high-tech clusters to grow.

The good practices that guided universities' first steps in implementing policies to support academic entrepreneurship were essentially drawn from the experience of leading American universities. However, there are significant disparities in universities. Therefore, the strategies that have worked for some do not lead others to similar results. In addition to their internal differences, universities are distinguished by the specific environment in which they operate.

The effectiveness of policies supporting academic entrepreneurship is impacted by the context, which varies in terms of business activities' development, national and regional innovation systems, public intellectual property policies, public funding, the level of industrial network growth, and the importance of innovation and R&D in the university environment.

Developed countries understand the economic and strategic benefits of academic entrepreneurship. In recent years many policy-makers around the world have been discussing issues related to this type of entrepreneurship and knowledge commercialization [29].

Several activities can make use of the university's expertise commercially, such as [30, 31]:

- R&D contracts,
- Patents,
- Licenses,
- Consulting,
- The creation of enterprises.

Among the previous activities, the creation of enterprises has emerged as a key element in commercializing the university's knowledge.

Despite the progress in its domestic promotion, the Algerian economy remains strongly dependent on the oil industry, highly vulnerable to external shocks, and unemployment growth. As a way to promote and maintain large-scale growth, Algeria must facilitate economic diversification and promote structural reforms.

The high level of uncertainty that has affected the business climate in the country has earned a lot of attention to the entrepreneurship over the past decade [7]. To enhance the business and economic development, and improve the entrepreneurship ecosystem, there is a need for a transformation of the university research results into commercial outcomes. This is possible through academic entrepreneurship.

Besides their two traditional missions (the production of new knowledge through research activities and the dissemination of this knowledge through teaching and publication activities), Algerian universities are called upon to pursue a third mission: transfer of this knowledge to the industrial world through economic development activities.

The awareness of students and researchers of the idea that creating business offers career opportunities seems to have a positive effect in reducing the gap between the academic and the industrial world [32], but it is not enough to go through all the stages of creating an academic spin-off in Algeria. This creation is supported by internal policies, by a set of external actors, and by the efforts of articulation between internal and external mechanisms.

To understand the scope of academic entrepreneurship in Algeria, it is important to place it in the broader context of relations between universities and enterprises.

While this strategy will be very successful for entrepreneurial orientation in universities, Algerian universities face significant challenges in terms of achieving their strategies and surviving today's shifting and developing environment. The presence of a broad distance between what happened in practice and what exists in reality is one of the major challenges.

**Table 1** Factors influencing academic entrepreneurship in Algeria

| Factors | |
|---|---|
| Large dependency on the oil industry | Large dependency on the oil industry |
| Organizational | Resources (human, physical and commercial) and capabilities (networks, etc.) |
| Financial | Difficulties securing bank loans or finding risk-capital, etc. |
| Technological | ICT infrastructure, internet connectivity, technology transfer, etc. |
| Others | Government policies, centralizations of incubators around the capital, accompaniment program, equipment, facilities, etc. |

It should also be mentioned that, despite efforts to lay the foundations of the information society and improve the ICT infrastructure and the core components of the digital economy, the country suffers from the digital gap problem due to several factors that have resulted in inadequate ICT implementation.

In Algerian universities, the challenges of academic entrepreneurship can be divided into financial, organizational, and technological factors, as illustrated in Table 1 [7].

Despite the importance of all these factors, this study focuses on analyzing the internal and external technological context and how it affects academic entrepreneurship among Algerian universities. This technological factor can deeply affect the innovative behavior of academic entrepreneurs.

It should be mentioned that in the previous studies on academic entrepreneurship, technological factor was not given enough attention, and identified as a main factor of interest for the success of the academic entrepreneurship process.

This study adopted this particular factor to examine its role in academic entrepreneurship within Algerian universities.

Literature-based on the study of the increasing tendency of academic entrepreneurship has underlined that technology transfer plays a significant role as a driver of the processes of knowledge valorization and intellectual property.

Such internal and external technical contexts can also constitute obstacles for entrepreneurial universities due to the inherent existence of Algeria's higher education system, lack of ICT infrastructure, existing technology parks, technological integration, equipment, etc.

Based on the above points, the following sections discuss the impacts that the technological context has on academic entrepreneurship in Algeria.

## 4 Methodology

### *4.1 Research Framework and Hypothesis*

Academic entrepreneurship may be adopted by individuals, operating independently or as part of a university system that can launch new projects or fostering innovation, within or outside the university via research centers, technology parks, etc. Academic entrepreneurship internally or externally carried out in and outside the university, includes innovation, business ventures, and strategic renewal.

Algeria is an entrepreneurship ecosystem that undergoes major changes and has enormous potentials that allow the development of academic entrepreneurship. To provide an appropriate theoretical structure within the current research, and analyze academic entrepreneurship in Algeria, we classified technological factors as follows (Fig. 2).

The conceptual model proposed to analyze the effect of digital context on academic entrepreneurship activities and the commercialization of the university knowledge includes five factors:

- ICT infrastructure: It covers all the devices, network connectivity, and technologies available at the university. This includes software, hardware, and website.
- Existing IT skills: Skills and knowledge related to the use, management, and technology development (software, applications, or devices).
- Technology parks: Open spaces related to a university to serve as business incubators.
- ICT sector: The ICT sector consists of industries whose main activities are linked to use, develop, product, and commercialize new technology.
- Technology transfer facilities: It includes all the facilities from authorities, universities, and institutions facilities to ensure that academic skills, knowledge, technologies, methods, etc., are available to a broader variety of users to develop new products s and services.

Mainly, it is presumed that all elements relevant to the Algerian technological background that affect academic entrepreneurship are classified into subcategories of internal (inside the universities) and external (outside the universities) influences. These subcategories will provide a clear description of the conditions and structures under which this research was carried out, for studying academic entrepreneurship. Therefore, we mainly hypothesize:

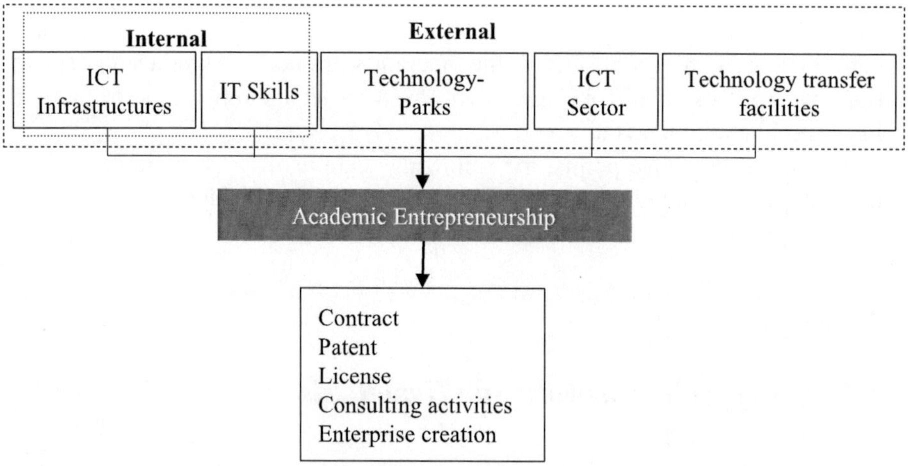


**Fig. 2** Conceptual model

*H1: There is a positive relationship between academic entrepreneurship activities and technological context.*

This hypothesis is included in the following sub-hypotheses:

*H1a: There is a positive relationship between academic entrepreneurship activities and ICT infrastructures.*
*H1b: There is a positive relationship between academic entrepreneurship activities and IT skills.*
*H1c: There is a positive relationship between academic entrepreneurship activities and technology parks.*
*H1d: There is a positive relationship between academic entrepreneurship activities and ICT sector.*
*H1e: There is a positive relationship between academic entrepreneurship activities and technology transfer facilities.*

## 4.2 Data Collection and Academic Entrepreneurs Profiles

To better analyze the research hypothesis and improve the accuracy of our research results, this study adopted a survey method to collect data, using Google forms. A questionnaire, containing 18 questions, was administered to 68 academics in Algerian universities were given a questionnaire.

The questionnaire was designed primarily to examine the Algerian academic entrepreneurship environment and to identify both the design characteristics of academic entrepreneurs and the effects of technological context on various academic entrepreneurship activities.

The questionnaire was first tested to ensure the simplicity, accuracy, and validity of survey content and measurement scales. Some elements have been updated based on the results of this study.

The questionnaire was conducted from November to December 2019, through an e-mail sent to 200 randomly sampled academics, of which 68 were successfully returned, resulting in a reaction rate of 34%.

The questionnaire grouped these five factors by adopting a 5 point Likert-type scale which ranges from '1- Strongly Disagree' to '5- Strongly Agree'. For other items, the questions are coded from 1 to 2 (yes or no).

The inter-item reliability of each variable has been assessed through Cronbach alpha (a measure of internal consistency or how closely connected a group of items is). Table 2 shows that the validity and the reliability criteria of all items were satisfied. The distribution of these variables is presented in Appendix 1.

Academics were tasked to select the most significant factors about academic entrepreneurship and technological context. To this end, the questionnaire was divided into two sections to capture the necessary information about academics including gender, age, qualification, etc., and the different information about the impact of the various factors on academic entrepreneurship activities.

**Table 2** Variables reliability score based on Cronbach's Alpha

| Variables | Means | St Dev | Cronbach's Alpha |
|---|---|---|---|
| ICT infrastructures | 1.93 | 0.739 | 0.754 |
| IT skills | 3.00 | 1.007 | 0.688 |
| Technology-parks | 1.59 | 0.604 | 0.726 |
| ICT sector | 2.32 | 0.969 | 0.698 |
| Technology transfer facilities | 1.85 | 0.833 | 0.713 |

**Table 3** Academics characteristics details (%)

| Characteristics | | University | University Centers | Higher Institutions |
|---|---|---|---|---|
| Gender | *M* | 53.8 | 71.4 | 68.5 |
| | *F* | 46.2 | 28.6 | 31.5 |
| Age | *<35* | 9.1 | 22.4 | 7.9 |
| | *<35–40>* | 25.8 | 47.3 | 14.3 |
| | *<40–45>* | 34.3 | 15.1 | 33.9 |
| | *<45–50>* | 22.6 | 8.8 | 28.7 |
| | *>50* | 8.2 | 6.4 | 15.2 |
| Status | Permanent | 82.3 | 61.4 | 86.7 |
| | Contract | 17.7 | 38.6 | 13.3 |
| Qualification | Ph.D. | 69.4 | 54 | 76.8 |
| | Master | 27.3 | 34.8 | 17.5 |
| | Other | 3.3 | 11.2 | 5.7 |
| Designation | Ph.D. student | 9.5 | 6.4 | 10.2 |
| | Assistant | 21.9 | 32.7 | 19 |
| | Lecturer | 41.2 | 51.6 | 39.7 |
| | Professor | 27.4 | 9.3 | 31.1 |

The characteristics of the 68 academics are summarized in Table 3.

The results show that the surveyed academics in universities (69.4%), university centers (54%), and higher institutions (76.8%) have a Ph.D. degree. They are predominantly male with more than 53% for the three categories. Results illustrate also that among 68 academics the majority is permanent and have between 35 and 50 years old. As shown in Table 3, more than 39% of academics in the three categories are lecturers.

# 5 Results and Discussion

## *5.1 Hypothesis Analysis*

After detailing the sample's general characteristics, we will in the following present the results of the analysis of questions related to the subject of this paper and organized the research hypothesis.

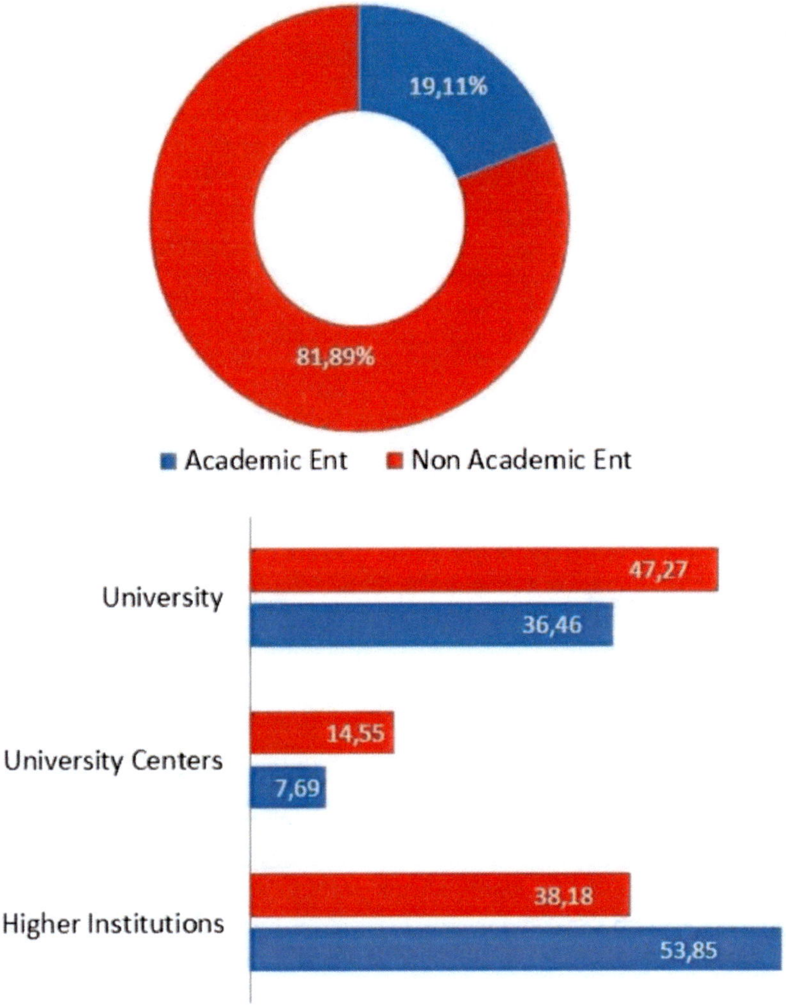


**Fig. 3** Distribution by academic entrepreneurship consideration

Before evaluating the effect of the different technological context elements, we need to provide some information about the consideration of entrepreneurship activities by Algerian academics. For this, some questions have been addressed to the 68 academics to divide them into two categories: academic entrepreneur and nonacademic entrepreneur. The results of this analysis are illustrated in Fig. 3.

Regarding academic entrepreneurship consideration, results indicate that only 13 academics have confirmed that they are academic entrepreneurs. This presents only 19.11% of our sample. Among them, 38.46% from universities, 53.85% from higher institutions, and 14.55 only from university centers.

By analyzing the collected data to identify the characteristics of academic entrepreneurs, we can see (Table 4) that among 13 academic entrepreneurs all of them have a Ph.D. degree and are permanent.

**Table 4** Characteristics of the 13 academic entrepreneurs

| Characteristics | | Academic Entrepreneurs | % |
|---|---|---|---|
| Gender | *M* | 9 | 69.23 |
| | *F* | 4 | 30.77 |
| Age | *<35–40>* | 4 | 30.76 |
| | *<40–45>* | 7 | 53.84 |
| | *<45–50>* | 1 | 7.70 |
| | *>50* | 1 | 7.70 |
| Designation | Lecturer | 7 | 53.84 |
| | Professor | 6 | 46.16 |

We can notice also that 69.23% are male. Also, more than 53% are lecturers, the rest are professors. But, no assistant or a Ph.D. student is registered as an academic entrepreneur.

The results in Table 4 also show that academics under 35 have not engaged in any entrepreneurial activity contrary to those who are over 35 years old.

We notice that there is a relationship between a researcher's age and the realization of academic entrepreneurship activities. In other words, more experienced academics may participate in entrepreneurial activities [33] related to, for example, consultancy activities because of their visibility and reputation they gained throughout their careers, in addition to research funding and patent filing initiatives.

Concerning the gender, results affirm partially the result of previous research studies, as men have more advantage in academic entrepreneurship [33, 34], and [35], but their percentage is still higher among non-academic entrepreneurs (63.63%). The academic qualification and designation elements should also be considered, since from the 13 entrepreneurs, lecturer and professors are the most oriented entrepreneurial academics.

It should be noticed that due to the time and energy devoted to the family, women have a great difficulty in pursuing an academic entrepreneurship career [36, 37].

Ultimately, Fig. 4 confirms the results of previous studies and stands out the academic entrepreneurship practices in the applied Sciences field comparing to other fields.

But, *in which entrepreneurial university activities have these 13 university entrepreneurs participated?*

The answer is that Algerian academics are not very experienced in academic entrepreneurship, as they are primarily focused on other methods to transfer and exploit their knowledge.

The results detailed in Table 5 confirm that academic entrepreneurship is not a common concept in Algerian universities.

Like in the other developing countries, Algerian universities typically contain bureaucratic units based mostly on the protection of technology and other forms of transfer such as consulting activities (84.61%) and patents (76.92%), then on R&D contracts (7.69%) or the creation of enterprises (15.38%), even if many authors have shown the relevance of enterprise creation in improving innovation and job creation [20, 38].

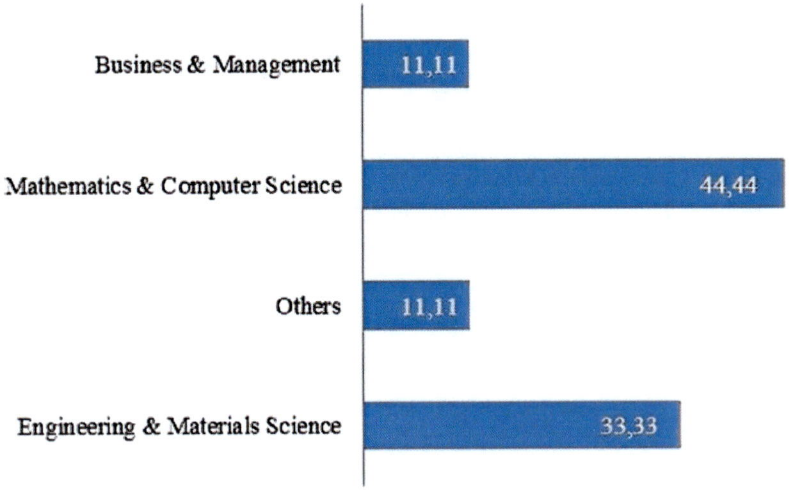


**Fig. 4** Distribution of academic entrepreneurs by fields

**Table 5** Academic entrepreneurship consideration by activities

| Activities | Nbr | % |
|---|---|---|
| R&D contract | 1 | 7.69 |
| Patent | 10 | 76.92 |
| License | 0 | 0.00 |
| Consulting | 11 | 84.61 |
| Enterprise creation | 2 | 15.38 |

This conducted academics to focus more on commercializing their knowledge through consulting and patents rather than in other academic entrepreneurship activities.

Nevertheless, the interest in encouraging entrepreneurial performance at universities has recently been growing. Since universities are committed to designing policies, devote resources to promote the creation of enterprises, through their Entrepreneurship House (*Dar El Mokawalatiya*), and why not improving R&D activities.

By analyzing the correlations, we found some moderately low (positive or negative) relationships between academic entrepreneurship and the different technological factors selected.

As shown in Table 6, the Pearson's coefficient for the correlation between each factor and academic entrepreneurship was positive, expect the technology parks (0.165).

The results of the correlation analysis, which varies from positive to negative correlation (see the measurement scale in Fig. 5), the values of the coefficient were rather medium; that means that there is not a strong relationship between the factors and entrepreneurial university. The strongest correlation is with the Technology transfer Facilities (0.230), followed by ICT Sector (0.163).

**Table 6** Assessment of correlation and the results of sub-hypotheses validity

| Factors | Correlation | Effect | Sub-hypothesis | Results |
|---|---|---|---|---|
| ICT infrastructures | 0.002 | Positive<br>Very weak | $H1_a$ | Accepted |
| IT skills | 0.037 | Positive<br>Very weak | $H1_b$ | Accepted |
| Technology-parks | −0.165 | Negative<br>Weak | $H1_c$ | Rejected |
| ICT sector | 0.163 | Positive<br>Weak | $H1_d$ | Accepted |
| Technology transfer facilities | 0.230 | Positive<br>Medium | H1e | Accepted |

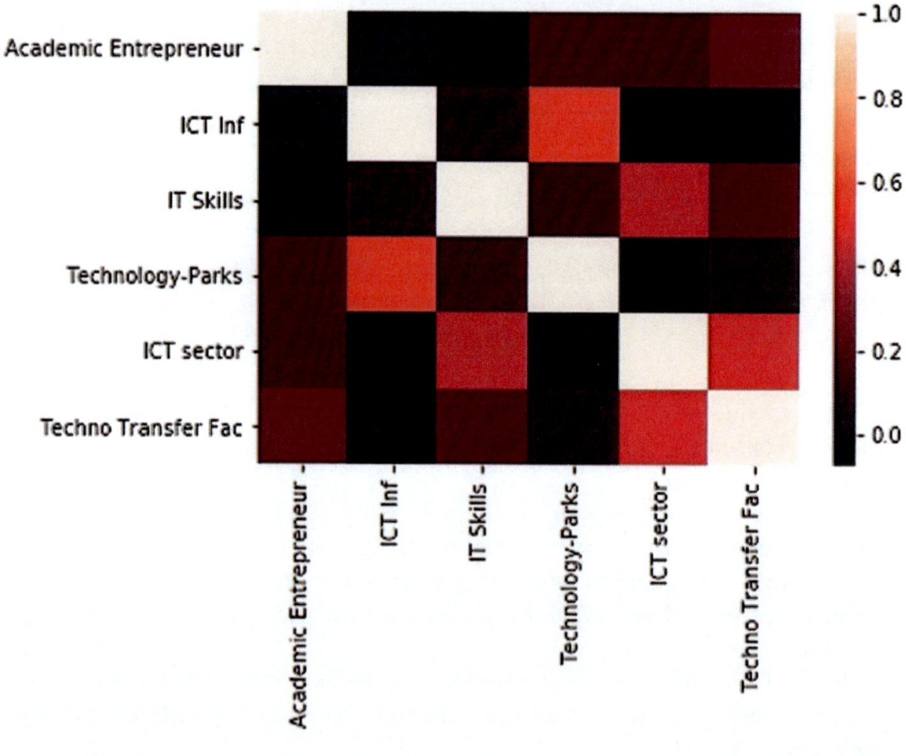


**Fig. 5** Correlation matrix plot

However, the other variables, internal and external ICT infrastructures and IT Skills, are weakly correlated with academic entrepreneurship (0.002) and (0.037) respectively.

The results in Table 6 show that four of the relationships supposed in the previous section are acceptable. This is to say, that there is a positive relationship between academic entrepreneurship and availability ICT infrastructures, the availability of

IT skills, the level of development of ICT sector, and the facilities that academics find to transfer their knowledge and developed technology.

Moreover, the negative value of technology parks cannot support the sub hypothesis H1c.

The correlation analysis conducted to test the sub hypotheses showed that each technological factor influences the academic entrepreneurship activities in Algeria. However, the relationship between academic entrepreneurship and the five selected factors of the Algerian technological context is not strong. The correlations scores indicate a low to moderate the relationship between entrepreneurial university activities and the technological context.

Results indicate that the Algerian universities must improve their internal ICT infrastructure and gain adequate technical and systems competencies to increase academic entrepreneurship. Therefore, academics must be more innovative and entrepreneurial.

## 5.2 *Discussion*

Based on the results drawn from the test of the sub hypotheses, descriptive statistics, and correlations analysis, the paper provided the adequacy of the technological context to measure the effect of its different factors (internal and external) in stimulating the level of academic entrepreneurship activities in Algerian universities.

To analyze deeply the relationship between the five selected factors and the university entrepreneurship, cluster analysis was used to identify clusters of Algerian academic entrepreneurs with similar characteristics; with the condition of the highly differentiating of the identified clusters. Besides, concerning the variables which have been applied for the creation of the cluster, each group of Algerian academic entrepreneur is homogeneous.

This supervised multivariate algorithm determines whether certain distinct clusters of academic entrepreneurs in Algerian universities that meet the following two criteria can be defined when some relevant characteristics are taken into consideration.

As we have quantitative variables, the number of clusters is not known, and the number of cases is not too large (68), for this study, a cluster-hierarchical approach is an effective statistical method.

It should be noticed that, to classify clusters, the Ward method and the Euclidean distance have been applied since these two methods are the most widely used methods for developing such analysis [39].

The clusters were labeled based on the entrepreneurial university activities and the academic entrepreneurs' profiles. These activities are defined as activities that have at least $n > 0$ links of at least a given strength according to the underlying similarity measure.

Since academic entrepreneurship activities are linked with enormous academic characteristics in the same cluster, based on the similarity measure used for the

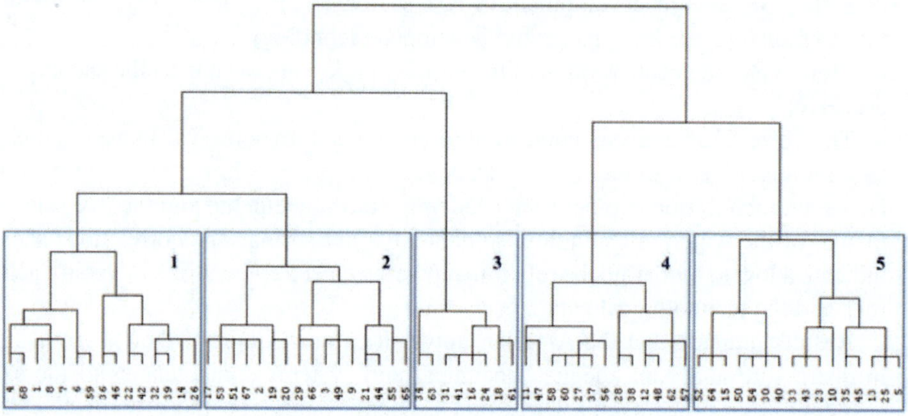


**Fig. 6** Dendrogram: Clusters in Algerian academic entrepreneurs

clustering, they are expected to form the very cognitive nodes of the academic entrepreneur they represent.

Five clusters, or group of an academic entrepreneur, may be distinguished as shown in the dendrogram (see Fig. 6).

A summary of these clusters is detailed in Table 7 which provides a crucial overview based on their characteristics in each cluster, to better examine the profile of the 13 academic entrepreneurs.

The third cluster is not included, because it contains any academic entrepreneurs. The results indicated also that Patents and Consulting activities are the most affected academic entrepreneurship activities by the technological context in Algeria.

We can say also, that the academic entrepreneurs that belong to Cluster 1 have realized one R&D contract, produced four Patents, and participated in five consulting activities. In Cluster 2, academic entrepreneurs have realized also four patents and four consulting strategic activities, counter to Cluster 4 and Cluster 5 where we find only one patent and one consulting activity for each.

Concerning the creation of enterprise, the two academic entrepreneurs that have created companies through academic spinoffs, belong to Cluster 1 and Cluster 2. This is to say that these two clusters gather the different academic entrepreneurship activities realized by academics.

## 6 Conclusion

The social and economic development of modern economies is increasingly shaped by a change in previously industrial relations to a knowledge-intensive service, information, and communication society. In this process of dematerializing economic activities and the associated rise of a knowledge economy, academics are becoming increasingly important.

**Table 7** Clusters of Algerian academic entrepreneurs

| Characteristics | | Clusters | | | |
|---|---|---|---|---|---|
| | | Cluster 1 | Cluster 2 | Cluster 4 | Cluster 5 |
| Gender | *M* | 5 | 3 | – | 1 |
| | *F* | – | 2 | 1 | 1 |
| Age | *<35* | – | – | – | – |
| | *<35–40>* | 1 | 1 | 1 | 1 |
| | *<40–45>* | 4 | 2 | – | 1 |
| | *<45–50>* | – | 1 | – | – |
| | *>50* | – | 1 | – | – |
| Status | Permanent | 5 | 5 | 1 | 2 |
| | Contract | – | – | – | – |
| Qualification | Ph.D. | 5 | 5 | 1 | 2 |
| | Master | – | – | – | – |
| | Other | – | – | – | – |
| Designation | Ph.D. student | – | – | – | – |
| | Assistant | – | – | – | – |
| | Lecturer | 4 | 4 | 1 | – |
| | Professor | 1 | 1 | – | 2 |
| Academic entrepreneurship activities | R&D contract | 1 | – | – | – |
| | Patent | 4 | 4 | 1 | 1 |
| | License | – | – | – | – |
| | Consulting | 5 | 4 | 1 | 1 |
| | Enterprise creation | 1 | 1 | – | – |

Little has been said on the effect of technological context on academic entrepreneurship in the literature. In this paper, the review analysis of the previous studies indicated that most studies did not consider this context. The analysis shows that there were just a few research studies that have discussed the influence of other factors (organizational, human resources, culture, etc., [40] on technology transferor or the academic spinoffs.

This paper, focused on the digitalization strategies to support academic entrepreneurship implemented by universities that operate in an environment that does not have the maturity of *Silicon Valley*. This environment, that we consider not mature, is characterized by an ecosystem that is poorly designed to support entrepreneurial activities.

According to Youtie and Shapira [41], in immature environments, the development of academic entrepreneurship requires a contribution from universities to the construction of this environment. The university then functions as a *knowledge hub* and its policy must be driven by the development of rich interactions and connections between the academic, business, and financial world. The university thus contributes to the development of an ecosystem favorable to economic entrepreneurship.

The rationale for the interest in academic entrepreneurship usually focuses on the economic benefits of the scientific and technological knowledge being commercialized [42]. Examples from Stanford University and Silicon Valley in the United States and University of Cambridge in the United Kingdom are also kept as emulative models.

The current study brought together the components of the Algerian technological context including ICT infrastructures, IT skills, ICT sector, technology parks, and Technology transfer facilities, to analyze their influence on the practice of the various entrepreneurial university activities by academics. Comparing to previous research, the distinctive feature of this analysis is broad of elements able to influence academic entrepreneurship activities in the Algerian technological context.

The study showed the importance of developing the technological context of academic entrepreneurship by its diverse elements' influence in stimulating the activities of academic entrepreneurship of Algerian universities.

As a consequence, we learned more about the characteristics of Algerian academics entrepreneurs and the entrepreneurial activities that they practice. Also, the results demonstrate that the Algerian university system should have flexible structures for establishing and achieving the task of academic entrepreneurship. Moreover, IT competencies in such an environment will foster academics entrepreneurial practices learning in the entrepreneurship ecosystem.

The structure of this study broadened also the analysis of the integration of academic entrepreneurship perspectives that cover the different activities occurring within and outside the university.

The good news is that Algerian universities and other public research institutions can create companies that can promote growth and secure prosperity. But, their importance to the development of cutting-edge technologies for high-tech products must increase further.

The results of this research can help to define the basic elements which must be available for academics that are interested in promoting the performance of entrepreneurial. Such elements can be used to encourage academic authorities to act on the future orientations of Algerian universities in aspects relating to the development of spinoffs, patents, R&D development, etc.

For academic entrepreneurship to be an effective policy in the economic sector, some factors need to be taken into account. Especially at two levels: 'What' and 'How' - which are mutually dependent and intertwined. Both, and especially the latter, need to be pursued in the short or long term with a view to the more adequate promotion of academic entrepreneurship activities. These factors include:

- Transfer culture within the research institutions, which appreciates exploitation activities such as start-ups and which offers scientists incentives and motivation.
- The transparency of start-up processes and clear rules on the use of patents. Besides, successful transfer facilities should not only be a point of contact for those interested but should also actively scout ideas and technology in their research facilities.
- Adopt policies that consolidate the values of ICT-backed academic entrepreneurship activities. This is based on the development of an integrated plan to formu-

late goals and future choices that are central to activating integration into the digital economy.
- Use ICT tools to encourage entrepreneurial skills among academics, and embark on the creation of start-ups dedicated to ICT. To this end, a technological innovation center, meeting international standards and intended for start-ups and academic entrepreneurs with innovative solutions, must be set up. A lively academic start-up culture, in turn, is an important success factor for creating employment opportunities for young academics - and also contributes to an economic and social dynamic from which education systems and (labor) markets can benefit.
- Move towards a more selective support model that must be part of a strategy of seeking increased efficiency.
- Develop the relationship between universities and companies in various sectors, to promote research-based on real business development needs and promote the exploitation of research results in the science system.
- Implement strategies that aim to increase the tendency of academics to utilize to strengthen the transfer of knowledge and technology and the innovative power in general.

These orientations will help to disseminate the research results of Algerian academics to being used by industry and society, so these results will not be limited to the academic environment. This strategy may also help to create new opportunities for Algerian academics for career trajectories, including entrepreneurship. Therefore, positive results on the technological and socio-economic development indicators of the country could be anticipated.

To make valuable contributions to job creation, innovation, and regional prosperity, academic entrepreneurs must be able to convert their technologies into marketable products. Universities and public research institutions should not only think entrepreneurially but also act.

Despite its contribution and results, our analysis is not without limitations. Firstly, our sample may not permit to generalize entirely the characteristics of all academic entrepreneurs, and the technological elements that can stimulate and improve the different entrepreneurial activities they practice. Secondly, regarding the priorities of this study, and due to constraint of time, we probably missed some important aspects of academic entrepreneurship activities in Algerian universities.

Similar research studies are, therefore, suggested in different Algerian universities, considering the other aspects. This can make a major contribution to the gap between academic and real-world practices. A potential research line should examine also how academic entrepreneurs' goals should not only suit the strategic purposes of universities but also with the opportunities offered by these universities and the entire Algerian authorities. Finally, the causality of the relations between the variables studied in this research should be deeply analyzed.

**Acknowledgments** The authors would like to gratefully and sincerely thank all respondents; without the enthusiastic participation of 68 academics this study would never have come into being. Thank you to all of you!

# Appendices

## *Appendix 1: Kernel Density Estimates (KDE) of Variables*

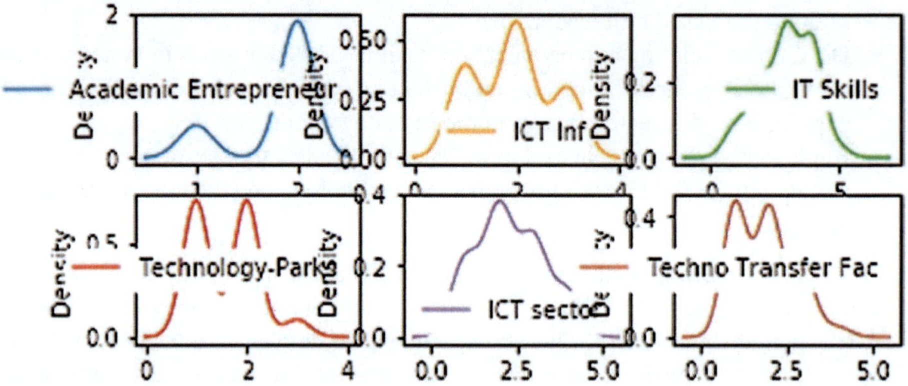


# References

1. B.R. Clark, Delineating the character of the entrepreneurial university. High Educ. Pol. **17**(4), 355–370 (2004)
2. B.R. Clark, The entrepreneurial university: Demand and response. Tert. Educ. Manag. **4**(1), 5–16 (1998)
3. H. Etzkowitz, The evolution of the entrepreneurial university. Int. J. Technol. Glob. **1**(1), 64–77 (2004)
4. B. Cohen, J.E. Amorós, L. Lundyd, The generative potential of emerging technology to support startups and new ecosystems. Bus. Horiz. **60**(6), 741–745 (2017)
5. S. Nambisan, Digital entrepreneurship: Toward a digital technology perspective of entrepreneurship. Entrep. Theory Pract. **41**(6), 1029–1055 (2017)
6. V. Ramaswamy, K. Ozcan, What is co-creation? An interactional creation framework and its implications for value creation. J. Bus. Res. **84**, 196–205 (2018)
7. S. Sedkaoui, An empirical analysis of the Algerian entrepreneurship ecosystem: Entrepreneurship ecosystem in Algeria, in *Handbook of Research on Ethics, Entrepreneurship, and Governance in Higher Education*, ed. by S. Nair, J. Saiz-Álvarez, (Hershey, PA, IGI Global, 2019), pp. 476–497
8. G. Dalmarco, W. Hulsink, G.V. Blois, Creating entrepreneurial universities in an emerging economy: Evidence from Brazil. Technol. Forecast. Soc. Chang. **135**, 99–111 (2018)
9. H. Etzkowitz, Innovation lodestar: The entrepreneurial university in a stellar knowledge firmament. Technol. Forecast. Soc. Chang. **123**, 122–129 (2016)

10. G. Secundo, S. Elena-Perez, Ž. Martinaitis, K.H. Leitner, An intellectual capital framework to measure third mission activities. Technol. Forecast. Soc. Chang. **123**, 229–239 (2017)
11. F.T. Rothaermel, D.S.D. Agung, L. Jiang, University Entrepreneurship: A taxonomy of the literature. Ind. Corp. Chang. **16**(4), 691–791 (2007)
12. Statistical Information Report, Ministry of Industry and Mines. General Management of Strategic Intelligence, Studies and Information Systems, n°31, November 2019
13. J.A. Schumpeter, *History of Economic Analysis* (Oxford University Press, New York, 1954)
14. J.A. Schumpeter, *The Theory of Economic Development: An Inquiry into Profits, Capital, Credit, Interest, and the Business Cycle* (Transaction Publishers, New Brunswick, NJ, 1934)
15. D.L. Birch, *Job Creation in America: How our Smallest Companies Put the Most People to Work* (The Free Press, New York, 1987)
16. D.J. Storey, S. Johnson, *Are Small Firms the Answer to Unemployment?* (Employment Institute, 1987)
17. P.D. Reynolds, New firms: Societal contribution versus survival potential. J. Bus. Ventur. **2**(3), 231–246 (1987)
18. S. Shane, S. Venkataraman, The promise of entrepreneurship as a field of research. Acad. Manag. Rev. **25**(1), 217–226 (2000)
19. Z.J. Acs, C. Armington, Employment growth and entrepreneurial activity in cities. Reg. Stud. **38**(8), 911–927 (2004)
20. M. Guerrero, J.A. Cunningham, D. Urbano, The economic impact of entrepreneurial universities' activities: An exploratory study of the United Kingdom. Res. Policy **44**(3), 748–764 (2015)
21. H. Etzkowitz, The norms of entrepreneurial science: Cognitive effects of the new university-industry linkages. Res. Policy **27**(8), 823–833 (1998)
22. T.E. Stuart, W.W. Ding, When do scientists become entrepreneurs? The social structural antecedents of commercial activity in the academic life sciences. Am. J. Sociol. **112**(1), 97–144 (2006)
23. M. Fontes, The process of transformation of scientific and technological knowledge into economic value conducted by biotechnology spin-off. Technovation **25**, 339–347 (2005)
24. M. Fontes, C. Sousa, Types of proximity in knowledge access by science-based start-ups. Eur. J. Innov. Manag. **19**(3), 298–316 (2016)
25. S. Shane, Academic entrepreneurship: University spinoffs and wealth creation, in *New Horizons in Entrepreneurship*, ed. by S. Venkataraman, (Edward Elgar Publishing, Inc, Cheltenham, UK, 2004)
26. H. Etzkowitz, A. Webster, C. Gebbardt, B.R. Cantisano Terra, The future of the university of the future: Evolution of ivory tower to entrepreneurial paradigm. Res. Policy **29**(2), 313–330 (2000)
27. B.R. Martin, The changing social contract for science and the evolution of the University. Science and innovation, in *Rethinking the Rationales for Public Funding*, ed. by A. Geuna, A. J. Salter, W. E. Steinmueller, E. Edward, (Cheltenham, 2003), pp. 448–471
28. P. Mustar, M. Wright, Convergence or path dependency in policies to foster the creation of university spin-off firms? A comparison of France and the United Kingdom. J. Technol. Transfer. **3**(1), 42–65 (2010)
29. V. Ratten, C.S. Marques, V. Braga, Knowledge, learning and innovation: research into cross-sector collaboration, in *Knowledge, Learning and Innovation*, ed. by V. Ratten, V. Braga, C. S. Marques, (Springer, Cham, 2018), pp. 1–4
30. A. Davari, A. Emami, V. Ramadani, S. Taherkhani, Factors influencing academic entrepreneurship: A case-based study. J. Sci. Technol. Pol. Manage. **9**(3), 284–295 (2018)
31. A. Ustundag, S. Ugurlu, M. Serdar, Evaluating the performance of technology transfer offices. J. Enterp. Inf. Manag. **24**(4), 322–337 (2011)
32. M. Wright, E. Piva, S. Mosey, A. Lockett, Academic entrepreneurship and business schools. J. Technol. Transfer. **34**, 560–587 (2009)
33. M. Abreu, V. Grinevich, The nature of academic entrepreneurship in the UK: Widening the focus on entrepreneurial activities. Res. Policy **42**(2), 408–422 (2013)

34. B. Clarysse, V. Tartari, A. Salter, The impact of entrepreneurial capacity, experience, and organizational support on academic entrepreneurship. Res. Policy **40**(8), 1084–1093 (2011)
35. C. Haeussler, J.A. Colyvas, Breaking the ivory tower: Academic entrepreneurship in the life sciences in UK and Germany. Res. Policy **40**(1), 41–54 (2011)
36. B. Probert, I just couldn't fit in': Gender and unequal outcomes in academic careers. Gend. Work. Organ. **12**(1), 50–72 (2005)
37. S. Winslow, Gender inequality and time allocations among academic faculty. Gend. Soc. **24**(6), 769–793 (2010)
38. P.P.I. Sanchez, M.C.J. Maldonado, R&D activity of university spin-offs: Comparative analysis through the measurement of their economic impact. Int. J. Innov. Learn. **18**(1), 45–64 (2015)
39. S. Sedkaoui, M. Khelfaoui, *Sharing Economy and Big Data Analytics* (ISTE-Wiley, London, 2020)
40. M.M. Fischer, P. Nijkamp, Entrepreneurship and regional development. Working Paper: Serie Research Memoranda. VU University Amsterdam, Faculty of Economics, Business Administration and Econometrics. Amsterdam (2009)
41. J. Youtie, P. Shapira, Building an innovation hub: A case study of the transformation of university roles in regional technological and economic development. Res. Policy **37**, 1188–1204 (2008)
42. D.J. Storey, B.S. Tether, New technology-based firms in the European Union: An introduction. Res. Policy **26**(9), 933–946 (1998)

# E-Learning: MOOC User Intention Analysis Using TAM and TTF with Social Motivation Factor and MOOC Features

**Yohannes Kurniawan, Sevenpri Candra, and Lucky Yudhistira Tungka**

**Abstract** The purpose of this article is to analyze user's intention toward massive open online course (MOOC) using technology acceptance model (TAM) and task-technology fit (TTF) with social motivation factors and MOOC features. In this research, IDCourserians (MOOC community in Indonesia) became its subject. The data for this research were collected from 183 respondents who already used MOOC and processed with SEM method using WarpPls 6.0. The results of this research show that variables inside TAM interplay with each other significantly. Individual technology fit does not have any effect toward perceived usefulness and perceived ease of use. Meanwhile, TTF has a significant and positive effect toward perceived usefulness and perceived ease of use. And the openness variable has a positive and significant effect toward perceived ease of use but does not have any effect toward perceived usefulness. The unique finding of this research is reputation does not have any effect toward perceived usefulness of an MOOC. Social recognition does not have any effect toward perceived usefulness but has a positive and significant effect toward continuance intention to use MOOC. The social influence does not have any effect toward perceived usefulness but has a positive and significant effect toward user's attitude in using MOOC.

**Keywords** E-learning · MOOC · Technology acceptance model · Task-technology fit · Social motivation · MOOC features

Y. Kurniawan (✉) · L. Y. Tungka
Information Systems Department, School of Information Systems, Bina Nusantara University, Jakarta, Indonesia
e-mail: ykurniawan@binus.edu

S. Candra
Management Department, BINUS Business School Undergraduate Program, Bina Nusantara University, Jakarta, Indonesia
e-mail: seven@binus.ac.id



E. Eilu et al. (eds.), *Digital Literacy and Socio-Cultural Acceptance of ICT in Developing Countries*, https://doi.org/10.1007/978-3-030-61089-0_7

# 1 Introduction

An MOOC (massive open online course) is an online college site where participants can access for free or by way of payment. It first appeared in 2008, when Stephen Downes and George Siemens made the first online course using the name "Connectivism and Connective Knowledge/2008" [1]. Then 4 years later in 2012, MOOC grew even more using sites called edX and Coursera which are the most popular MOOC sites in the world. Coursera is a private institution, while edX is a nonprofit institution created by Harvard University and MIT. In November 2012, MOOC sites such as edX, Coursera, Stanford University, and Udacity managed to attract participants, so MOOC users at that time reached more than 1,900,241 students in 196 countries that were affiliated and were looking for just one program on the Coursera site [2]. The MOOC has increased its development in recent years. However, in 2015, Bartolome and Steffens said that it was observed that MOOC has highly increased its dropout ratio. Based on the average, MOOC students who successfully completed their courses are less than 10% of the number of users [3]. In addition, one of the factors in the MOOC course can be successful because of curiosity and hope for progress or development in work [4].

In Indonesia, the development of MOOC began in 2014, where the compilation was run by six public and private universities launching an online lecture system. However, this year the system is only intended for students in the six universities to be able to take courses at other universities. Furthermore, in 2015, finally several universities and institutions began looking for MOOC by providing an MOOC platform, among them are Ciputra University issuing Ciputra UCEO, UGM Faculty of Social and Political Sciences issuing FOCUS, and Open University issuing MOOCs. But despite the phenomenon that exists with various MOOCs in Indonesia, the MOOC approval ratio is still relatively low. According to reports received, the approval ratio only reached 10% of the number of students enrolled in the MOOC course. In 2016, Firmansyah and Timmis [5] conducted research by making IDCourserians from Jakarta, Indonesia, as research subjects. IDCourserians is an Indonesian Coursera community that offers to facilitate Courserians meetings and as a place to share experiences and knowledge between Courserians. IDCourserians often hold face-to-face meetings with the aim that learning materials received at Coursera can be more easily obtained. This was triggered by the MOOC platform that talked more about while the Indonesian people were less proficient in English compared to Indonesian.

Besides the language, the problem of motivation and understanding of the material is also a barrier for students in Indonesia in completing the MOOC course. From the results of the agreed study, participants agreed that they did not like to study MOOC alone. They are happier if they have friends or classmates which show that meeting social needs is related to the motivational drivers of students [5]. Another problem can be seen in a survey conducted in Indonesia by DailySocial.id in 2017 with the title "MOOC in Indonesia Survey 2017." This survey was conducted with the aim of finding out the characteristics of MOOC users in Indonesia. From the survey results obtained, it is proven that MOOC users in Indonesia are familiar with MOOC but have not been able to maximize the level of use of MOOC. This arises in the statement that MOOC with a low level of bandwidth needs to be used because of the problem of Internet access by users from Indonesia, while the Internet in Indonesia is difficult both in terms of price and quality [6].

In the research that has been done, the technology acceptance model (TAM) method is emphasized for use because TAM is easy to apply and is a cross-cultural method. TAM measures the level of user convenience in using a technology, the level of usefulness or function of the technology itself, as well as the behavior of the users of that technology. The factor used in TAM to make measurements is the perceived ease of use which in this case is interpreted as the level of ease of someone in using a technology. Then the next factor is perceived usefulness in this case interpreted as the level of usefulness or function of the technology. Then the attitude and behavioral intention which is the behavior of the users themselves in using MOOC is very influential on increasing the use of MOOC by individuals for the future [7]. In addition to using the TAM model, technology-task fit (TTF) is also one of the models used to take measurements in the above research. The TTF model is used to measure the suitability of users with technology and tasks carried out by each individual. If an information system can make it easier for users to do tasks more easily and efficiently, then the user will increasingly adopt the information system. If interpreted in the context of MOOC, students will increasingly increase the use of MOOC if the technology in the MOOC can facilitate the process of learning activities (tasks) carried out, where in this case the technology in MOOC meets the needs or requirements of students as MOOC users [8].

MOOC itself as part of e-learning has certain features that are needed. The main characteristic of MOOC is openness, namely, the level of ease of access to the MOOC platform in terms of access to text materials, videos, and website itself. The measurement of openness is limited not only to the material but also to the ease of users to access the MOOC platform. It was said in previous research that free access is one of the main things in the openness feature of MOOC. In addition to the openness feature, another important feature that serves as a benchmark is reputation. It was said in previous research that reputation is one of the measures that can influence one's interest or one's behavior toward the use of a particular technology. In MOOC, reputation is one of the factors that influence an individual's use of the MOOC. If the reputation of an MOOC is increasingly well known, it will increasingly influence individuals to use the MOOC platform [7].

The previous studies found that in the e-learning process, the student's behavior could be influenced by other students. This factor is called social motivation. Social motivation itself is divided into two main measures, namely, social recognition and social influence. Social recognition itself is a social recognition of one's achievements or titles. In MOOC, social recognition can be interpreted as an acknowledgment of the ability or achievement of someone achieved in the MOOC, while social influence is a social influence or the influence of society. In relation to MOOC, students influence each other to use or not to use an MOOC platform [7]. The MOOC in Indonesia has begun to develop, but if you look at the data above, there

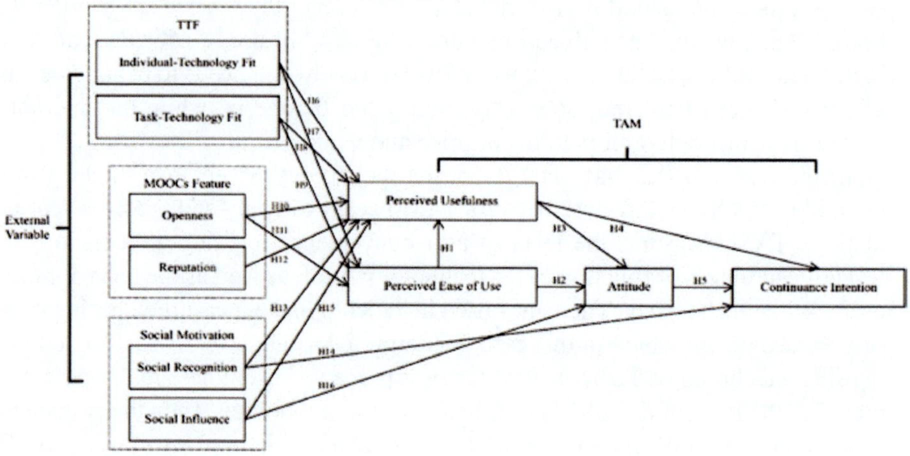


**Fig. 1** The research model

are so many research models that have the potential to be done in Indonesia but have not been done in previous studies. The previous research [6] focused on the characteristics of Indonesian MOOC users including interest in the type of lessons and level of knowledge of Indonesian MOOC users regarding the MOOC itself. Then research conducted by Firmansyah and Timmis [5] found that problems in Indonesia were motivations, different languages and cultures, and infrastructures that were not ready. So this paper will try to do research using other models as a research framework. This research will integrate the TAM model, TTF, MOOC features, and social aspects by making Indonesian MOOC users as research subjects. It is hoped that through this research, a solution or model that is suitable for MOOC for users in Indonesian geographical coverage can be found. It is also hoped that the results of this research can bring Indonesian technology forward to a better digital era. Figure 1 shows the research model of this research.

## 2 The Methods

The method used is the method of data collection with questionnaire method with a number of samples using Slovin's theory. Data collection technique snowball sampling is also used. The samples in this study were respondents who knew MOOC and had used MOOC. The measurement scale in this study was the Likert scale. The data analysis method uses SEM-PLS with WarpPls 6.0 software. In this study, the data collection process was carried out for 5 months last year using the online questionnaire method and combining data collection techniques with the snowball method. This questionnaire was made in the form of Google forms and then distributed online to the MOOC community which has a Facebook and WhatsApp group, IDCourserians. In addition to the MOOC community, the questionnaire was also distributed to people who used MOOC, and then the researchers asked for help from respondents to distribute questionnaires to other people or prospective respondents who also used MOOC. This method continues to be done as the longer the respondent will increase in number as the snowball (snowball) keeps rolling, the better.

# 3 The Results

The following are the results of the characteristics of MOOC users in Indonesia (Table 1 shows the respondents' profiles):

Operationalization of variables is the translation of the variables in the study to measure these variables. In this study, there are 16 variables that will be examined for its effect. Researchers will describe the indicators of each variable to be used as a benchmark in preparing statements. These can be seen in Tables 2, 3, and 4.

**Table 1** The profile of respondents

| Questions | Options | Answers |
|---|---|---|
| Gender | Male | 117 |
| | Female | 66 |
| Age | 20–25 | 102 |
| Education level | Bachelor's degree | 109 |
| Type of work | Employee | 88 |
| | College student | 65 |
| | Student | 14 |
| Domicile | DKI Jakarta | 98 |
| | Jawa Barat | 23 |
| | Banten | 14 |
| Average MOOC usage in a week | <3 h | 91 |
| The average respondent who has completed MOOC (until the final stage of the final exam) | Ever | 99 |
| Respondents' preferences in learning MOOC (in completing the MOOC course from the beginning to the final exam) | 4–6 weeks | 83 |
| What MOOC platform do you use? | Coursera | 94 |
| | EdX | 61 |
| | Udacity | 44 |
| What is the basis for you to choose a course in MOOC? | Interesting material | 142 |
| | Quality teachers | 87 |
| | The course completion time is relatively short | 67 |
| What type of assignment is more in line with your preferences in doing MOOC? | Multiple choice | 147 |
| From the platform below, what platform do you know as MOOC from Indonesia? | IndonesiaX | 84 |
| | CiputraUCEO | 41 |
| | MOOC Universitas Terbuka | 41 |

**Table 2** Operationalization of variables (Part 1)

| Variable | Dimension | Indicator | Statement of questionnaire |
|---|---|---|---|
| Perceived usefulness (PU) | | *Subjective assessment* | I believe that using MOOC can improve my learning performance |
| | | *Enhance job performance* | Using MOOC can increase the effectiveness of my learning |
| | | *Achieving learning goals* | Using MOOC makes it easy for me to understand a subject |
| Perceive ease of use (PEOU) | | *Free of effort* | I think MOOC is easy to use |
| | | *Ease of acquiring skills* | In my opinion, MOOC makes me more competent |
| | | *Ease of understanding* | For me, the interaction with MOOC is clear and easy to understand |
| *Attitude* toward using MOOCs (ATT) | | *Positive feeling* | I believe that using MOOC is a good thing |
| | | *Advisable* | I would advise other people to use MOOC |
| | | *Satisfaction* | I am satisfied with using MOOC |
| Continuance intention (CI) to use | | *Intention to use in the future* | I will continue to use MOOC in the future |
| | | *Increase use in the future* | I will increase my use of MOOC in the future |
| MOOC features (MF) | *Openness* (OP) | *Flexibility* | I can easily and freely take any courses without conditions (subscribe, payment, minimum GPA, etc.) |
| | | *Free access* | I can freely access and use course material for free |
| | | *Change education practices* | I can use the course material that I learned through MOOC in my work |
| | | *Remixing* | I feel free to combine course material with one another to produce something new |

Source: Adapted from [7]

**Table 3** Operationalization of variables (Part 2)

| Variable | Dimension | Indicator | Statement of questionnaire |
|---|---|---|---|
| MOOC features (MF) | *Reputation* (RP) | *Individual satisfaction* | For me, the reputable MOOC platform will be able to offer interesting courses for me |
| | | *Individual confirmation* | For me the university as an MOOC partner will be an important factor in ensuring good quality of study |
| | | *Individual confirmation* | MOOC tends to provide courses given by professors from universities with high reputation |
| | | *Excellence of institution* | MOOC courses are offered by goo/class/favorite universities |

(continued)

**Table 3** (continued)

| Variable | Dimension | Indicator | Statement of questionnaire |
|---|---|---|---|
| Social motivation (SM) | *Social recognition* (SR) | *Realizing people's own abilities and skills* | In my opinion, it is important if MOOC is adopted as worker training by employers |
| | | *Relationships with others in society* | In my opinion, it is important if the quality of MOOC is appreciated and accepted by others |
| | | *Institutional support* | In my opinion, it is important if MOOC credits are recognized by the university |
| | *Social influence* (SI) | *User perceives that others explicitly encourage their participation in MOOC* | Other participants' opinions about MOOC encouraged me to use MOOC |
| | | *User perceives that others explicitly approve their participation in MOOC* | The other participants' opinions about MOOC influenced the level of my use of MOOC |
| | | *User perceives that others explicitly condition their use of MOOC* | The other participants' opinions about MOOC conditioned me to use MOOC |

Source: Adapted from [7]

**Table 4** The operationalization of variables (Part 3)

| Variable | Dimension | Indicator | Statement of questionnaire |
|---|---|---|---|
| Individual technology fit (ITF) | | *Individual willingness* | I can complete courses in MOOC according to my own wishes (without coercion) |
| | | *Individual participation* | I actively participate in discussions and evaluations in the MOOC |
| | | *Individual performance* | I tried to get an award as the best student in the MOOC course |
| Task-technology fit (TTF) | | *Fit to requirements* | The course at MOOC provides material that matches the study criteria I want |
| | | *Fit with educational practice* | In my opinion, the MOOC material that I received is in accordance with the lessons I learned |
| | | *Tool easy to understand* | I can understand how to use the MOOC platform easily |
| | | *Suitable with task* | In my opinion, MOOC is appropriate in helping me complete online courses |

Source: Adapted from [7]

The following are the results of the path coefficients and p-values after processing the data using WarpPls 6.0 software (Table 5).

This study tries to integrate TAM, TTF, social motivational factors, and MOOC features to test the determinants of MOOC users' continuous intentions. The data collected comes from anyone who knows MOOC and has used MOOC before. After an empirical analysis, there are 12 hypotheses out of 16 hypotheses accepted

**Table 5** The results of hypothesis test

| No | Hypothesis | *p*-values | Path coefficients | Decision | Significance/not |
|---|---|---|---|---|---|
| H1 | PEOU--->PU | <0.001 | 0.330 | Accepted | Significance |
| H2 | PEOU--->ATU | <0.001 | 0.480 | Accepted | Significance |
| H3 | PU--->ATU | 0.003 | 0.201 | Accepted | Significance |
| H4 | PU--->CITU | 0.020 | 0.149 | Accepted | Significance |
| H5 | ATU-->CITU | <0.001 | 0.383 | Accepted | Significance |
| H6 | ITT--->PU | 0.421 | 0.015 | Rejected | Not |
| H7 | ITT--->PEOU | 0.306 | 0.037 | Rejected | Not |
| H8 | TTF--->PU | 0.002 | 0.209 | Accepted | Significance |
| H9 | TTF--->PEOU | <0.001 | 0.456 | Accepted | Significance |
| H10 | OP--->PU | 0.212 | 0.059 | Rejected | Not |
| H11 | OP--->PEOU | <0.001 | 0.304 | Accepted | Significance |
| H12 | RP--->PU | 0.134 | −0.081 | Rejected | Not |
| H13 | SR--->PU | 0.071 | 0.106 | Rejected | Not |
| H14 | SR--->CITU | <0.001 | 0.280 | Accepted | Significance |
| H15 | SI--->PU | 0.127 | 0.083 | Rejected | Not |
| H16 | SI--->ATU | 0.029 | 0.137 | Accepted | Significance |

in this study. The findings in this study found that the more someone has an easy perception (ease of use) in using MOOC, the level of usefulness (usefulness) of the MOOC will also increase. This statement is in accordance with previous research conducted by Abdullah and Ward [9]. If someone feels the ease of using MOOC, it will affect the attitude or behavior in using the MOOC. These results are relevant to previous research conducted by Yang and Su [10].

The relevant results regarding the effect of perceived usefulness and its effect on attitude were also found in previous studies. The research conducted by Wu and Chen [7] also proves that if someone feels the benefit of using MOOC, it will affect the person's behavior. In the case of MOOC, it was also found that perceived usefulness influences a person's continuing intention to use MOOC, where these results are also relevant to the findings of previous findings [11]. And to utilize TAM in MOOC, the way that organizations can do this is to add practical exercises. Practical exercises are added in order to produce good benefits for students [10].

In MOOC itself, one's attitude will be a strong determining factor in influencing continuance intention. This means that the user's attitude toward an MOOC will largely determine whether the user will continue to use MOOC or not. This study demonstrates that the effect of attitude on continuance intention is positive and significant according to the theory put forward by Davis et al. [11]. MOOC provider organizations can benefit by improving the technology of the MOOC. In this research, it was found that the effect of task technology fit will affect the perception of the ease of an MOOC and the perception of the usefulness of an MOOC, so it is recommended for organizations to develop MOOC technology by focusing on the characteristics of the MOOC and the features of the MOOC. These results are relevant to previous findings made by Khan et al. [8].

Based on the findings in this study, the feature in MOOC that can be developed by organizations is openness. Organizations can increase the intention of users to continue to use MOOC by maximizing the MOOC feature of openness. Openness features can be improved by providing free access and more course options as recommended in previous studies by Alraimi et al. [4]. Then in addition to that, the organization can also benefit with another step that is utilizing social motivational factors. In the findings of this study, social recognition has a significant effect on user intentions to continue using MOOC. It is therefore recommended for MOOC provider organizations to focus on introducing MOOC. The introduction of MOOC is carried out jointly by providers, partner institutions, and organizations so that social penetration can be made to increase community interest in using MOOC. Using digital media can make MOOC more attractive by inspiring using positive opinions and "power users." "Power users" can provide a positive picture through their experience to increase community knowledge and the benefits of using MOOC [8].

In addition, it can be seen in Fig. 2 the preference in choosing a course; the top three answers show the consideration of students in choosing an MOOC course. MOOC provider organizations can take advantage by utilizing these results. The top three answers are interesting material, quality instructors, and relatively short course completion time. In presenting courses, MOOC providers need to present interesting material, as well as courses taught by qualified instructors and relatively short course completion time. Through these three steps, the MOOC provider organization will be able to attract users to choose an MOOC course.

In addition to the recommendations above, also included are recommendations that were developed personally after conducting an analysis of the existing MOOC website. The recommendations given are features and user interface recommendations. The following are some examples of features in MOOC, and their comparison with other MOOC can be seen in Table 6.

It can be seen in the table above on platform features Coursera, IndonesiaX, and Udacity have web and mobile platforms. Only Open edX does not have a mobile platform. Based on the search engine features, both Coursera and Udacity have search engine features on web and mobile platforms. Likewise, Open edX has search engine features on web platforms, but IndonesiaX does not have search engine features on both platforms. Then downloading the material can be done by

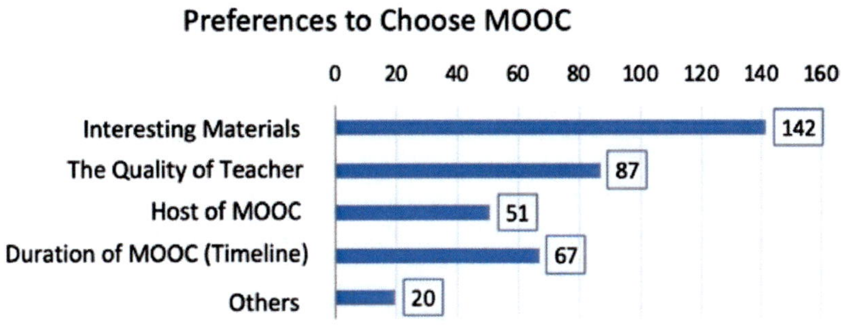


**Fig. 2** The preferences to choose MOOC

**Table 6** The feature comparison between several MOOC providers

| Features | MOOC sites | | | | | | | |
|---|---|---|---|---|---|---|---|---|
| | Coursera | | Open edX | | IndonesiaX | | Udacity | |
| | Web | Mobile | Web | Mobile | Web | Mobile | Web | Mobile |
| Platform | √ | √ | √ | x | √ | √ | √ | √ |
| Search engine | √ | √ | √ | – | x | x | √ | √ |
| Download (material) | √ | √ | √ | – | √ | √ | √ | √ |
| Notification | x | √ | x | – | x | √ | x | √ |
| Free access | Request +/– 24 days | | x | – | √ | √ | x | x |

the four MOOC sites. For notification features, Coursera, IndonesiaX, and Udacity can only provide notifications via a mobile application but not on the web. Free access to the material is done by IndonesiaX and Coursera while not on the other two sites. However, Coursera can get access to free material by submitting an application that will be processed for approximately 24 days. By looking at the comparison of features above, a combination of features is performed to improve an MOOC site. Some recommendations that need to be considered for MOOC providers to improve the features of an MOOC website can be seen in Table 7. The following are recommended features for MOOC providers in developing an MOOC website.

The level of ease and usefulness is vital in e-learning, in this research, and especially in MOOC. With ease and good usability, it will increase the intention of users to continue to use an MOOC. Some of the findings in this research are believed to be a good step in the development of the MOOC site in Indonesia. The researcher realizes that there is an opportunity to conduct further research to include various other aspects in the research model which currently cannot be included in the study due to time and place limitations. An aspect that can be included in future research is increasing the number of respondents in order to get better data. In addition, the addition of variables into models such as theory of planned behavior (TPB) can examine actual user behavior so that the use of a research object will actually be measurable. The addition of other variables, namely, self-determination theory (SDT), can also add quality to research. SDT can measure user's motivation variables better by considering internal and external factors that affect one's motivation. By adding this variable to the research, it is expected that research results can be better so that it can maximize the development of an MOOC. The researchers believe that the addition of other aspects such as TPB and SDT as well as other aspects such as economics in future research will produce theoretical and practical implications that are useful for industries and communities who are interested in e-learning, especially in this case of MOOC. Based on some user interface (UI) references from MOOC that already exist, the authors make a combination of UI designs to be a recommendation for MOOC providers. Figure 3 recommends UI designs in the web.

It can be seen in Fig. 3 the UI in the recommended Home menu that displays sliders for recommendations on various course topics such as "Business," "Coding," and "Social Media" in the form of sliders that will change each time period. On the

**Table 7** The recommended features for MOOC

| Features | Description | Indicators |
|---|---|---|
| Access and course content without pay | The open educational community focuses MOOC to present MOOC with access to course content without paying. By providing free access and free course content, it will facilitate student access to an MOOC site. This is done because not all online course users have sufficient funds to register. Some online course users tend to think twice before spending money to access content that they don't already know will be useful or not. Therefore it is better to provide content access for free, and payment will be made if the user wants to take the course certificate | Free access, ease of use |
| Download content for offline learning | Downloading content for offline learning allows the user to first download the course material available to be learned at any time according to the time of the user and anywhere without the need for an Internet connection. This is done because not all users are able to connect to good Internet services at all times. By providing content download feature, users can learn anytime and anywhere without having to be connected to the Internet | Flexible, ease of use |
| Access without prerequisites | Access without prerequisites will make it easier for users to register and attend an MOOC course. By not providing the prerequisites for registering on an MOOC site, it will open up opportunities for anyone to enter the site and learn from the courses available in the MOOC. Access without these prerequisites needs to be given because not all online course users meet certain requirements such as GPA, degrees, and other academic requirements. MOOC users also consist of people who register because they have some academic limitations such as not being able to continue to a higher level, so they seek knowledge through MOOC | Flexible, ease of use |
| Notifications for reminders | This notification is done to remind users of online courses in completing the course being undertaken. Online course users tend to forget to complete the course regularly so it needs to be reminded. By using notifications on the mobile application or by sending email via the web to the user's email, it will present a social motivational factor to participate and encourage users to use MOOC | Individual participation, social impulse (other users) |
| Awarding the best students | The awarding of the best students refers to the awarding of students in the course who have the best grades. The award given can be a badge or the title of best student in the course. This is done because users tend to want to get social recognition from the community and also appreciation from others. So with the presence of this feature, the user will strive to become the best student so that it can increase the use of MOOC | Social recognition, appreciation |
| The search feature is based on the relationship between the keywords searched with the title | The use of features on the MOOC search engine will make it easier for users to find related courses. For example, if a user types the letter "b," then a course recommendation will come out such as "business," "big data," "business analysis," "business strategy," and "all words that have the prefix" b and "that matches the user's interests and profile. This is done because users tend to want to get ease in doing searches that are in accordance with their interests. Using features on the search engine MOOC site is a good step in making it easier for users to find courses that match the user's interests | Ease |

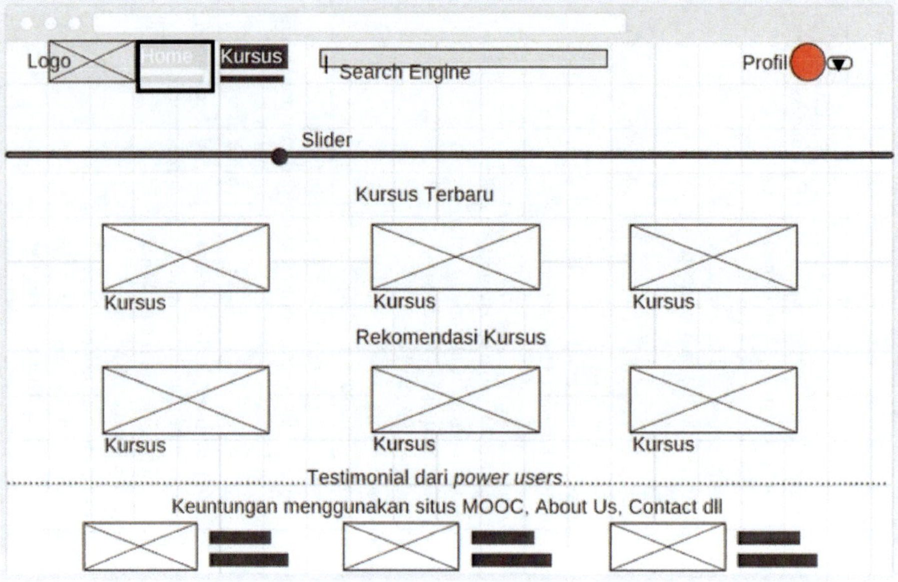


**Fig. 3** User interface Home on the web

Home menu also displays the latest courses and course recommendations according to user interests. The search engines are also included in the Home menu. In addition, testimonials from power users and information about the MOOC site were also displayed.

In Fig. 4, it can be seen that the course menu displays the courses that the user has taken, along with the name of the course and the progress that the user has passed. Besides that, the most popular courses are also displayed based on the categories that the users are interested in. Filters for courses are also displayed to make it easier for users to search for courses.

The UI in Fig. 5 will be displayed if the user clicks on one of the courses being followed by the user. Then the user will direct to the My Course page which has three sub-menus, namely, the course menu, discussion, and progress. On this page, there is an explanation of the courses taken, course schedule, materials, videos, tests, and final exam schedule of the course. On the right, there are course materials and transcripts that are presented in the list form each week. If the list is clicked, it will direct the user to the Material UI page.

Figure 6 is a UI recommendation for course material. On this page on the left, there are course materials and transcripts that are presented in a list every week, to make it easy for users to move to the material in accordance with the desired week. The previous and next buttons are also included to help users replace weekly material. Video material is equipped with subtitles to facilitate users' reading if they have difficulty in hearing. Material in the form of videos and transcripts also allows for download by users. The "Discussion" sub-menu will direct users to the Discussion page, while the "Progress" sub-menu will direct users to the Progress page.

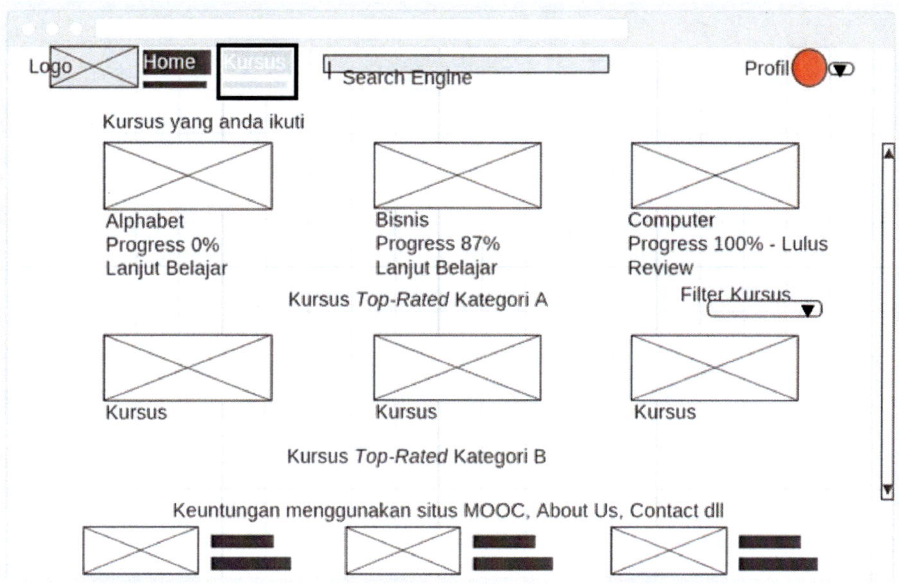


**Fig. 4** User interface course on the web

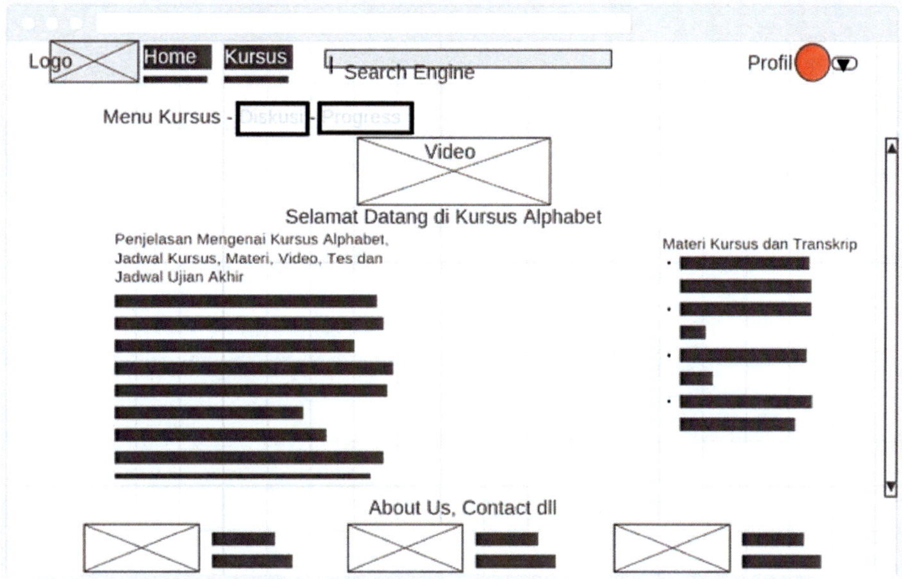


**Fig. 5** User interface My Course on the web

In Fig. 7, it can be seen that in the Discussion page, there are thread lists for discussion on the left-hand side, to make it easier for users to choose the thread where the user wants to discuss. When clicked, it will appear as shown on the right, in the form of other user comments and user comments in discussing discussions related to the thread.

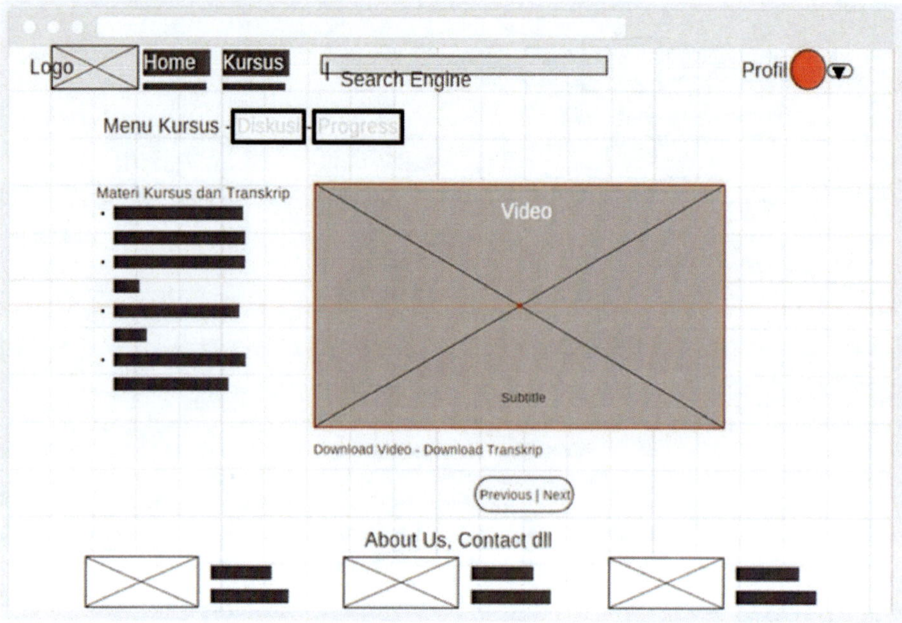


Fig. 6 User interface for download material on the web

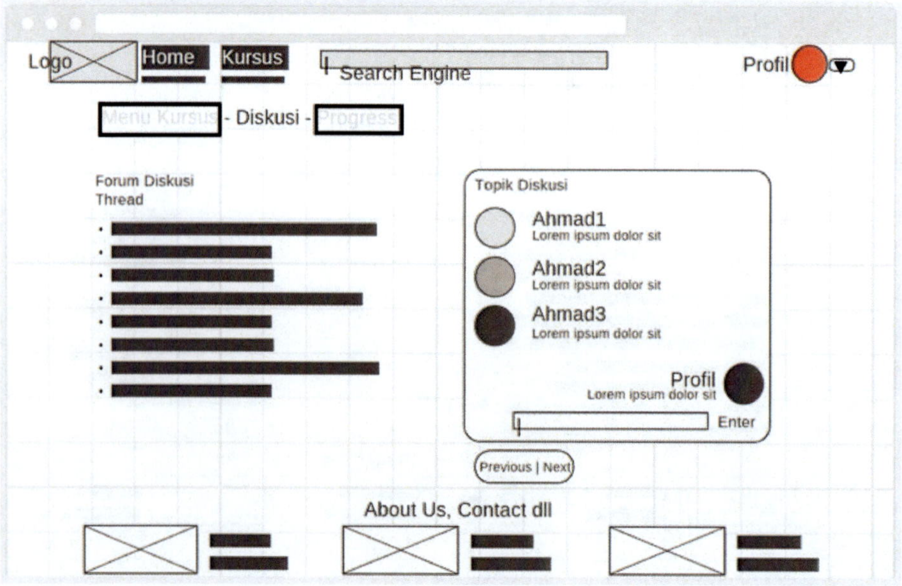


Fig. 7 User interface Discussion Forum on the web

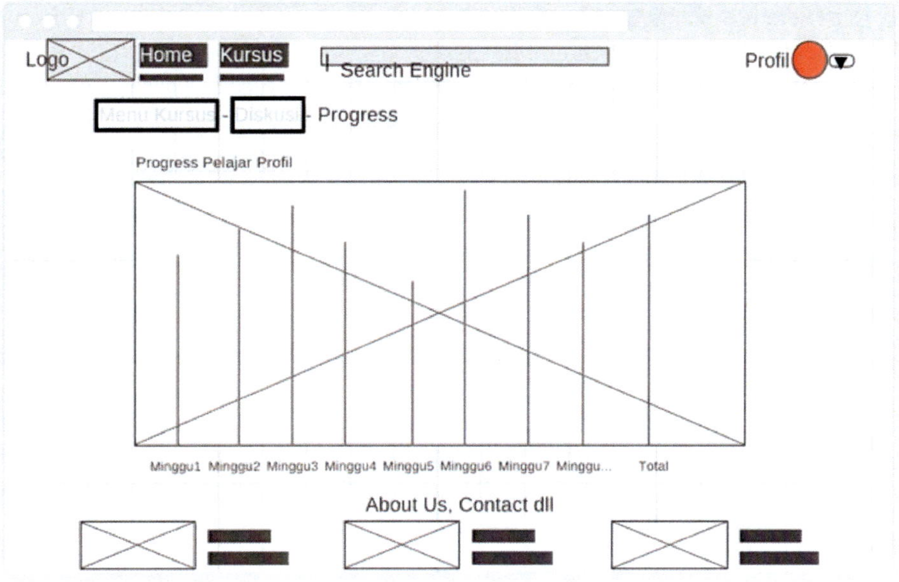


**Fig. 8** User interface Progress of Learning on the web

In Fig. 8, it can be seen that in the UI Progress recommendation, there is a learning progress from the user. User learning progress is displayed in graphical form and divided into time periods each week. On this page, the user can see the learning progress that has been obtained and the weight of lessons that have been learned by the user.

## 4 Discussion

MOOC in Indonesia is growing rapidly. This was marked by the presence of MOOC providers such as IndonesiaX, Open University MOOCs, FOCUS Fisipol UGM, and other MOOC sites. The growth of MOOC is also marked by the presence of MOOC provider fans on various social media such as Facebook, Twitter, Instagram, and YouTube (KampusUNJ.COM, 2016). The rapid development of MOOC in Indonesia has become a new phenomenon that can be utilized to improve public education. The changes in user behavior toward MOOC are caused by many factors. Based on the results of research, it was found that some of these factors are technology, social motivation, and features in the MOOC. This ultimately affects user intentions to use MOOC, or not to use MOOC.

Based on the results of research conducted, it has found several conclusions as follows:

1. There is a positive and significant effect between perceived ease of use on the perceived usefulness of an MOOC.
2. There is a positive and significant effect between perceived ease of use on attitude in the use of MOOC.
3. There is a positive and significant effect between perceived usefulness on attitude in the use of MOOC.
4. There is a positive and significant influence between the perceived usefulness of continuance intention to use MOOC.
5. There is a positive and significant effect between attitudes toward continuance intention in the use of MOOC.
6. There is no positive and significant influence between individual technologies fit on the perceived usefulness of an MOOC.
7. There is no positive and significant influence between individual technologies fit on the perceived ease of use of an MOOC.
8. There is a positive and significant effect between task technologies fit on the perceived usefulness of an MOOC.
9. There is a positive and significant effect between task technologies fit on the perceived ease of use of an MOOC.
10. There is no positive and significant effect between openness on the perceived usefulness of an MOOC.
11. There is a positive and significant effect between openness on the perceived ease of use of an MOOC.
12. There is no positive and significant effect between reputations on the perceived usefulness of an MOOC.
13. There is no significant effect between social recognitions on the perceived usefulness of an MOOC.
14. There is a positive and significant effect between social recognitions of continuance intention to use MOOC.
15. There is no significant effect between the perceived usefulness of an MOOC.
16. There is a positive and significant influence between social influences on attitude to use MOOC.

Based on observations from the results of the study, the user intentions to use MOOC are influenced by the ease and usefulness of the MOOC site. Thus, for profit or nonprofit organizations that provide MOOC sites, there is a need to optimize MOOC by presenting features that make it easy for users to use MOOC and increase the usefulness of the MOOC itself. The features included in the form of searching according to keywords, free access, access without prerequisites, and downloads for offline learning. In addition to the ease and level of usability, other factors such as social motivation also affect user intentions to use MOOC; thus features that need to be included to influence user intentions using social motivational factors are notifications as reminders and awards for the best students on the MOOC site.

## References

1. Y.E. Desiany, Academia. Edu. Retrieved January 10, 2018, from https://www.academia.edu/9765258/Masa_depan_MOOC_Massive_Open_Online_Course_Di_Indonesia (2014)
2. P. Adamopoulos, What Makes a Great MOOC?, in *An Interdisciplinary Analysis of Student Retention in Online Courses. Thirty Fourth International Conference on Information Systems, Milan* (2013), pp. 1–21
3. A. Bartolome-Pina, K. Steffens, Are MOOCs promising learning environments? Comun. Med. Educ. Res. J. **22**(44), 91–99 (2015)
4. K.M. Alraimi, H. Zo, A.P. Ciganek, Understanding the MOOCs continuance: The role of openness and *reputation*. Comput. Educ. **80**, 28–38 (2015)
5. M. Firmansyah, S. Timmis, Making MOOCs meaningful and locally relevant? Investigating IDCourserians—An independent, collaborative, community hub in Indonesia. Res. Pract. Technol. Enhanc. Learn. **11**(1) (2016)
6. DailySocial.id: MOOC in Indonesia Survey 2017 (2017)
7. B. Wu, X. Chen, Continuance intention to use MOOCs: Integrating the technology acceptance model (TAM) and task technology fit (TTF) model. Comput. Hum. Behav. **67**, 221–232 (2017)
8. I.U. Khan, Z. Hameed, Y. Yu, T. Islam, Z. Sheikh, S.U. Khan, Predicting the acceptance of MOOCs in a developing country: Application of task-technology fit model, social motivation, and self-determination theory. Telematics Inform. **35**(4), 964–978 (2018)
9. F. Abdullah, R. Ward, Developing a General Extended Technology Acceptance Model for E-Learning (GETAMEL) by analysing commonly used external factors. Comput. Hum. Behav. **56**, 238–256 (2016)
10. H.-H. Yang, C.-H. Su, Learner behaviour in a MOOC practice-oriented course: In empirical study integrating TAM and TPB. Int. Rev. Res. Open Distrib. Learn. **18**(5), 35–63 (2017)
11. F.D. Davis, R.P. Bagozzi, P.R. Warshaw, User acceptance of computer technology: A comparison of two theoretical models. Manag. Sci. **35**(8), 982–1003 (1989)

# Employees' Acceptance and Adoption of Online Communication Tools as Part of an E-Government Strategy in the Public Sector in Kenya

**Paul Waititu and Charmaine Du Plessis**

**Abstract** This chapter explores and describes the perceptions of employees, in one public sector organisation, of using online communication tools (OCTs) as a strategy for adopting e-Government. The findings address the paucity of research when it comes to using technology in government-to-employee (G2E) services in the public sector. Using methodological triangulation, a single study case design was employed.

The results indicate that employees need to form part of a consultation process to determine their training needs and preferences, prior to implementing any technologies. A lack of training and consultation results in negative perceptions and a low adoption rate when it comes to technologies (such as OCTs), which minimise the likelihood of these tools being used for G2E purposes. The study findings could be valuable for other public organisations in Africa, by assisting them to follow the right processes when implementing any innovation. Moreover, several theoretical guidelines are proposed to enhance internal communication and public services within the context of e-Government.

**Keywords** E-Government · Employees · Internal communication · Online communication tools · Public sector organisation

## 1 Introduction

E-Government cannot be regarded as a new phenomenon, as its implementation has been ongoing for several decades now in many countries the world over [13]. Dukić and Bertović [19] and Solinthone and Rumyantseva [60] define e-Government as

P. Waititu (✉)
Department of Health Services, County Government of Nakuru, Nakuru, Kenya

C. Du Plessis
Department of Communication Science, University of South Africa, Pretoria, South Africa



E. Eilu et al. (eds.), *Digital Literacy and Socio-Cultural Acceptance of ICT in Developing Countries*, https://doi.org/10.1007/978-3-030-61089-0_8

"the use of technology to enhance the access to and delivery of government services to benefit citizens, business partners and employees". It is considered a central aspect of public service management in enhancing the quality of service delivery in public sector organisations. To date, most related literature has centred around four types of e-Government services, namely, (1) government to citizen (G2C), (2) government to government (G2G), (3) government to business (G2B) and (4) government to employee (G2E).

In particular, most countries consider e-Government to be an innovative strategy, which is suitable for addressing economic programmes, as well as social and political agendas, and promoting transparency, accountability, good governance and democracy [10, 19, 35]. According to [7, 34], this strategy is also used with the aim of curbing corruption, promoting efficiency and enhancing quality service delivery in the public sector. It is characterised by systemic and drastic changes which are aimed at transforming existing structures, cultures, values and practices, using information and communication technologies (ICTs) though an assortment of online communication tools (OCTs), to provide quality government services. E-Government is credited with improving public service delivery, besides "overseeing" the implementation of new innovations in ICT systems within the public sector. It is relied on to bring about efficient e-services, enhance interactive communication and facilitate the sharing of information between public sector organisations, their employees and the public [7, 13, 25, 35, 51, 56, 61].

The recent United Nations (UN) [62] survey identifies Denmark, Australia and the Republic of Korea as the leading countries in providing e-Government services in the world, with an E-Government Development Index (EGDI) score of 0.9 and above. Other countries with very high EGDI score include the United Kingdom, Sweden, Singapore, Finland, New Zealand, France and Japan. The reason for this success is that those countries have seriously enhanced the use of digital platforms in all government communication and engagements. However, Africa countries still lag behind when compared with the rest of world, with only Mauritius, Seychelles, South Africa, Tunisia, Morocco and Ghana being placed in the high-EGDI-level category. This scenario is attributed to poor internet connectivity, prohibitive cost of access and inadequate skills in ICT. According to [34, 56]), the stakeholders of different public sector organisations have diverse needs. As stated by [34], the needs of employees require more attention in terms of training and managerial support; hence, G2E services should be provided via multiple channels through appropriate systems and tools. G2E services, according to [32], include "online interactions that are facilitated by the use of a set of communication tools between employees and government units for the purpose of giving the employee instant access to information in regard to topics such as compensation, learning and training opportunities", among other services. Importantly, [9] note that most e-Government channels and services are citizen-centric. Employees' perceptions of the innovations being implemented in their respective organisations do, however, differ from those of other stakeholders. It is for this reason that if e-Government is to be effective, G2E services should receive as much emphasis as other stakeholders do.

One of the expected effects of G2E services is that they will enhance interactive online communication among employees, resulting in better customer service.

According to several authors [5, 32–35, 49, 56, 57, 61], this can be done by using ICT resources and various OCTs such as the World Wide Web (WWW), websites, social media, e-mail, local area networks and mobile phone services such as SMS.

The objective of this chapter is to elucidate employees' acceptance and adoption of ICTs within the context of e-Government in Kenya and in particular G2E. A public sector organisation in Kenya is used as an example of employees' perceptions of the use of OCTs in G2E services as part of an e-Government strategy which seeks to improve public services. This is in a bid to address the following research question:

- What perceptions do employees at a public sector organisation in Kenya hold about their acceptance and adoption of online communication tools as part of the e-Government strategy?

## *1.1 Theoretical Background and Significance of the Study*

The chapter draws on the insights provided by the relevant aspects of four theories, namely, those of the diffusion of innovation, social penetration, social exchange and relational perception [4, 11, 15–17, 31, 58, 66]. The use of these four theories serves to evaluate the study, by providing useful insight into public sector employees' perceptions in adopting OCTs for G2E services in Kenya.

Until recently, precisely because G2E services were considered part of G2G and G2C services [9, 60], they were overlooked by researchers, thereby creating gaps in the research in terms of their impact on efficient public service delivery. Filling this gap has become necessary, since there currently is paucity of research which explores the determinants prompting public sector employees in developing countries (especially in Africa) to adopt or ignore OCTs for e-Government services. Indeed, more evidence on G2E would encourage the robust adoption and use of OCTs among public sector employees as part of their routine duties, hence the need for this study.

Here, the focus is on the relevance, acceptance and adoption of OCTs for e-Government services on the part of public sector employees, in a bid to enhance service delivery. This study is premised on the supposition that such employees' lack of consultation, involvement and training on e-Government affect their acceptance and use of OCTs for e-Government, thereby impairing the G2E services they are expected to deliver.

## *1.2 E-Government in Kenya*

The implementation of e-Government in Kenya is linked to the country's national strategy for 2008–2030, popularly known as Vision 2030 [28, 35, 49], a long-term development plan which identifies ICTs as important drivers undergirding the country's economic, social and political pillars. Since the advent and launch of e-Government in Kenya in 2004, an evident rise in the demand for virtual communication and services has been experienced in various public sector organisations [35, 51, 57].

Currently, e-Government is used in almost all public sector organisations, be they government or quasi-governmental agencies. It incorporates online systems using OCTs to provide public services to citizens and other stakeholders alike. Some of the e-Government services offered for G2E include human resources, procurement, finance, budget and email. Moreover, the use of e-Government is recognised as being able to provide business opportunities to private organisations and individuals willing to act as ICT and e-Government service providers, to enhance access to online government services.

Due to the complexity of e-Government, which is a recent innovation in Kenya, its implementation and operations call for serious consideration. In 2015, Kenya ranked 86th out of 143 countries worldwide, with a networked readiness index of 3.8 [21, 35] far ahead of many other African countries. However, in the [62] survey, Kenya came 122nd out of 193 member states, with an EGDI of 0.4541 and an Online Service Index (OSI) of 0.6250, in respect of the implementation and use of e-Government. This is an indication that Kenya is still sluggish in terms of the implementation and adoption of e-Government, and much needs to be done to improve this scenario, to prevent the initiative from failing.

Many e-Government enterprises are reported to have failed in other countries, due to certain adoption factors being lacking, which hampered its implementation [2, 3, 5, 13, 22, 32, 38, 40, 57]. In this chapter, adoption factors encompass any aspects which provide, motivate or support a conducive milieu for the use of OCTs for G2E services. Included in those factors are sound infrastructure, the planned implementation of innovations, an accommodative culture, appropriate training, stakeholder consultation and involvement, trust, online security, organisational support, the ease of use of an innovation and users' acceptance of an innovation, among others. According to certain authors [46, 51, 57, 62], the implementation of e-Government in Kenya faces various challenges, including weak infrastructure, poor implementation of innovations, the alienation of stakeholders during implementation, inadequate training on the use of ICTs among government employees, entrenched graft, the digital divide, eroded trust in the technology employed by both citizens and employees and resistance from public sector employees. It is imperative that public sector organisations in Kenya address those challenges, to prevent the e-Government initiative from collapsing.

One of the key edifices of e-Government is sound and up-to-date infrastructure [4, 10, 29, 33]. In fact, ICT infrastructure forms the backbone of an e-Government enterprise in any country; hence, contemporary ICT infrastructure is a prerequisite for its successful implementation. In fact, up-to-date ICT infrastructure supports all the systems of an e-Government enterprise by providing the required milieu in terms of, for instance, accessibility, content development, data management and data transfer and security. Arguably, the failure of most e-Government initiatives is due to weak infrastructure, and in this regard, Kenya is no exception [38, 40, 46, 57,

64]. The situation can only be remedied if modern and appropriate ICT platforms are made available and are considered acceptable and easy to use by public sector employees through a comprehensive e-Government strategy.

## *1.3 E-Government Strategy in Kenya*

According to [32, 33], the uniform implementation of e-Government across all states is in no way possible, due to diverse environments and unique status of each country. That uniqueness should, however, be considered as the basis for planning and implementing e-Government, well before the system is launched, to avoid unnecessary challenges. One of the determining factors for the success of any e-Government initiative is the formulation and implementation of a well-thought-out strategy.

Although not a new concept, e-Government strategy is a relatively recent notion within the ICT purview in Africa, and especially in Kenya. Riany et al. [57] deem an e-Government strategy to entail "a plan for e-Government systems and their supporting infrastructure which maximises the ability of management to achieve organisational objectives". Scholars such as [10, 49] posit that an e-Government strategy is a critical success factor in any country wishing to transform its public sector. Such transformation can be achieved by identifying units, structures and systems that need to be integrated and finding ingenious ways to achieve this, to allow interactions between various stakeholders (both internal and external to the organisation). E-Government strategy has been hailed for being able to remove various barriers, such as high organisational operating costs and bureaucracy, which are synonymous with the traditional offline public service and hinder citizens from accessing efficient, quality services.

In 2004, the Government of Kenya formulated an institutional framework by launching its e-Government strategy [27, 49, 51, 64]. This strategy outlines how government expects to modernise its services and orient them to achieve results, be more efficient and become citizen-centric. The main aim is to guide the migration of government services from manual operations to various online systems, so as to enhance transparency, accountability and good governance. The strategy also provides a framework for how citizens and private businesses can access government information and services through OCTs.

Initially, most e-Government services in developing countries (Kenya included) were offered through stand-alone systems and fragmented OCTs within their respective organisations [13, 64]. This means that organisational and inter-organisational information was exchanged manually, due to a lack of interoperability between those systems and tools. Moreover, the ICT service providers for e-Government were regulated though several agencies. A strategy, which envisaged harmonising the various e-Government ICT enterprises (which had been uncoordinated and were working in isolation) into an integrated system, would thus eliminate the wastage of public resources and avoid confusion among users [10, 49, 64].

This situation was addressed through the establishment of the ICT Authority, which merged several public ICT regulatory agencies. Other institutional structures that came into being because of the e-Government strategy include the Ministry of Information and Communications and the ICT board, while internet connectivity was provided to urban areas across Kenya. Efforts at establishing connectivity are currently being realised through the implementation of the National Optic Fibre Backbone (NOFBI) project, which aims to provide internet connectivity to all 47 counties in Kenya, as a way of improving government service delivery at the grass-roots level [41, 45].

In 2006, a National ICT Sector roadmap for the period 2008–2012 was launched, with the aim of upgrading Kenya's national competitiveness and developing knowledge-based culture through the "Digitalised Villages" project [49]. The constitution of Kenya [26] and the Kenya National ICT Master Plan 2014–2017 [29] further provided a foundation for the establishment of an e-Government legal framework to address political and managerial structures at both the national level and across devolved units [26, 29, 49]. In addition, the e-Government strategy has enabled government to roll out various online systems to provide myriad online public services, including information pertaining to government human resource management, an e-Registry for business registration, integrated financial management, land administration, road safety and traffic management, e-tax, online immigration services and education management.

In 2013, the Kenyan government launched several "one-stop shop" citizens' service hubs dubbed "Huduma centres" [29], which offer national government services in a single location. The aim of the centres is to provide government services and information to those citizens who cannot access e-Government services due to the digital divide and a lack of prerequisite ICT skills or for any other reason. Currently, at least one Huduma centre has been established in each of Kenya's 47 counties. For these centres to offer optimum services, however, public service employees ought to be involved from the onset and must be adequately trained in e-Government. This is to enable them to acquire the necessary knowledge, attitude and skills for implementing G2E and to modify their perceptions of the adoption and use of the available OCTs across all e-Government services.

According to [29, 51, 57], the Kenyan government planned to enhance its e-Government strategy through the use of OCTs as channels for G2E, besides offering other services within the various public sector organisations.

## 1.4 Government to Employees (G2E)

Several arguments have been put forward in the literature regarding the acceptance of, or resistance to, new technologies which depend on employees' perceptions of both their benefits and ease of use. For example, [39, 42] argue that there is a need to ease employees' anxiety and modify their perceptions about change (in this case, innovations) – a perspective which is supported by [4, 34, 55, 61]. Employees' perceptions are crucial in enhancing their trust in the use of new media technologies in G2E services and subsequently applying them in customer services [4, 22, 39, 61, 66]. Currently, many public service employees do not seriously consider OCTs to be important tools for enhancing G2E services. As [6] point out, most government employees tend to use OCTs only to perform basic routine tasks.

In essence, employees' training in the use of OCTs, as well as their consultation and involvement prior to and during the implementation of any technological innovation, is an important theoretical requirement for using OCTs in the context of e-Government [3, 4, 22, 34, 36, 39, 44, 57]. According to [50], employees perceive G2E services as "better" if they themselves are involved in the implementation process. The aforementioned authors emphasise the importance of consultation to obtain employees' approval and develop their intention to adopt innovations. Notably, [32, 33] point out that integrating OCTs into employees' working routine systems and implementing appropriate training programmes are seldom standard practice in public service organisations. Moreover, although employees are not always consulted in respect of the implementation of e-Government initiatives, governments expect them to actively utilise their tools to facilitate service delivery. Consultation, in this study, refers to consultative participation which, according to [50], occurs when an organisation consults its stakeholders before arriving at a significant decision. On the other hand, involvement occurs when stakeholders can actively participate in the organisation's decision-making process.

It is reported that a few public sector organisations have adopted a proactive approach to the implementation of digital innovations used for e-Government, such as using OCTs as an initiative for enhancing G2E services [38]. Thus far, the prevalence of this approach is not widespread in public sector organisations in Kenya, but if adopted, it could add significant value to the use of OCTs for G2E services. Therefore, it is prudent for public sector organisations to consult with staff and allow those employees to interactively participate in the implementation of innovations, besides training them on their use. Some of the OCTs that are applicable to G2E are highlighted below.

## 1.5 Online Communication Tools Available in the Public Sector in Kenya for Internal Communication

Chmielecki [14] describes internal communication as a controlling function that coordinates all formal and informal communications occurring within an organisation. Additionally, [13, 68] observe that internal communications affect all functions within an organisation and any ineffectual communication is thus a precursor to problems within that entity.

OCTs (which include web-based and other online communication channels) are used in providing e-Government services and represent increasingly important methods for employees to communicate through G2E. Several OCTs that are commonly used for G2E services are mentioned below, and they include the internet, WWW, intranet, extranet and e-mail. Besides the organisation's system, most of these platforms can also be accessed via smartphones.

### 1.5.1 Internet

Some scholars (see [24, 27]) state that the influence exerted by the internet cannot be compared to that of any other ICT innovation. The internet has revolutionised the technological world exponentially over the past few decades, by providing online connectivity via numerous networks, and its impact is felt in virtually all spheres of life [6, 25, 53]. One of the most notable features of the internet is the WWW, which [8] consider to its mainstay: it is through the WWW that online connectivity and the numerous networks and services associated with the internet are provided. Moreover, the WWW has enabled the creation of online social networks or virtual societies to address varying issues [12, 67].

### 1.5.2 Intranet and Extranet

Another OCT that is popular in organisations is the intranet [59]. This online connection within organisations is used to facilitate internal communications among employees. If this network is connected to serve external stakeholders or associates such as suppliers, consultants or contractors, it is referred to as the extranet.

### 1.5.3 E-Mail

E-mail, which is the digital version of the ordinary mailbox, has been in existence since the early 1960s, and its use intensified when it was adopted by the internet in the early 1970s. It is still preferred and widely used in most organisations, as it is deemed secure and confidential [20, 37]. E-mail is used to pass on information in the organisation, using various document formats. Due to its asynchronous nature, employees can access information even then they do not have a regular internet connection.

### 1.5.4 Social Media

Besides e-mail, the use of social media has become prevalent in the public sector organisation. Social media is a recent internet-based innovation (Web 2.0) that allows users to communicate and share various types of content electronically [1, 18, 37]. It enables the establishment of assorted virtual networks and communities in which members can create and share ideas, opinions and information, using various social media tools. The contents include topics in a variety of formats, such as documents, videos and photos. Moreover, the contents are accessible through a computer and other web-enabled mobile devices such as tablets or smartphones.

Many public sector organisations are now adopting social media tools to engage with their stakeholders, upon which they are regarded as enterprise social media (ESM) [23, 52]. In this context, ESMs can be described as web-based platforms which employees use to communicate in G2E and other e-Government services, using social media tools. Some of the most commonly used ESM tools include corporate blogs, microblogging tools (e.g. Twitter), instant messaging (e.g. WhatsApp, Telegram) and social networking tools (e.g. Facebook, Instagram and LinkedIn).

## 2 Method

### *2.1 Research Design*

This study adopted a single case design, as proposed by the seminal author [65], and worked from the assumption that employees are not adequately consulted, involved or properly trained on OCTs before e-Government is implemented in their respective organisations.

Kumar [43] identifies a unit of analysis as an individual person or object from whom/which data is obtained. In this case, individual employees served as units of analysis.

### *2.2 Sampling*

The study's target population comprised 10,252 permanent employees of a public sector organisation in Kenya. The sample, drawn purposively from the accessible population of 115 employees at both middle- and top-level management in one regional office, consisted of those employees who had admission rights to the organisation's OCT system and whose duties required the use of OCTs in their day-to-day work. The employees who participated in the study were selected from the ten strata (departments) in the organisation. Of those employees, 30 were targeted to complete the online questionnaire; another 30 were selected to participate in the focus group discussions (FGDs), while ten top managers were included in the in-depth interviews. The fairly small sample did not compromise the validity of the findings, as the issue of generalisation was not envisaged. A pilot study was also done to assist the researchers in evaluating how the measuring instruments were understood. Moreover, the researchers conducted a case study and applied triangulation (mixed methods approach) in the methodology, to analyse and interpret the data.

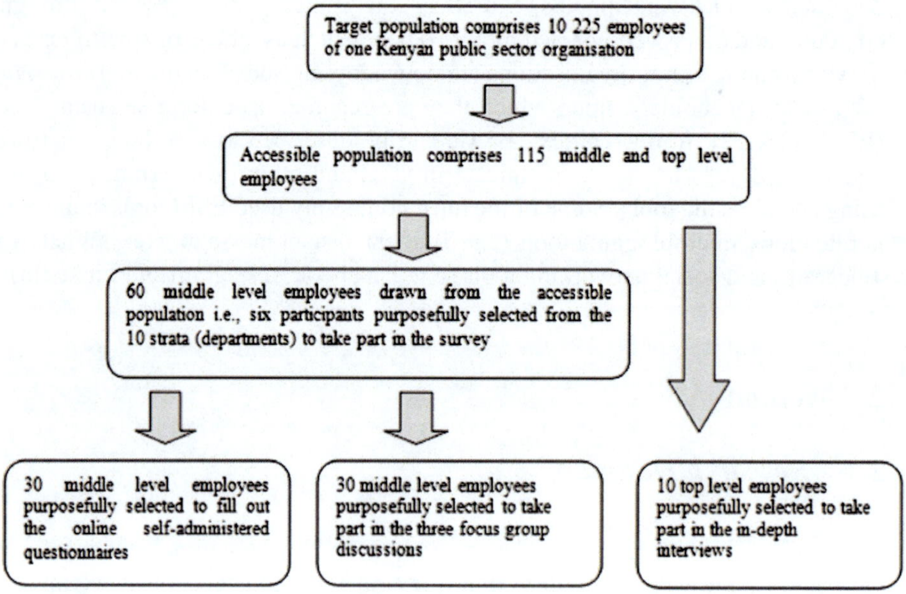


**Fig. 1** The sampling process followed for this study. (Source: Authors' exposition)

The study used a non-probability sample similar to that of a study by [2]. A probability-type sample was not deemed appropriate here, since not all employees use OCTs. The sampling process followed is summarised in Fig. 1.

## 2.3 *Data Collection Instruments*

Evidence was collected from multiple sources, to better understand the case under study. Data collected through multiple sources assists the researcher in examining the research problem from different perspectives, besides allowing him/her to view the phenomenon in a holistic manner that leads to similar conclusions [63].

In an attempt to answer the research question, a range of quantitative and qualitative methods and techniques (online self-administered questionnaire for the survey, interview schedule for the in-depth interviews and moderator's guide for the FGDs) was used to collect primary data from the public sector organisation. Various items corresponding to the literature were incorporated into the three measurement instruments, to test employees' perceptions. In addition, methodological triangulation was used to increase the validity of the findings, while a pretest was done on all the questions in the measurement instruments prior to the study (see [30, 54]).

The final step was to analyse the findings from the three data sources and compare them in order to establish how employees perceived OCTs within the context of G2E.

## 3 Results

The data analysis and findings were guided by the research question and connected to the theoretical supposition of OCTs and their use for G2E services, as elucidated below.

### *3.1 Quantitative Data*

A 90 per cent response rate was achieved, with 27 online self-administered questionnaires being electronically completed and returned. The collected quantitative data were verified and analysed using the computer software SPSS (version 25.0), to create descriptive statistics.

The researchers applied various scales to assess employees' perceptions of the use of OCTs for G2E services, in line with the research question. The scales were constructed on five different hierarchical response levels. The online self-administered questionnaire requested the respondents to simply indicate, on the scale, their perceptions of various aspects related to the use of OCTs in G2E. However, this study highlights only the three most salient issues that became evident in the findings.

As regards the survey findings on employees' perceptions of them being consulted by management prior to the implementation of OCTs, the results show that approximately 66 per cent ($n = 18$) felt there had been no such consultation. Given such limited consultation between employees and management when implementing an online communication system, it is fair to deduce that this affected employees' perceptions of the meaningful use of OCTs for G2E services. The concept of consultation is aligned to the adoption factors of trust and commitment [3, 42, 66]. Organisations can increase the level of trust and commitment by consulting employees when setting up new communication innovations such as OCTs. As [50] indicate, consultation reduces users' apathy and enhances their intention to adopt innovations in the organisation. Figure 2 portrays employees' perceptions of their consultation on the implementation of innovations.

As regards whether employees were adequately trained in the use OCTs for G2E services, more than half of the respondents (63%) ($n = 17$) had negative perceptions. This indicates that they deemed themselves to be insufficiently skilled at using OCTs for G2E services, which might be a factor in the slow adoption thereof. This finding is linked to the adoption factors of trust, commitment and management support [3, 4, 22, 32, 33, 36, 66]. By contrast, trained employees are likely to trust OCT tools and commit to using them for G2E services. According to [32, 33], there is an urgent need for employees to be trained in the use of new technologies and to receive periodic updates or undergo upskilling on OCTs and platforms, in order to keep abreast of rapidly changing technological developments. Figure 3 depicts employees' perceptions of their training in the use of OCTs for G2E services.

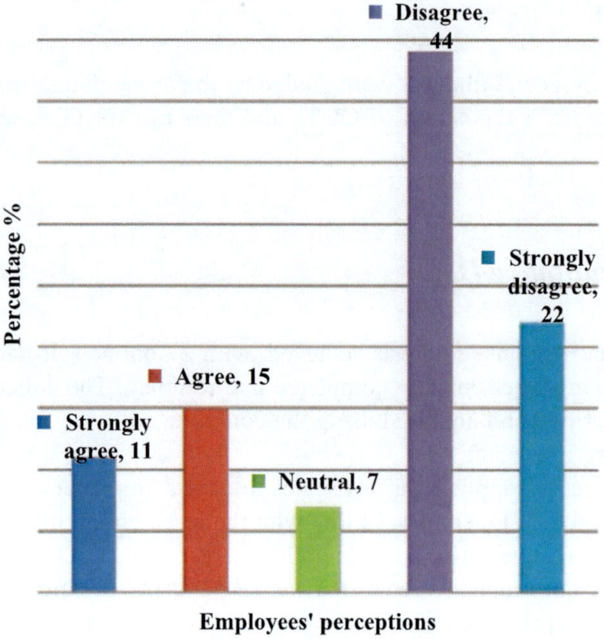


**Fig. 2** Employees' perceptions on their consultation in the implementation of OCTs. (Source: Authors' exposition)

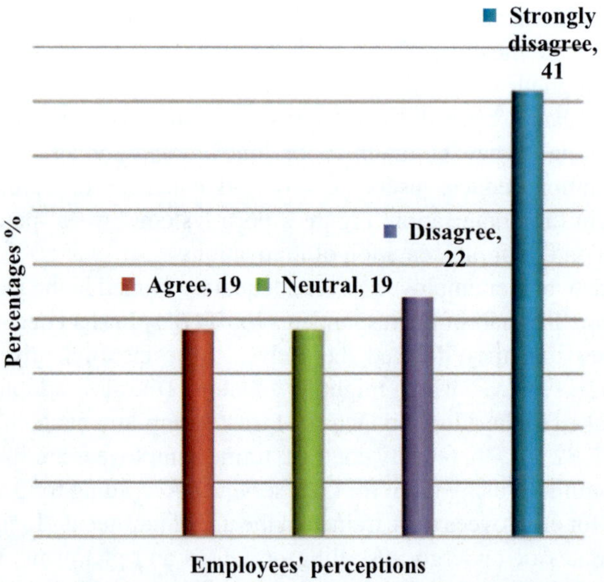


**Fig. 3** Employees' perceptions on their training in the use of OCTs for G2E services. (Source: Authors' exposition)

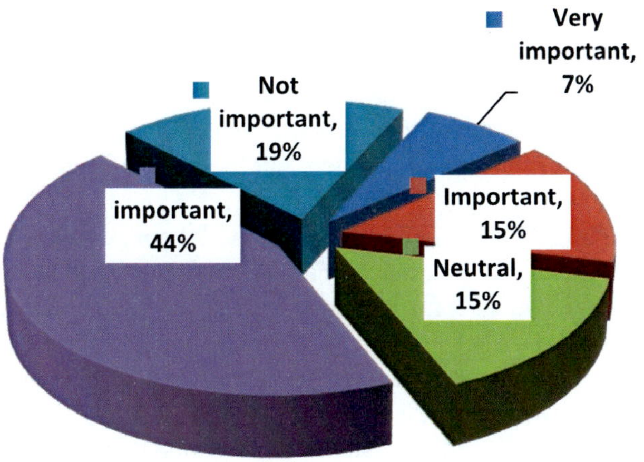


**Fig. 4** Employees' perceptions on the role of e-Government in the implementation of OCTs for G2E

Another salient issue was whether employees understood the role of e-Government in enhancing their use of OCTs for G2E services. According to the findings, most employees (63%; $n = 17$) felt that the existence of e-Government did not influence their involvement in the implementation of OCTs, yet it is important as they are the system's end users. This finding is linked to the adoption factors of trust, commitment and involvement, which are essential in forming employees' perceptions on the use of OCTs for G2E services [36, 66]. Equally, [38] argue that employees' perceptions of whether to adopt or reject an innovation depend on their involvement in its implementation. Figure 4 depicts employee perceptions of the role of e-Government in promoting the use of OCTs for G2E services.

## 3.2 *Qualitative Data*

The approach for analysing qualitative data, advocated by [47] and reinforced by [48], was used on the qualitative data obtained from the interviews and focus groups. Data were manually coded and organised into several categories, and common themes to draw meaning from them were identified in a bid to answer the research question.

### 3.2.1 Focus Groups

Three FGDs were conducted, each consisting of ten middle-level employees with a 100 per cent attendance rate. Several themes emerged from the coding, the most noticeable of which included employee training, involvement and management support in OCTs.

One theme which emerged from the FGDs was a lack of employee training and proficiency or improvement in using OCTs, especially for G2E services. One participant remarked: "We (employees) need to know [...] and [to be] trained [i]n the use of online tools, as some of us are still green". Another commented: "To me, OCTs can be used [for] communication, but there is [a] need for training [in] how to use [them]. I would emphasise [...] training...". A third participant observed: "As we support the use of OCTs to provide services [...] our skills need to be enhanced". As indicated by [22, 32, 33], these findings are related to the adoption factors of training. Faulkner et al. [22] further indicate there has been no serious impact on the use of OCTs for G2E in some public sector organisations, since the implementation of e-Government programmes in many developing nations does not afford employee training the seriousness which it deserves.

Another theme expressed the issue of organisational management support in the implementation of OCTs for G2E services. As these participants' comments indicated: "Regarding e-Government for employees, its support is not adequate [...]" and "Employees are not supported enough and are not involved on the use of online system[s] in the organisation". According to [38], inadequate management support acts as an obstacle to the adoption of OCTs. As the study's findings show, most participants concurred that despite having access to ICT on different digital platforms, the management was not implementing technologies in the appropriate manner. They hold the view that management support was inadequate as regards the implementation and that this affected their adoption of OCTs for the purposes of providing G2E services. As [4, 32, 38, 39] posit, that deficiency in management support (in respect of the implementation of OCT platforms) impedes the adoption factors of training, involvement, trust and commitment. Consequently, this negatively affects employees' perceptions and curtails their willingness to adopt such tools – a situation that is detrimental to G2E services as well as internal communication.

### 3.2.2 In-Depth Interviews

Similar sentiments emerged from the in-depth interviews with the ten top managers, whose insights provided crucial additional information, as they play a considerable role in the implementation of OCTs. Nine of the envisaged ten managers were interviewed (i.e. 90% coverage). Some of the dominant themes identified included consultation, participation and training.

There was a common feeling among the top managers that consultation was being hampered by existing top-down processes in the organisation's internal communication. One manager commented: "Bureaucracy in the organisation can be resolved by promoting open online communication".

Despite the issue of employees' non-involvement in the implementation of e-Government being prominent, all was not lost. As one manager commented: "The use of OCTs is providing a new horizon where the management and employees account to each other through online interactions". Moreover, other top managers also attributed employees' negative attitudes towards the implementation of OCTs for G2E to inadequate training. In this regard, a manager commented: "Training, skills and motivation in the use of OCTs need to be improved [...] for internal services, we will definitely have a very sound organisation".

These findings are associated with the adoption factors of trust, commitment and involvement, which are crucial if employees' perceptions are to be modified to the extent that they readily adopt OCTs for G2E [36, 66]. As [32, 33] indicate, most employees in the public sector do not possess the necessary knowledge, skills and motivation to effectively do so. What is needed is prior training for employees, before they can be expected to embrace OCT platforms in their internal communications [3, 32, 33, 44].

## 4 Discussion

With regard to the research question, the findings indicate that employees in the public sector are not always consulted prior to new technologies such as e-Government (on which OCTs are anchored) being "imposed" on them. This situation might cause apathy among employees, thus impeding the adoption of OCTs for G2E services. The findings echo those of a similar study carried out by [39], which reported that a lack of consultation contributed to employees' poor perceptions of the use of innovations intended for internal communication. To this end, [9] advise public sector organisations not only to concentrate on their citizens but also to extend their influence over their employees by motivating them through continuous dialogue and consultation.

Employees' involvement in the implementation of innovations is another major issue. It is apparent that, in this case, the respondents were not optimally involved in the implementation of new technologies such as e-Government. This finding is supported by other studies [36, 39], which point out that the implementation of innovations and their subsequent adoption are hindered by an organisation's failure to make employees an active part of these processes.

The other pertinent issue deals with training and the acquisition of essential skills by managers and employees alike, in the use of OCTs for G2E services. It is apparent that the study respondents did not possess the requisite knowledge, skills or attitudes necessary to optimally use OCTs when delivering G2E services. This confirms the findings of [36, 39], who assert that it is important for organisations to regularly engage with, and built the capacity of, their managers and employees, to ensure that they all keep abreast of rapidly changing technologies (especially new developments in the field of OCTs).

The overall findings of this study revealed that employees in the selected public sector organisation were not consulted or involved in the implementation of OCT platforms earmarked for the provision of G2E services – neither during the planning nor the implementation stages. Moreover, most employees did not possess the necessary skills for applying OCTs and providing the subsequent online G2E services.

Despite the popularity of e-Government, it is evident that its implementation has not enhanced the adoption of OCTs for G2E services, contrary to expectations. This situation is likely to contribute a scenario which sees employees making minimal use of such tools in their internal interactions.

## 4.1 Conclusion

The findings of this study elucidate that despite the implementation of e-Government and subsequent OCTs in public sector organisations in Kenya, employees' perceptions of their adoption within the context of G2E seem to be negative and their actual use minimal. This might be attributed to a lack of consultation, on the part of management, prior to implementing innovations, employees' inadequate training in the use of such tools and their exclusion from decisions around the implementation of novel technologies. The fact that e-Government in Kenya is still in its infancy may also have contributed to employees' apathy towards the adoption of OCTs co-opted for the purposes of delivering G2E services.

Specifically, employees had become apathetic about using OCTs for G2E services, thus adversely affecting the potential of such platforms to provide and improve inclusive e-Government services in the organisation. The situation should be cause for concern to bureaucrats, if their intention is to employ OCTs for purposes of streamlining the provision of e-Government services. Hence, there is a need for public sector organisations to adequately train all their employees, in addition to consulting and involving them when implementing new innovations. To remedy this situation, two recommendations can be made, as outlined below:

- To promote G2E, an organisation's management should consult, build capacity and involve all employees in implementing the use of the various OCTs to enhance G2E services.
- Proper mechanisms must be put in place for implementing e-Government, to promote the use of OCTs for delivering services from government to employees.

## 4.2 Further Research

Due to constant technological innovation especially in public sector organisations, there is a need to conduct future research on the topic broached in this study. Further research is recommended, specifically to assess the relationship between demographic variables such as age, gender and level of academic attainment. Studying the adoption of OCTs for G2E services from employees' diverse perspectives in public sector organisations will help the latter to determine the most appropriate OCTs for wide-ranging purposes.

## 4.3 Limitations of this Study

The study's limitations are acknowledged as being the following:

- It is impossible to generalise the study findings to other public sector organisations, as the research was carried out in a single public sector organisation, with a small realised sample.
- A small, non-probability type of sample from a single regional office was used to represent the entire organisation; therefore it may be necessary to include a wider and larger sample, before drawing a final conclusion.

## 4.4 Theoretical and Practical Implications of the Study

The main contribution of this study was to propose theoretical guidelines to assist public sector organisations in Africa to adopt and implement OCTs for e-Government services and to stimulate further debate on the topic. In terms of practical implications, the study findings provide various suggestions to not only strengthen but also promote the use of online tools among public sector employees. Since OCTs are pivotal for G2E service delivery and other areas of e-Government in public sector organisations, it is prudent to focus more attention on this area.

# References

1. J. Abbas, J. Aman, M. Nurunnabi, S. Bano, The impact of social media on learning behavior for sustainable education: Evidence of students from selected universities in Pakistan. Sustainability **11**(6), 1–23 (2019)
2. H. Alenezi, A. Tarhini, R. Masa'deh, A. Alalwan, N. Al-Qirim, Factors affecting the adoption of e-Government in Kuwait: A qualitative study. Electron. J. e-Govern. **15**(2), 84–102 (2017)
3. B.N. Al-Jenaibi, The new electronic government: Are the federal authorities ready to use e-government? Int. J. Know. Soc. Res. **6**(3), 45–74 (2015)
4. W. Alrawabdeh, E-government diffusion in Jordan: Employees' perceptions toward electronic government in Jordan. Am. J. Appl. Sci. **14**(1), 124–133 (2017)
5. A.S.M. Al-rawahna, S. Chen, C. Hung, The barriers of e-Government success: An empirical study from Jordan. Int. J. Manag. Publ. Sect. Inform. Commun. Technol. **9**(2), 1–18 (2018)
6. E. Apăvăloaie, The impact of the internet on the business environment. Proc Econ. Fin. **15**(14), 951–958 (2014)
7. E. Asorwoe, Can e-Government mitigate administrative corruption? An empirical study into the potential role of e-Government in eradicating administrative corruption in sub-Saharan Africa. Global J. Multidiscipl. Soc. Sci. **3**(4), 41–47 (2014)
8. S. Balan, P. Ponmuthuramalingam, A study on world wide web. Inform. Ret. Web Search Tech. **4**(3), 3532–3535 (2016)
9. A. Baležentis, G. Paražinskaitė, The benchmarking of the government-to-employee (G2E) technology development: theoretical aspects of the model construction. Socialinės Technologijos (Soc. Technol.) **2**(1), 53–66 (2012)

10. E.A. Botchway, E.O. Yeboah-Boateng, T.E. Kwofie, Benefits of e-governance implementation in physical infrastructure development at the local government level. Electron. Govern. Int. J. **12**(4), 395–405 (2016)
11. A. Carpenter, K. Greene, Social penetration theory, in *International Encyclopedia of Interpersonal Communication*, ed. by C. R. Berger, M. E. Roloff, 1st edn., (Wiley, Hoboken, 2016), pp. 1–5
12. D. Castro, R. Atkinson, Beyond internet universalism: A framework for addressing cross-border internet policy (Information Technology and Innovation Foundation, Washington, DC, 2014). http://www2.itif.org/2014-crossborder-Internet-policy.pdf
13. J. Chipeta, A review of e-Government development in Africa: A case of Zambia. J. e-Govern. Stud. Best Pract., 1–13 (2018)
14. M. Chmielecki, 2015 Factors influencing effectiveness of internal communication. J. Manag. Bus. Admin. **23**(2), 24–38 (2018). Central Europe, Sciendo.
15. J.W. Dearing, J.G. Cox, Diffusion of innovations theory: principles and practice. Health Aff. **37**(2), 183–190 (2018)
16. M. Dibra, Rogers' theory on diffusion of innovation: the most appropriate theoretical model in the study of factors influencing the integration of sustainability in tourism businesses. Procedia Soc. Behav. Sci. **195**, 1453–1462 (2015)
17. J. Dijkstra, Social exchange: relations and networks. Soc. Netw. Anal. Min. **5**(1), 1–3 (2015)
18. A. Dodokh, M.A. Al-Maaitah, Impact of social media usage on organisational performance in the Jordanian Dead Sea cosmetic sector. Eur. J. Bus. Manag. **11**(2), 75–91 (2019)
19. D. Dukić, B. Bertović, How do government employees use information and communication technologies? Tech. J. **9**(2), 142–146 (2015)
20. M. Dunahee, H. Lebo, *The World Internet Project International Report*, 6th edn. (University of Southern California, Los Angeles, 2016)
21. S. Dutta, T. Geiger, B. Lanvin, Global Information Technology Report: ICTs for inclusive growth (2015). http://www3.weforum.org/docs/WEF_Global_IT_Report_2015.pdf
22. N. Faulkner, B. Jorgensen, G. Koufariotis, Can behavioural interventions increase citizens' use of e-Government? Evidence from a quasi-experimental trial. Gov. Inf. Q. **36**(1), 61–68 (2019)
23. J. Fu, S. Sawang, Y. Sun, Enterprise social media adoption: its impact on social capital in work and job satisfaction. Sustainability **11**(16), 6–17 (2019)
24. Z. Gaál, L. Szabó, N. Obermayer-Kovács, Personal knowledge sharing: Web 2.0 role through the lens of generations, in *Proceedings of 15th European Conference on Knowledge Management*, 4–5 September, Santarem, Portugal (2014), pp. 365–370
25. J.M. Gomez-Reynoso, R. Sandoval-Almazan, The adoption of e-Government services in Mexico: A citizens' perception analysis. Am. J. Ind. Bus. Manag. **3**(6A), 12–22 (2013)
26. Government of Kenya (GoK), *Constitution of Kenya* (National Council for Law Reporting, Nairobi, 2010)
27. Government of Kenya, e-Government Strategy: The Strategic Framework, Administrative Structure, Training and Standardisation Framework (Cabinet Office, Office of the President, Nairobi, 2004)
28. Government of Kenya, *Kenya Vision 2030* (2007). http://nesc.go.ke
29. Government of Kenya, The Kenya National ICT Master Plan, 2014–2017 (Ministry of Information and Communication, Nairobi, 2013)
30. R. Heale, D. Forbes, Understanding triangulation in research. Evid. Based Nurs. **16**(4), 98–98 (2013)
31. J. Holthausen, Scientific review of the social exchange theory and its contribution to solving purchasers' decision-making issues, in *Presented at the 1st IBA Bachelor Thesis Conference*, 27 June, Enschede, The Netherlands (2013)
32. O.A. Ibrahim, N.H. Zakaria, Towards a model of e-Government services adoption among employees in developing countries. J. Inform. Syst. Res. Innov. **8**, 66–74 (2014)
33. O.A. Ibrahim, N.H. Zakaria, Towards the development of an adoption model for e-Government services in developing countries, in *Pacific Asia conference on information systems, PACIS 2015 proceedings* (2015), p. 137

34. O.A. Ibrahim, N.H. Zakaria, E-government services in developing countries: A success adoption model from employees' perspective. J. Theor. Appl. Inf. Technol. **94**(2), 382–396 (2016)
35. E.N. Imbamba, N. Kimile, A review of the status of e-Government implementation in Kenya. Reg. J. Inform. Knowl. Manag. **2**(2), 14–28 (2017)
36. D.W. Irawanto, Employee participation in decision-making: evidence from a state-owned enterprise in Indonesia. Management **20**(1), 159–172 (2015)
37. A.I. Iskandar, A.F. Arham, N. Shohaime, The use of social media, e-mail and instant messaging as the predictors of an employee's work performance. J. Acad. UiTM Negeri Sembilan **5**(1), 127–136 (2017)
38. P. Joshi, S. Islam, e-Government maturity model for sustainable e-Government services, from the perspective of developing countries. Sustainability **10**(6), 2–28 (2018)
39. K.K. Katsaros, A.N. Tsirikas, S.N. Bani, Exploring employees' perceptions, job-related attitudes and characteristics during a planned organisational change. Int. J. Bus. Sci. Appl. Manag. **9**(1), 36–50 (2014)
40. S.T. Khaemba, G.M. Muketha, N. Matoke, Factors affecting citizen readiness for e-Government systems in Kenya. J. Res. Eng. Appl. Sci. **2**(2), 59¬67 (2017)
41. M.M. Kiminza, S. Were, Factors affecting successful implementation of fiber optic cable projects in Kenya: a case of Nairobi city county. Devel. Country Stud. **6**(6), 52–59 (2016)
42. M. Komodromos, Employees' perceptions of trust, fairness, and the management of change in three private universities in Cyprus. J. Hum. Resour. Manag. Labor Stud. **2**(2), 35–54 (2014)
43. S. Kumar, Understanding different issues of unit of analysis in a business research. J. General Manag. Res. **5**(2), 70–82 (2018)
44. J.B. Lee, A. Gregory, G.A. Porumbescu, Engendering inclusive e-Government use through citizen IT training programs. Gov. Inf. Q. **36**, 69–76 (2019)
45. M. Lemlem, M. Moronge, Determinants of optimal implementation of fibre optic projects in Kenya: a case of Liquid Telecom Kenya. Strategic J. Bus. Change Manag. **4**(49), 912–932 (2017)
46. T. Mawela, N.M. Ochara, H. Twinomurinzi, e-Government implementation: a reflection on South African municipalities. S. Afr. Comp. J. **29**(1), 147–171 (2017)
47. M.B. Miles, A.M. Huberman, J. Saldana, *Qualitative Data Analysis: A Methods Sourcebook* (Sage, London, 2014)
48. M.B. Miles, A.M. Huberman, *Qualitative Data Analysis: An Expanded Sourcebook* (Sage, Thousand Oaks, 1994)
49. A.N. Mungai, e-Government strategy implementation and performance of the public sector in Kenya. Int. Acad. J. Hum. Resour. Bus. Admin. **2**(3), 301–338 (2017)
50. C. Naidoo, R.R. Ramphal, The factors that affect public participation for effective municipal service delivery: a case of ward committees. S. Afr. J. Ind. Eng. **29**(4), 82–93 (2018)
51. K. Okong'o, M. Kyobe, Empirical examination of e-Government in developing countries and its value in Kenya's public service. Electron. J. Inform. Syst. Eval. **21**(1), 35–45 (2018)
52. N. Oostervink, M. Agterberg, M. Huysman, Knowledge sharing on enterprise social media: practices to cope with institutional complexity. J. Comput.-Mediat. Commun. **21**(2), 156–176 (2016)
53. M. Page, C. Firth, C. Rand, *Internet value chain: A study on the economics of the Internet* (GSMA, London, 2015). http://www.gsma.com/publicpolicy/wp-content/uploads/2016/05/GSMA_The-Internet-Value-Chain_WEB.pdf
54. G. Peersman, Overview: data collection and analysis methods in impact evaluation, in *Methodological briefs – Impact evaluation 10* (UNICEF Office of Research, Florence, 2014). https://www.unicef-irc.org/publications/pdf/brief_10_data_collection_analysis_eng.pdf
55. N.P. Rana, Y.K. Dwivedi, B. Lal, M.D. Williams, M. Clement, Citizens' adoption of an electronic government system: towards a unified view. Inf. Syst. Front. **19**, 549–568 (2017)
56. V.R. Rao, Improving government-to-employee (G2E) services through mobile technology – a survey. Int. J. Comput. Appl. **168**(6), 33–45 (2017)
57. G.K. Riany, S. Were, A. Kihara, Influence of e-Government strategy implementation on the performance of public service delivery in Kenya. Int. J. Strateg. Manag. **7**(2), 32–49 (2018)

58. E.M. Rogers, *Diffusion of Innovations*, 4th edn. (Free Press, New York, 1995)
59. D.R. Roshan, K.S. Rao, Intranet design strategies. Int. J. Electric. Comp. Eng. **6**(3), 936–944 (2016)
60. P. Solinthone, T. Rumyantseva, e-Government implementation. MATEC Web Conf. **79**, 1066, 1–11 (2016)
61. J.D. Twizeyimana, A. Andersson, The public value of e-Government – a literature review. Gov. Inf. Q. **36**, 167–178 (2019)
62. United Nations (UN). e-Government Survey: Gearing e-Government to Support Transformation towards Sustainable and Resilient Societies (UN, New York, 2018)
63. D. Wahyuni, The research design maze: Understanding paradigms, cases, methods and methodologies. JAMAR **10**(1), 69–80 (2012)
64. F.O. Wamoto, e-Government implementation in Kenya: An evaluation of factors hindering or promoting e-Government successful implementation. Int. J. Comp. Appl. Technol. Res. **4**(12), 906–915 (2015)
65. R.K. Yin, *Case Study Research: Design and Methods*, 5th edn. (Sage, Thousand Oaks, 2014)
66. H. Zahid, B.H. Din, Determinants of intention to adopt e-Government services in Pakistan: an imperative for sustainable development. Resources **8**(3), 1–25 (2019)
67. N. Zeitel-Bank, U. Tat, Social Media and Its Effects on Individuals and Social Systems: Human Capital without Borders – Knowledge and Learning for Quality of Life, in *Paper Presented at the Management, Knowledge and Learning International Conference, Pharos University, Portorož, Slovenia* (2015)
68. S. Zondi, N. Cassim, A.M. Karodia, Internal communication challenges and issues: A case study of Transnet freight rail business unit coal, Vryheid (KwaZulu-Natal Province, South Africa). Kuwait Chap. Arab. J. Bus. Manag. Rev. **4**(6), 105–145 (2015)

**Dr. Paul Waititu** is a Communication and Public Health Specialist in the Department of Health Services at the County Government of Nakuru in Kenya. He holds a PhD in organisational communication from the University of South Africa (Unisa), South Africa, and a diploma in Public Health from the Kenya Medical Training College, Kenya. His research areas of interest include online communication, knowledge-sharing and health communication.

*Prof. Charmaine du Plessis* is a full Professor in the Department of Communication Science, University of South Africa, South Africa. She holds a PhD in marketing communication and a postgraduate diploma in marketing management from the University of South Africa (Unisa). In addition, she completed a certificate in brand management at the University of Cape Town. She was rated by the National Research Foundation in South Africa for the impact of her research in her field of specialisation, namely, brand communication with a focus on online communication in various contexts. She has published and presented widely both locally and internationally.

# Human-Computer Interaction System for Improving Digital Literacy Among Speech- and Hearing-Impaired in Developing Countries

**Suriya Sundaramoorthy and Balaji Muthazhagan**

**Abstract** This chapter discusses the implementation of a human-computer interaction system for deaf and mute people in developing countries. The system does not have a dependency on active Internet connection, consumes low-resolution input, and at the same time comprehends the sign language used by deaf and mute people and translates it accordingly for normal people to understand. Such a system shows scope for improving digital literacy among deaf and mute people because of increased understanding and communication. The system is implemented as a three-layer system using image processing and convolutional neural networks: first layer to identify the hand in an image frame, second layer to remove superfluous information, and third layer to classify the gestures.

**Keywords** Sign language · Hand gesture recognition · Human-computer interaction (HCI) · Convolutional neural networks (CNN) · Feature extraction

## 1 Introduction

About 5% of the entire world's population, which accounts to about 470 million people, is said to have some form of hearing impairment [1]. There are three types in which we define the degree of hearing impairments: mild, moderate, and profound. People who have mild hearing impairments are not affected drastically and can use speech to communicate, whereas people having medium and profound hearing impairments find it extremely hard to do their daily activities. They rely on

S. Sundaramoorthy (✉)
Department of Computer Science and Engineering, PSG College of Technology, Coimbatore, Tamil Nadu, India
e-mail: ss.cse@psgtech.ac.in

B. Muthazhagan
Arcesium, Bangalore, India



E. Eilu et al. (eds.), *Digital Literacy and Socio-Cultural Acceptance of ICT in Developing Countries*, https://doi.org/10.1007/978-3-030-61089-0_9

channels such as helper pages, lip reading, by using gestures, or through sign language [2]. Mute people also use to rely largely on sign language as a form of communication. Even though there exists a multitude of such channels, there are some difficulties which are encountered by deaf and mute people when conversing with a normal person. There is a lack of awareness among people about deaf and mute culture [3]. Not everyone is specialized in sign languages, and even if they are specialized in one, there are more than 300 sign languages to pick from [4]. This makes it difficult for the conversation to take place. In developing countries, this problem is more prevalent because there are fewer number of specialized sign translators.

Digital literacy is defined as a person's ability to compose content using digital platforms. The evaluation metric used to evaluate this composition is the correctness of grammar, the structure of the composition, the use of good typing skills, and the creative ability to produce content [5]. With the advent of the Internet on almost every handheld device, there has been a shift of digital literacy from stand-alone computers to mobile devices. For deaf and mute people, the ability to create and consume quality video and audio content is obstructed because of their impairments. Firstly, they must learn how to interact with these devices, and secondly, they must efficiently capture their thought process into texts, audio, or video for others to consume. Using a translator as a mediatory to transform sign languages into understandable context and vice versa can solve this problem. For example, although a video can be uploaded to YouTube by a deaf-mute person, there needs to be a translator in between to translate that video into a common language over sign language to garner more views and for efficacious communication.

In developing countries, the deaf and mute people have lower access to novel technologies and network architectures [6]. Also, the Internet speeds as tabulated in Table 1 are much lower than developed countries [7]. Most of the images and video will be captured on mobile devices which have a lower resolution. To bridge these gaps, in this chapter, we propose a system which does not have a dependency on active Internet connection, consumes low-resolution input, and at the same time comprehends the sign language used by deaf and mute people and translates it accordingly for normal people to understand.

## 2 Related Works

The research and development of sign language recognition systems have been around since the late 1980s. Tamura et al. [8] and T. Simon et al. [35] proposed a system in 1988 in which they used the three-dimensional extraction of features such as shape, location, and movement and had them converted into two-dimensional features which were used in classifying the sign language frame. In 1995, Starner et al. [9] proposed a system based on Hidden Markov Model (HMM) in which they captured hand shapes and orientation using a colored glove and classified about 40 words. A principal component analysis (PCA)-based algorithm was proposed by Sawant et al. [10] which converted gestures into text and subsequently voice.

**Table 1** Average Internet speeds lower than 15 megabits/second [7]

| | Average Internet speed (Mbps) |
|---|---|
| Country | April 2020 |
| Haiti | 12.17 |
| Philippines | 12.09 |
| Botswana | 11.77 |
| Zimbabwe | 11.3 |
| El Salvador | 11.23 |
| Ghana | 11.02 |
| Uganda | 10.59 |
| Tajikistan | 10.38 |
| Zambia | 10.32 |
| Tanzania | 10.13 |
| Bangladesh | 9.96 |
| Iraq | 9.9 |
| India | 9.81 |
| Uzbekistan | 9.68 |
| Libya | 8.44 |
| Algeria | 8.25 |
| Rwanda | 8.22 |
| Sudan | 7.09 |
| Venezuela | 6.07 |
| Afghanistan | 6.02 |

Eigenvalues and eigenvectors were used as the features. The system they developed correctly classified 26 words from the Indian Sign Language (ISL).

A multitude of machine learning algorithms have been proposed to gesture and image recognition. Murthy et al. [11] used a supervised feed-forward network for classifying ten gestures. The system they had come up with had an accuracy of 89%. A similar work was proposed by Bhowmick et al. [12], in which they proposed a multilayer perceptron artificial neural network-based hand gesture recognition which was able to correctly classify eight alphabets and achieved an accuracy of 89.05%. A multisensor system was proposed by Molchanov et al. [13] in which they were able to identify various gestures of the driver's hand. They used the raw information obtained from radar and different sensors and calibrated it. This was passed down to a CNN which identified and classified as one of the proposed ten gestures with an accuracy of 94.1%.

Lin et al. [14] proposed a CNN model which used a skin mask filter using Gaussian mixture in which the non-skin color components were removed and then passed on to the CNN. The system achieved an accuracy of 95.94% and was able to successfully classify seven hand gestures. Static hand gesture recognition was proposed by Flores et al. [15] in which deep preprocessing was done. This preprocessing helped in removing the features which were invariant such as lighting conditions, noise, unwanted rotation, etc. The system they proposed was experimented on two different architectures, where the first architecture achieved an accuracy of 95.37%

and the second architecture was able to achieve an accuracy of 96%. It successfully classified 24 gestures. A max-pooling-based CNN was proposed by Nagi et al. [16] for gesture recognition which extracted the hand contour using color segmentation. The system developed was able to achieve an accuracy of 95% based on the experiment carried out on over 6000 sign language images. Yang and Zhu [17] applied CNN for classifying 40 daily vocabularies as part of Chinese Sign Languages. The system used AdaGrad and AdaDelta as the optimizer functions for learning the neural network.

Rioux-Maldague and Giguere [18] proposed a system in which Kinect-captured images are subjected to feature extraction and the identification of hand gesture was carried out. In the system, they proposed a threshold-based preprocessing, which is used to appropriately size the image and center it. The system was able to achieve a 99% precision on known users and 79% precision on unknown users. Huang et al. [19] also used Kinect to address sign language recognition using 3D CNN. The network extracts spatial temporal features which are later transformed to authentic features. The system they developed achieved an accuracy of 94.2% by successfully classifying 25 signs. A real-sense sign language recognition was proposed by Huang et al. [20] in which they captured 26 alphabet signs in 65,000 frames, split as 52,000 and 13,000 for training and testing, respectively. They were able to achieve an accuracy of 99% on real sense. Italian sign language gestures were recognized in a system proposed by Pigou et al. [21] which also used Kinect to record inputs and classify. They had a series of preprocessing steps which included the likes of background subtraction and used the NAG (Nesterov's Accelerated Gradient) descent optimizer. Their system achieved an accuracy of 91.7%. A sign language recognition system was proposed by Tang et al. [22] again through the input of Kinect sensor. The system encompassed transfer learning from LeNet. The testing they carried out was implemented using deep belief networks and convolutional neural networks. Their testing showed that deep belief networks were better than convolutional neural networks.

American Sign Language (ASL) recognition was done by Tushar et al. [23] in which they used batch normalization to achieve quicker training convergences and data overfitting was reduced using dropout. The model they proposed achieved an accuracy of 98.50%. Oyedotun and Khashman [24] used the dataset prepared by Thomas Moeslund to classify 24 ASL alphabets. Their system was implemented using stack denoising auto encoders which had an accuracy of 92.83%. Another implementation, proposed by Bheda and Radpour [25], used a CNN architecture consisting of three convolutional layers followed by max pooling layer, followed by a dropout and two groups of fully connected layers with the optimizer being stochastic gradient descent. This system was able to identify letters and digits in ASL with accuracy of 82.5% and 97%, respectively. An ASL dataset with a black background was used by Islam et al. [26] for a real-time recognition system. They used the convex hull algorithm for fingertip detection and then passed on to an artificial neural network which classified 37 signs with an accuracy of 94.32%. Hidden Markov Model (HMM) was used for the detection and identification of continuous signs by Koller et al. [27] in their implementation. They used dynamic programming-based approach to preprocess their data. A dual-stream CNN model was proposed by Kumar et al. [28] which uses joint distance topographic descriptor (JDTD) and joint angle topographic descriptor (JATD) as the input to the system. They crafted a dataset of over 50,000 Indian Sign Language videos using motion capture cameras. The system they proposed had an accuracy of 92.24%.

## 3 Proposed System

- The human-computer interaction system must ensure the following characteristics: the input being fed can be of lower resolutions, which means that the system must be able to process images of lower or blurry quality, since higher-resolution cameras are costlier.
- The system must not have a reliance on active Internet connection since developing countries as shown in Table 1 do not have accelerated Internet speeds.
- The system must be able to correctly translate the sign language used by deaf and mute people with a high accuracy.

Taking these characteristics into consideration, the system is broken down into three components (Fig. 1):

- Component 1 – Identification of the hand in the given frame

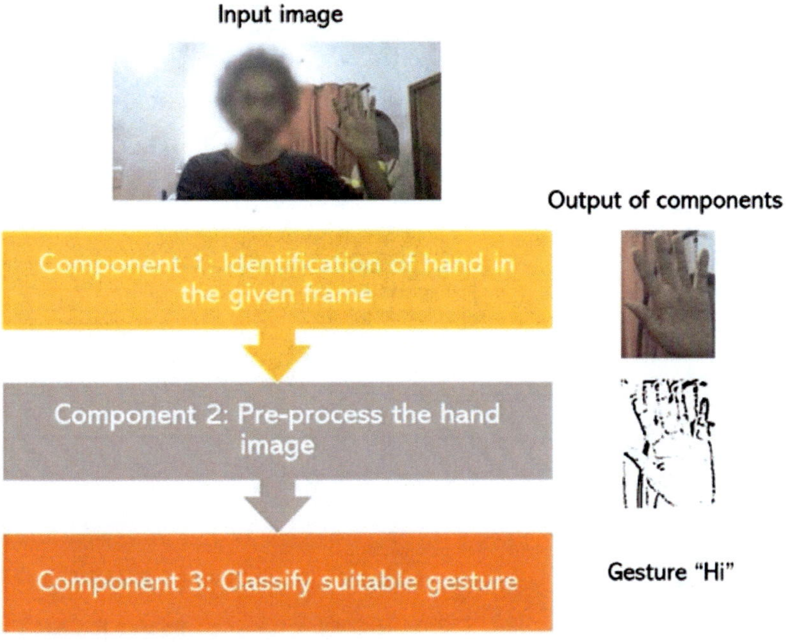


**Fig. 1** Components of the proposed system

- Component 2 – Preprocessing the hand image to remove noises, color components, and backgrounds
- Component 3 – Classifying the hand image as a suitable gesture

## 3.1 Component 1: Identification of the Hand in the Given Frame

Detecting hands in each image frame is an example of object identification where the object to be identified is hands. When it comes to object identification, we rely largely on three different possible algorithms to achieve this. The first is recurrent convolutional neural network (R-CNN). The architecture of R-CNN proposed by Girshick et al. [29] solves the object detection task in two levels. The first level is used to identify the bounding boxes, and the second level is used to pass the identified regions for classification. There are 2000 region proposals made in the first step which are used to classify the image. However, since this algorithm is broken up into two parts, it takes a lot of time to execute. This was improved by Girshick et al. [30] in another algorithm called the Fast R-CNN in which he introduced the concept of feature maps over region proposals. Although this decreased the time of execution, it still was not deemed completely end to end. This shortcoming was removed in Faster R-CNN [31] where a region proposal network was used instead of a selective search algorithm (Fig. 2).

To decrease the latency of the abovementioned algorithms, it is necessary to transform the two-step classification into a single-step classification. Algorithms

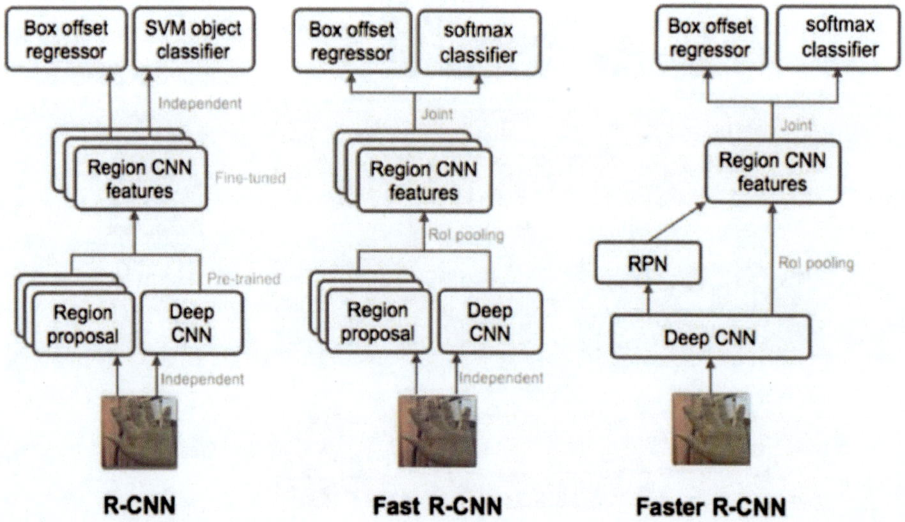


**Fig. 2** R-CNN architectures

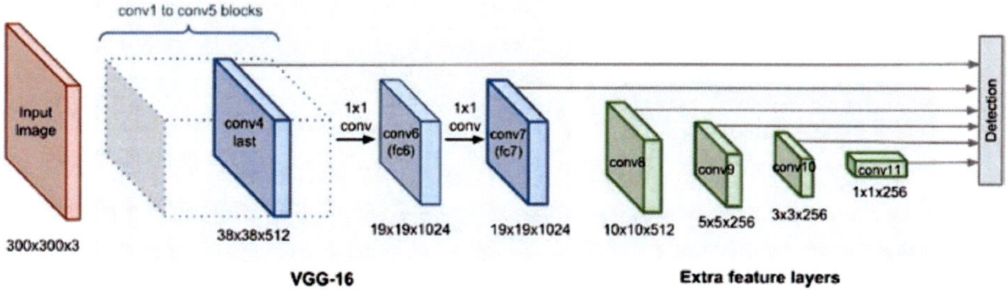


**Fig. 3** SSD architecture with VGG-16 backbone network

which run by this intuition, having the trade-off of a slightly lower accuracy, are SSD (single-shot detection) [32] and YOLO (you only look once) [33]. In single-shot detection, the network architecture comprises of two parts: a head and a backbone network. They are aligned in a pyramidal fashion so that all the convolutional layers have connections to the output layer. This is done because the objects to be detected have varying sizes of dimensions to cater to and a single dimension cannot be considered as the grid size. The concept of anchor boxes exists which corresponds to having a precise size and position with respect to the current cell under consideration. This is responsible for the feature maps being tiled in a convolutional manner. The anchor boxes are dynamically rescaled to coordinate with the size of the receptive field of the dimension being considered (Fig. 3).

Another popular architecture which is considered for object detection is YOLO [33]. In YOLO, the network which forms the backbone network is derived from GoogLeNet in which the convolution layers alter the inception module. The entire foundation architecture is called darknet. There are 24 convolutional layers which carry out the task of extracting features and 2 dense layers which are responsible for predicting the classes (Fig. 4).

An object is said to be detected in a cell if the center of the object is inside the cell. In that case, the cell is thereby held responsible for detecting the object. The tensor components which are predicted as a part of the YOLO algorithm are as follows:

- Bounding box coordinates: There are four values which define the bounding box coordinates – x of the center, y of the center, and height and width of the bounding box. These values fall in the range of 0–1 because of normalization.
- Confidence score: The measure which tells us how likely an object is present in the current cell is confidence score. It is defined as the combined product of the probability of finding an object and the Intersection over Union (IOU) which is a measure of how much the selected region aligns with the ground truth upon the total area covered by them.
- Probability of associated class: After identifying that an object is present, we need to proceed to calculate the measure as to which class the object belongs to. If there are *k* classes to choose from, there are *k* probabilities which are given as the output, and the maximum probability refers to the class most suitable for the object.

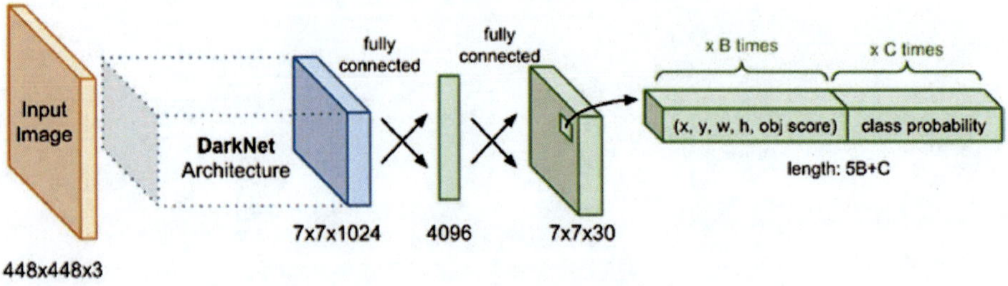


**Fig. 4** YOLO architecture

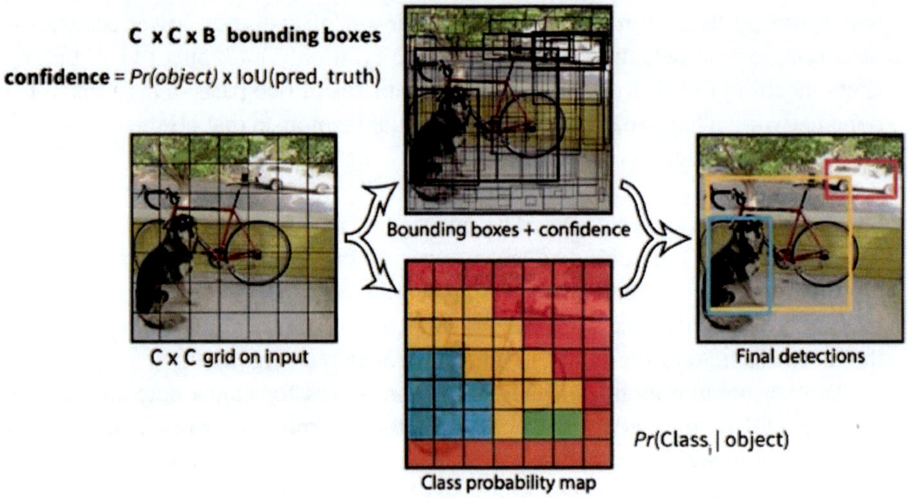


**Fig. 5** Output of the YOLO architecture

If we consider that an image is split into $C * C$ cells and $B$ bounding boxes are formed with $k$ classes to classify into, then the following scores are reported:

- 4 values which are of the bounding box
- 1 confidence level score pertaining to the bounding box
- $k$ probabilities for each class

Thus, there are a total of $C * C * (4B + 1B + K)$ values which are the output tensor (Fig. 5).

YOLOv3 [34] is based on the darknet architecture with an additional 53 layers. In this chapter, we trained the YOLOv3 model against the CMU Hands DB, EgoHands, and Oxford Hands Dataset for recognizing hand images with 80:10:10 for training/validation/testing ratio (Table 2).

The training time was around 11.5 h on a single GTX-1080 Ti graphics card. The RMSE values for training, validation, and test were 0.0713, 0.0991, and 0.0987, respectively. The component was able to successfully identify hand images in image frames (Figs. 6 and 7).

**Table 2** Countries having average Internet speeds lower than 15 megabits/second

| Dataset name | No of frames | Size |
|---|---|---|
| CMU Hands DB | 2758 | 588 MB |
| EgoHands | 4800 | 1.3 GB |
| Oxford Hands Dataset | 4170 | 240 MB |

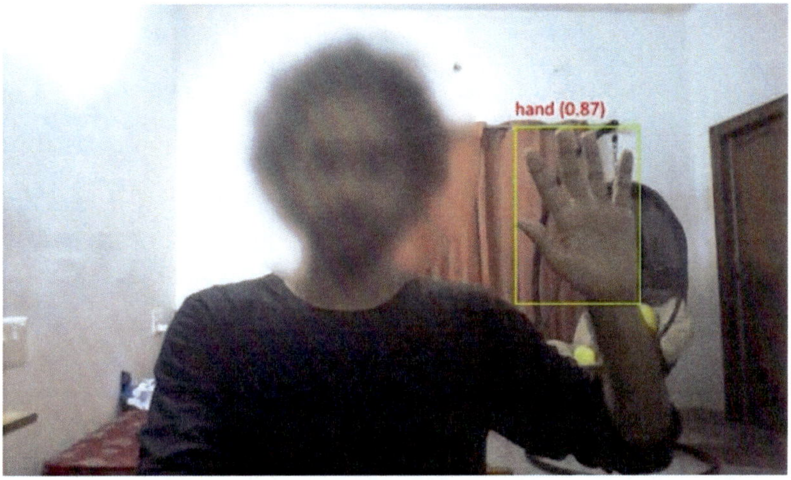


**Fig. 6** Output of component 1

## 3.2 *Component 2: Preprocessing the Hand Image*

The obtained hand image from component 1 has background noise, irrelevant features, and a wrong color space which will not yield proper results when passing down to the convolutional neural network. Thus, there is a need to preprocess the image such that only relevant features are passed down.

The first step in this process would be to convert the color image into gray scale since color spaces can be ignored and does not add additional information with respect to feature extraction (Fig. 8).

Post gray scale conversion, we need to reduce the image to a binary image: containing pixels having only two values, completely black or completely white. This process is referred to as thresholding and is one of the easiest known methods for segmentation. In this chapter, we consider a thresholding method called the Otsu thresholding [36]. The method first calculates a list of threshold values. Thereafter it iterates through all the values by evaluating the measure of spread for the corresponding pixels on either side of the value. The intensities which are lighter than the threshold are called the foreground, and the intensities which are darker than the threshold are called the background. The algorithm computes a threshold value which makes sure that the combined sum of background and foreground spread is minimum. Let us consider the help of a 6x6 image in Fig. 9 to explain this functioning. In this image, we consider that there are six levels of intensity. The histogram

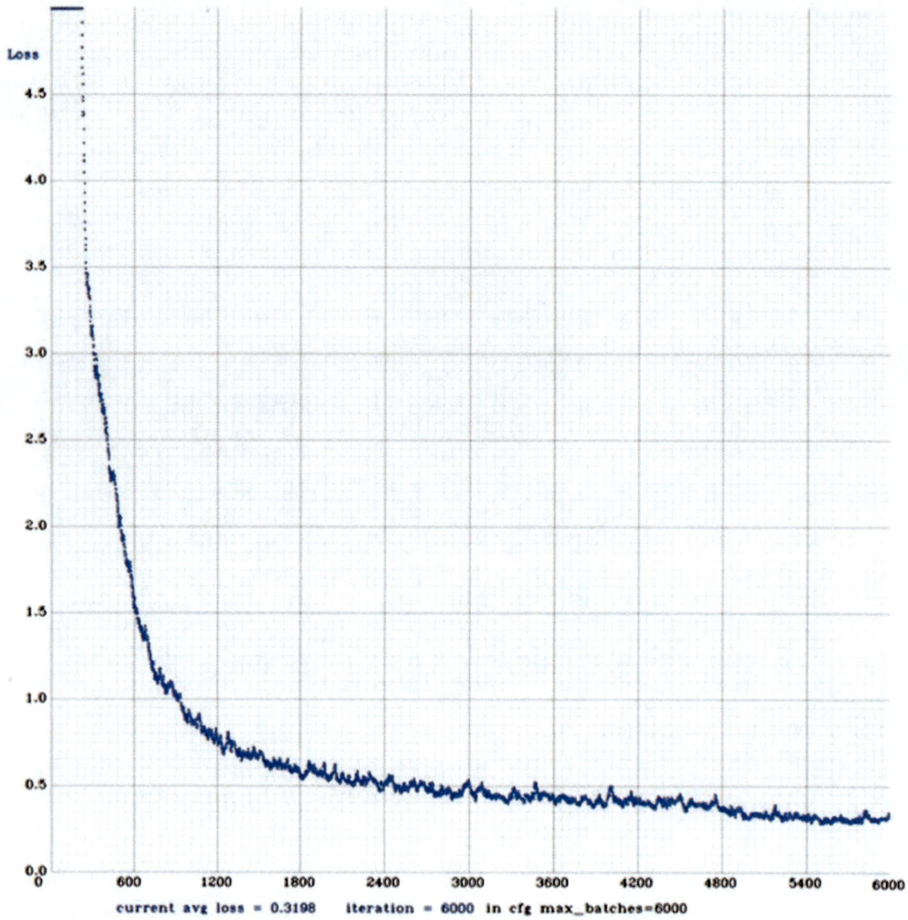


**Fig. 7** Training information for YOLOv3 on the considered hand datasets

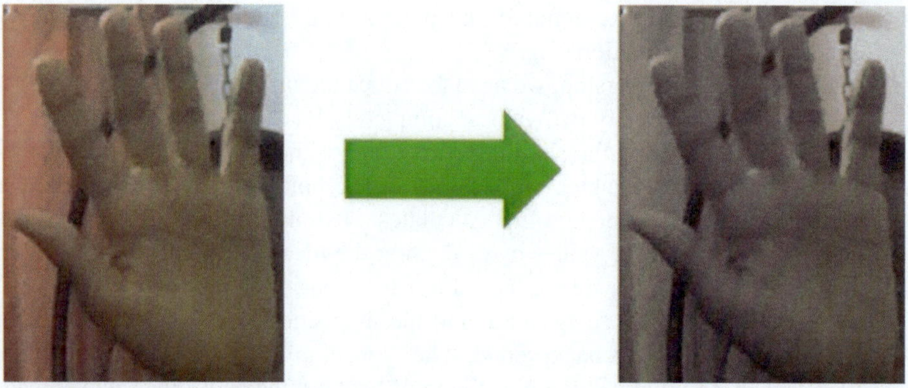

**Fig. 8** Gray scale conversion of hand images

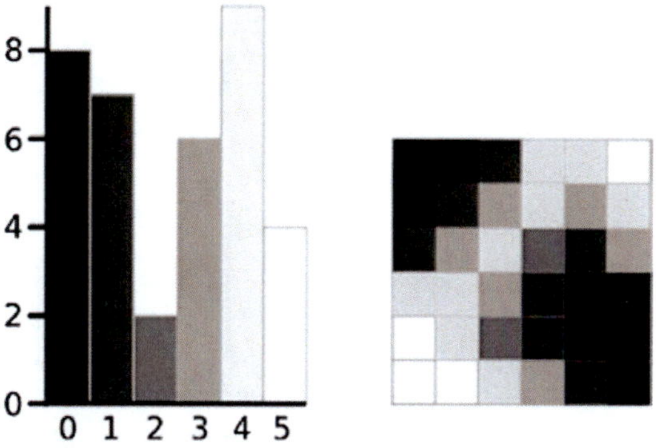


**Fig. 9** 6×6 image and its histogram

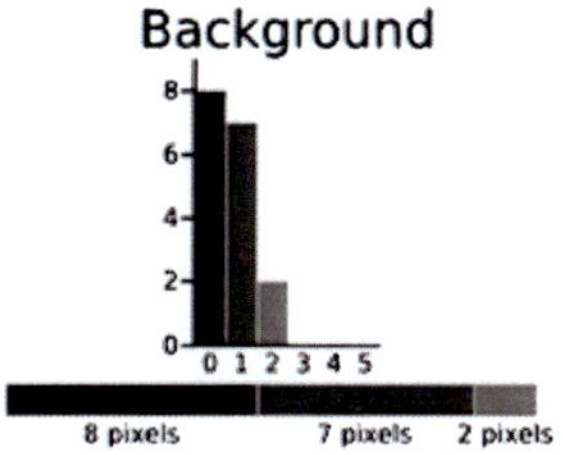


$$\text{Weight} \quad W_b = \frac{8+7+2}{36} = 0.4722$$

$$\text{Mean} \quad \mu_b = \frac{(0 \times 8) + (1 \times 7) + (2 \times 2)}{17} = 0.6471$$

$$\text{Variance} \quad \sigma_b^2 = \frac{((0-0.6471)^2 \times 8) + ((1-0.6471)^2 \times 7) + ((2-0.6471)^2 \times 2)}{17}$$

$$= \frac{(0.4187 \times 8) + (0.1246 \times 7) + (1.8304 \times 2)}{17}$$

$$= 0.4637$$

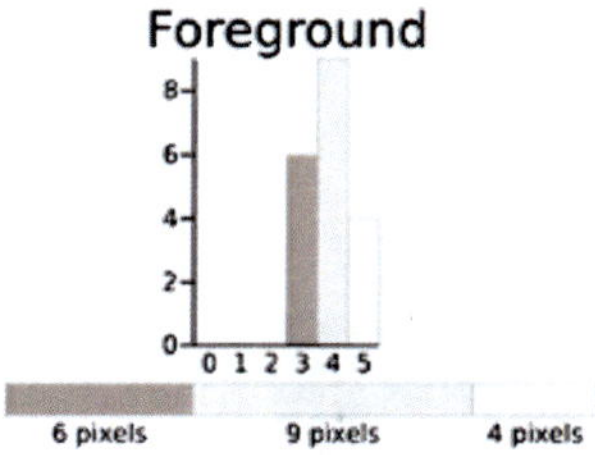


$$\text{Weight} \quad W_f = \frac{6+9+4}{36} = 0.5278$$

$$\text{Mean} \quad \mu_f = \frac{(3 \times 6) + (4 \times 9) + (5 \times 4)}{19} = 3.8947$$

$$\text{Variance} \quad \sigma_f^2 = \frac{((3-3.8947)^2 \times 6) + ((4-3.8947)^2 \times 9) + ((5-3.8947)^2 \times 4)}{19}$$

$$= \frac{(4.8033 \times 6) + (0.0997 \times 9) + (4.8864 \times 4)}{19}$$

$$= 0.5152$$

**Fig. 10** Distinction into background and foreground

is plotted showing the intensity of pixels and the number of pixels having that intensity. Post the plotting of the histogram, we iterate keeping each intensity value as the threshold and compute the background and the foreground. If we consider the threshold as 3, then the background and foreground along with their weight, mean, and variance calculations are shown in Fig. 10.

Post-differentiating it as background and threshold, we calculate a within-class variance which is the sum of the weights multiplied by their variances. For example, this will be 0.4722 $*$ 0.4637 + 0.5278 $*$ 0.5152 which is equal to 0.4909. We continue this process for all the possible intensities as shown in Fig. 11 and choose the threshold with the minimum within-class variance. In this example, we find that threshold 3 gives the minimum within-class variance. Pixels having greater intensities than that of 3 are made as the background, and pixels having intensity lower than or equal to 3 are made as the foreground (Fig. 12).

This same calculation is exposed via the OpenCV method cv.threshold with cv. THRESH_BINARY and cv.THRESH_OTSU as input parameters [37].The output of this component is an image which is binary and has the relevant features which can be used for classification (Fig. 13).

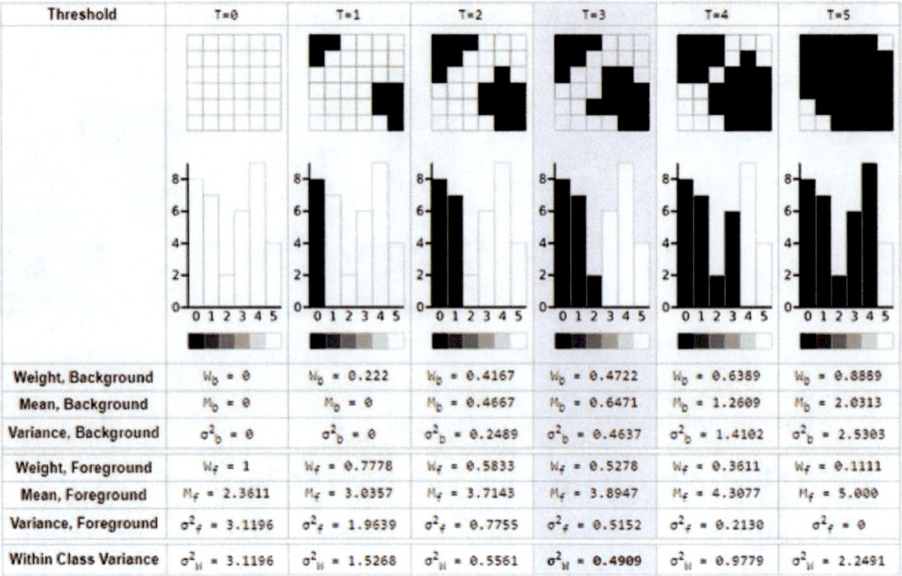


**Fig. 11** Within-class variance for all thresholds

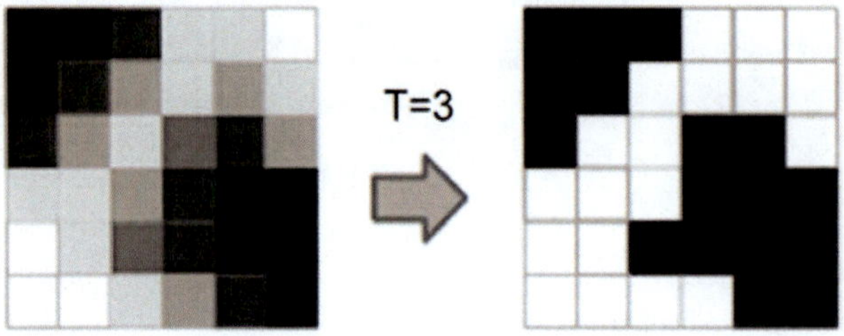


**Fig. 12** Final output of Otsu's threshold method

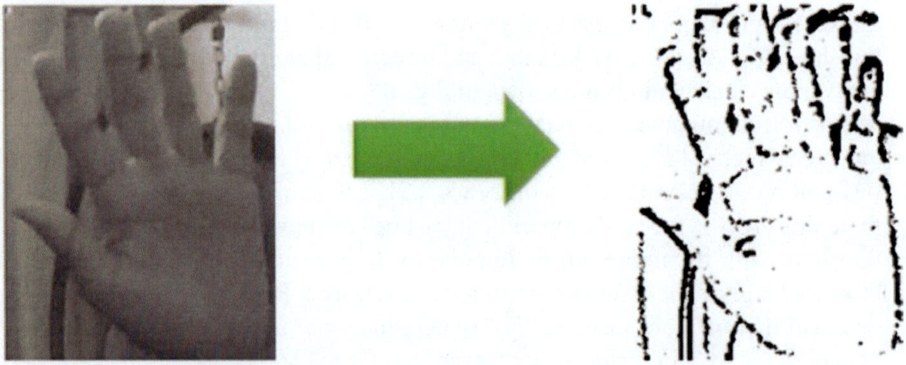

**Fig. 13** Application of Otsu's threshold on hand images

### *3.3 Component 3: Classification of Identified Gesture*

This component will take an input image containing the gesture/sign and will aim to classify the gesture. The literature survey showed that CNN models have a particularly good accuracy when it comes to classification based on images over image processing methods where handcrafted features are used. For us to adopt a CNN model, it is necessary that we get the right dataset and train on them. In this chapter, we adopt MNIST Sign Language dataset [38] and get the relevant training images by passing it down to component 2 to remove the superfluous features (Fig. 14).

The dataset contained all the letters excluding the character "z," and each letter was assigned a class from 0 to 24 alphabetically. There are 27455 in the training data folder and 7172 images in the test data folder.

The architecture of the framed convolutional neural network that was considered in this implementation is depicted in Fig. 15. There were three convolutional layers with max pooling layers alternated followed by a flatten layer, a dense layer, and a dropout layer. This was implemented using Keras with an epoch size of 20 and a batch size of 128 on the training data images. The model achieved an accuracy of 99.583% on the testing dataset as depicted in Fig. 16.

From the confusion matrix in Fig. 17, we can identify that majority of the wrongly classified gestures belonged to class 6 (character G). The system uses the trained weight file in an offline manner and does not need an active Internet connection to make predictions. The total size of the entire system was 112.13 MB.

## 4 Conclusion and Future Work

The chapter focused on the implementation of a human-computer interaction system which translates sign language gestures into a form understandable by people who are not well versed in sign languages. This system is largely targeted for deaf and mute people and can improve digital literacy among them by enabling them to communicate efficaciously. It was implemented as three components using image processing and convolutional neural networks. The first component identified the hand in an image frame. The second component was responsible for preprocessing and removed unnecessary features. The third component classified the image frame with gestures into text. The developed application consumed low memory, runs without the need of an active Internet connection, and works on low-resolution input which is highly effective in developing countries given the resource deficiency. The system achieved an accuracy of 99.583% on the MNIST sign language dataset. Future study should include a leaner system by using the same architecture for detecting hands and identifying the gesture. Also, custom datasets should be developed and trained specifically to the region since sign languages vary from one place to another.

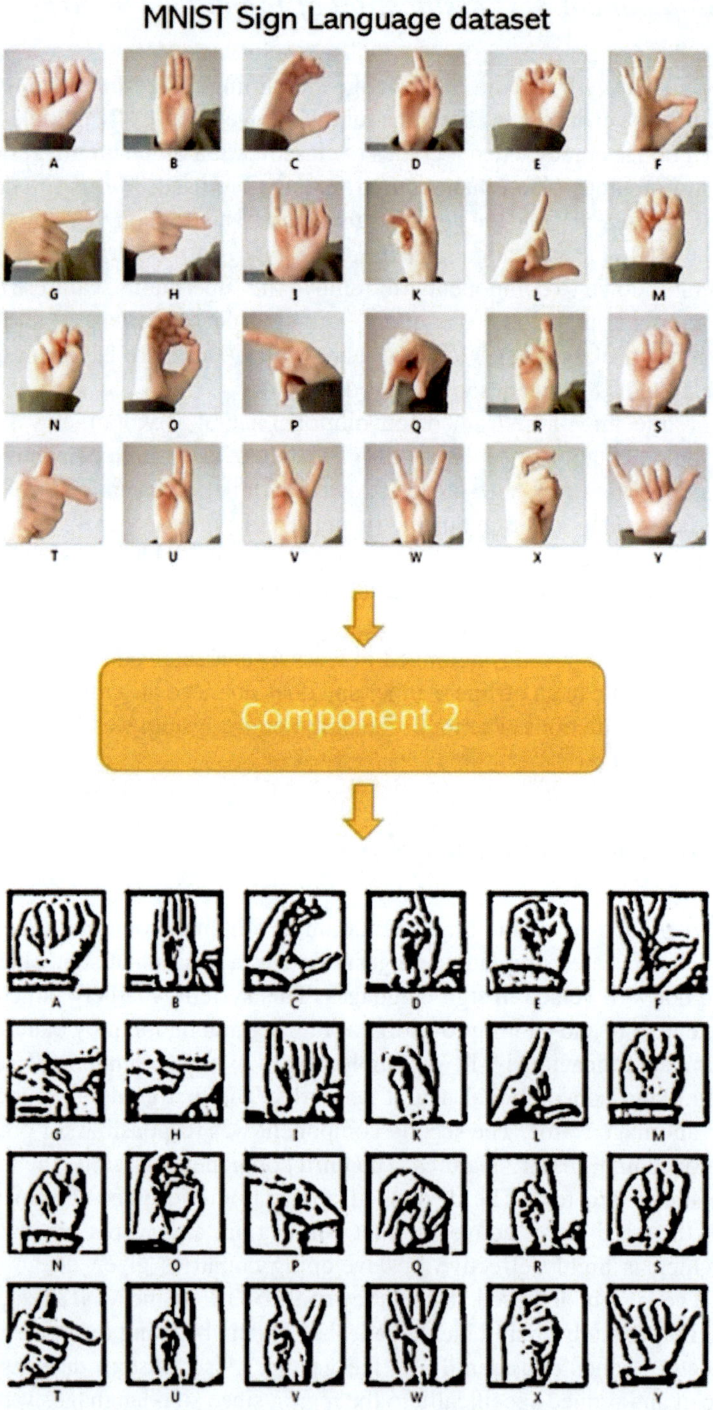


**Fig. 14** Transformation of MNIST Sign Language dataset using component 2

```
Model: "sequential_1"
_________________________________________________________________
Layer (type)                 Output Shape              Param #
=================================================================
conv2d_1 (Conv2D)            (None, 28, 28, 75)        750
_________________________________________________________________
max_pooling2d_1 (MaxPooling2 (None, 14, 14, 75)        0
_________________________________________________________________
conv2d_2 (Conv2D)            (None, 14, 14, 50)        33800
_________________________________________________________________
max_pooling2d_2 (MaxPooling2 (None, 7, 7, 50)          0
_________________________________________________________________
conv2d_3 (Conv2D)            (None, 7, 7, 25)          11275
_________________________________________________________________
max_pooling2d_3 (MaxPooling2 (None, 4, 4, 25)          0
_________________________________________________________________
flatten_1 (Flatten)          (None, 400)               0
_________________________________________________________________
dense_1 (Dense)              (None, 512)               205312
_________________________________________________________________
dropout_1 (Dropout)          (None, 512)               0
_________________________________________________________________
dense_2 (Dense)              (None, 24)                12312
=================================================================
Total params: 263,449
Trainable params: 263,449
Non-trainable params: 0
_________________________________________________________________
```

**Fig. 15** CNN architecture used in component 3

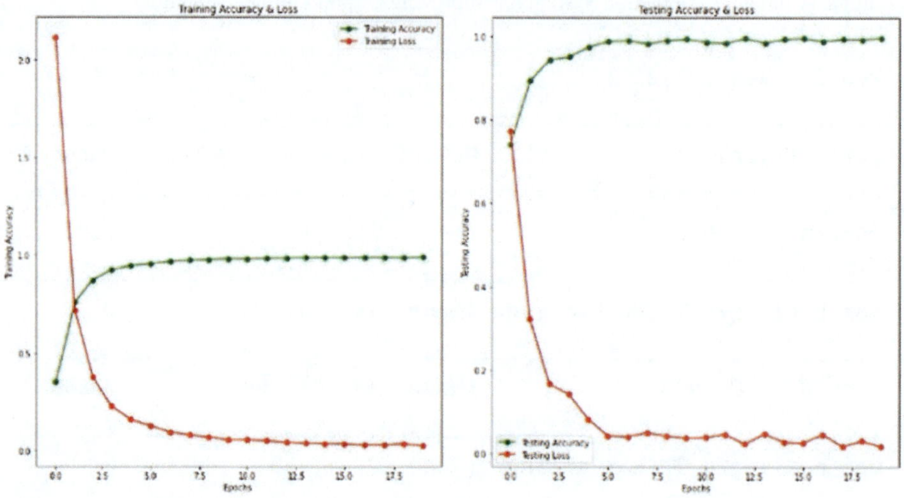


**Fig. 16** Training and testing accuracy

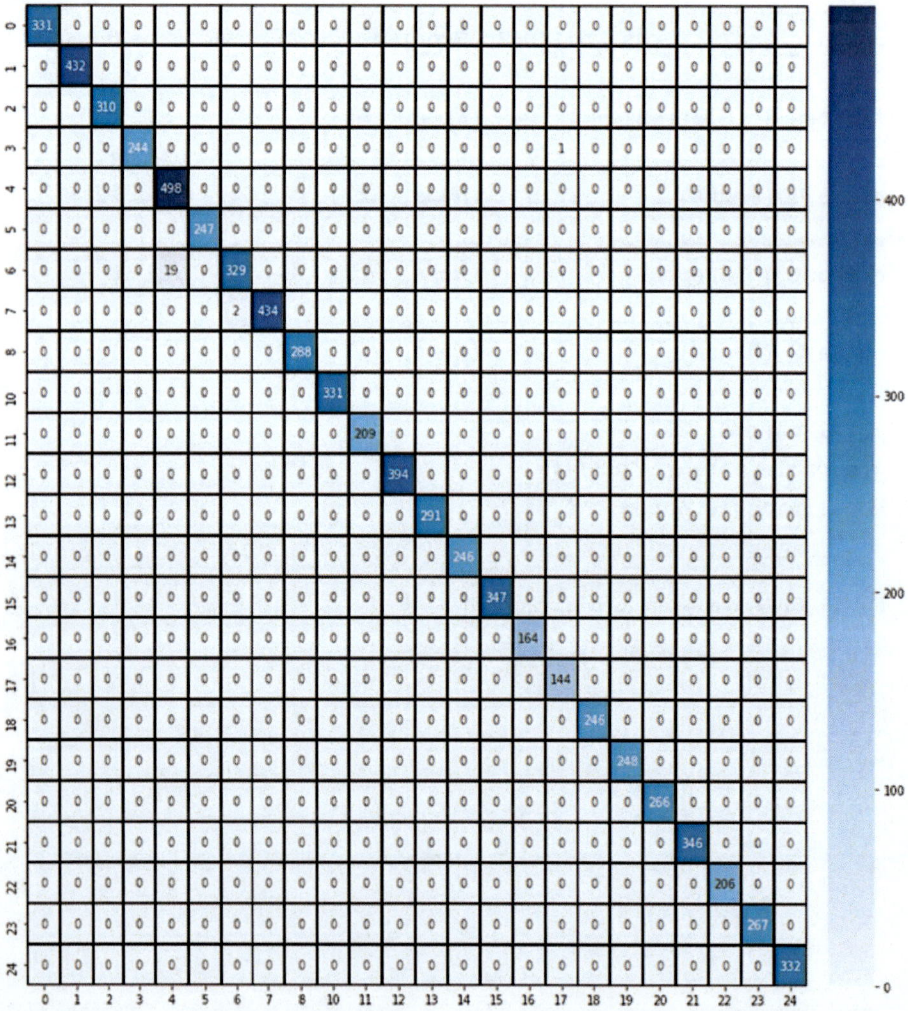


**Fig. 17** Confusion matrix of predicted letters

## References


1. Deafness and hearing loss, in *World Health Organization*, https://www.who.int/news-room/fact-sheets/detail/deafness-and-hearing-loss. Accessed 20 May 2020
2. Degrees of hearing loss, in *Hearing Health Foundation*, https://hearinghealth-foundation.org/degrees-of-hearing-loss. Accessed 20 May 2020
3. Assistive devices for people with hearing, voice, speech, or language disorders, in *National Institute of Deafness and Other Communication Disorders* (2019), https://www.nidcd.nih.gov/health/assistive-devices-people-hearing-voice-speech-or-language-disorders. Accessed 20 May 2020
4. List of sign languages, in *Wikipedia* (2020), https://en/wikipedia.org.wiki/List_sign_of_languages. Accessed 20 May 2020
5. What is digital literacy and why does it matter?, in *Renaissance* (2019), https://www/renaissance.com/2019/02/08/blog-digital-literacy-why-does-it-matter/. Accessed 27 May 2020
6. Technology in developing countries, in *Compassion International Blog* (2020). https://blog.compassion.com/technology-in-developing-countries/. Accessed 27 May 2020
7. Monthly comparisons of internet speeds from around the world, in *Speedtest Global Index*, https://www.speedtest.net/global-index. Accessed 27 May 2020
8. S. Tamura, S. Kawasaki, Recognition of sign language motion images. Pattern Recognition **21**, 343–353 (1988). https://doi.org/10.1016/0031-3203(88)90048-9
9. T. Starner, A. Pentland, Real-time American sign language recognition from video using hidden Markov models. Comput. Imag. Vision Mot. Based Recog, 227–243 (1997). https://doi.org/10.1007/978-94-015-8935-2_10
10. S.N. Sawant, M.S. Kumbhar, Real time sign language recognition using PCA, in *2014 IEEE International Conference on Advanced Communications, Control and Computing Technologies* (2014), https://doi.org/10.1109/icaccct.2014.7019333
11. G. Murthy, R. Jadon, Hand gesture recognition using neural networks, in *2010 IEEE 2nd International Advance Computing Conference (IACC)* (2010), https://doi.org/10.1109/iadcc.2010.5423024
12. S. Bhowmick, S. Kumar, A. Kumar, Hand gesture recognition of English alphabets using artificial neural network, in *2015 IEEE 2nd International Conference on Recent Trends in Information Systems (ReTIS)* (2015), https://doi.org/10.1109/retis.2015.7232913
13. P. Molchanov, S. Gupta, K. Kim, K. Pulli, Multi-sensor system for drivers hand-gesture recognition. in *2015 11th IEEE International Conference and Work-shops on Automatic Face and Gesture Recognition (FG)* (2015), https://doi.org/10.1109/fg.2015.7163132
14. H.-I. Lin, M.-H. Hsu, W.-K. Chen, Human hand gesture recognition using a convolution neural network, in *2014 IEEE International Conference on Automation Science and Engineering (CASE)* (2014). https://doi.org/10.1109/coase.2014.6899454
15. C.J.L. Flores, A.E.G. Cutipa, R.L. Enciso, Application of convolutional neural networks for static hand gestures recognition under different invariant features, in *2017 IEEE XXIV International Conference on Electronics, Electrical Engineering and Computing (INTERCON)* (2017), https://doi.org/10.1109/intercon.2017.8079727
16. J. Nagi, F. Ducatelle, G.A.D. Caro, et al. Max-pooling convolutional neural networks for vision-based hand gesture recognition, in *2011 IEEE International Conference on Signal and Image Processing Applications (ICSIPA)* (2011), https://doi.org/10.1109/icsipa.2011.6144164
17. S. Yang, Q. Zhu, Video-based Chinese sign language recognition using convolutional neural network, in *2017 IEEE 9th International Conference on Communication Software and Networks (ICCSN)* (2017), https://doi.org/10.1109/ic-csn.2017.8230247
18. L. Rioux-Maldague, P. Giguere, Sign language fingerspelling classification from depth and color images using a deep belief network, in *2014 Canadian Conference on Computer and Robot Vision* (2014), https://doi.org/10.1109/crv.2014.20
19. J. Huang, W. Zhou, H. Li, W. Li, Sign language recognition using 3D convolutional neural networks, in *2015 IEEE International Conference on Multi-media and Expo (ICME)* (2015, https://doi.org/10.1109/icme.2015.7177428

20. J. Huang, W. Zhou, H. Li, W. Li, Sign language recognition using real-sense, in *2015 IEEE China Summit and International Conference on Signal and Information Processing (ChinaSIP)*, (2015). https://doi.org/10.1109/chinasip.2015.7230384
21. L. Pigou, S. Dieleman, P.-J. Kindermans, B. Schrauwen, Sign language recognition using convolutional neural networks, in *Computer Vision – ECCV 2014 Workshops Lecture Notes in Computer Science*, (2015), pp. 572–578. https://doi.org/10.1007/978-3-319-16178-5_40
22. A. Tang, K. Lu, Y. Wang, et al., A real time I and posture recognition system using deep neural networks. ACM Trans. Intell. Syst. Technol. **6**, 1–23 (2015). https://doi.org/10.1145/2735952
23. A.K. Tushar, A. Ashiquzzaman, M.R. Slam, Faster convergence and reductions of overfitting in numerical and sign recognition using DCNN, in *2017 IEEE Region 10 Humanitarian Technology Conference (R10-HTC)*, (2017). https://doi.org/10.1109/r10-htc.20178289040
24. O.K. Oyedotun, A. Khashman, Deep learning in vision-based static hand gesture recognition. Neural Computing and Applications **28**, 3941–3951 (2016). https://doi.org/10.1007/s00521-016-2294-8
25. V. Bheda, D. Radpur, *Using deep convolutional networks for gesture recognition in American sign language* (2017), arXiv preprint arXiv:1710.16836
26. M.M. Islam, S. Siddiqua, J. Afran, Real time hand gesture recognition using different algorithms based on American sign language, in *2017 IEEE International Conference on Imaging, Vision & Pattern Recognition*, (2017). https://doi.org/10.1109/icivor.2017.780854(icIVPR)
27. O. Koller, S. Zargarn, J. Ney, R. Bovden, Deep sign: Enabling robust statistical continuous sign language recognitions via hybrid CNN-HMMs. Int. J. Comput. Vision **126**, 1311–1325 (2018). https://doi.org/10.1007/s11263-018-1121-
28. E.K. Kumar, P. Kishore, M.T.K. Kumar, D.A. Kumar, 3D sign language recognition with joint distance and angular coded color topographical descriptor on a 2 – stream CNN. Neurocomputing **372**, 40–54 (2020). https://doi.org/10.1016/j.neucom.2019.09.059
29. R. Girshick, J. Donahue, T. Darrell, J. Malik, Rich feature hierarchies for accurate object detection and semantic segmentation, in *2014 IEEE Conference on Computer Vison and Pattern Recognition*, (2014). https://doi.org/10.1109/cvpr.2014.81
30. R. Girshick, Fast R-CNN, in *2015 IEEE International Conference on Computer Vison (ICCV)*, (2015). https://doi.org/10.1109/iccv.2015.169
31. S. Ren, K. He, R. Girshick, J. Sun, Faster R-CNN: Towards real-time object detection with region proposal networks. IEEE Trans. Pattern Anal. Mach. Intell. **39**, 1137–1149 (2017). https://doi.org/10.1109/tpami.2016.2577031
32. W. Liu, D. Anguelov, D. Erhan, et al., SSD: Single shot multibox detector, in *Computer Vision – ECCV 2016 Lecture Notes in Computer Science*, (2016), pp. 21–37. https://doi.org/10.1007/078-3-319-46448-0_2
33. J. Redmon, S. Divvala, R. Girshick, A. Fahadi, You only look once unified, real-time object detection, in *2016 IEEE Conference on Computer Vision and Pattern Recognition (CVPR)*, (2016). https://doi.org/10.1109/cvpr.2016.91
34. J. Redmon, A. Farhadi, YOLOv3: An incremental improvement, in *Computer Vision and Pattern Recognition (cs.CV)* (2018), arXiv:1804.02767
35. T. Simon, H. Joo, I. Matthews, Y. Sheikh, Hand keypoint detection in single images using multiview bootstrapping, in *2017 IEEE Conference on Computer Vision and Pattern Recognition (CVPR)*, (2017). https://doi.org/10.1109/cvpr.2017.494
36. A. Greensted, The Labe book pages, in *Sitewide RSS,* http://www.labbook-pages.co.uk/software/ingProc/otsuThreshol.html. Accessed 19 May 2020
37. Image Thresholding, in *OpenCV*. https://docs.opencv.org/master/d7/d4d/tutorial_py_threshodling.html. Accessed 27 May 2020
38. Tecperson, Sign Language MNIST, in *Kaggle* (2017), https://www/com/datamunge/sign-lanuage-mnist. Accessed 20 May 2020

# Integrating Web-Based Learning in Uganda: Possibilities and Challenges

**Baguma Asuman, Md Shahadat Hossain Khan, and Hamisi Ramadhan Mubarak**

**Abstract** This chapter presents the possibilities of integrating web-based learning (WBL) in higher educational institutions (HEIs) of Uganda. In response to an increasing demand in using web to facilitate teaching and learning in both blended and online environment, HEIs are encouraging their faculty members to integrate and use these technologies in their teaching practices. However, previous studies have reported that this integration is still at its initial stages in the HEIs of Uganda. Studies have also highlighted that these initial stages, integration of WBL, are encountering a number of constraints that slow down their operation. The aim of this chapter is to identify/explore the major challenges faced by the HEI teachers in using WBL from a broader perspective and to find out the possible remedial measures for integrating WBL in the HEIs of Uganda and other developing countries.

**Keywords** Web-based learning · ICT · Barriers · Online learning · Integration

## 1 Introduction

With the global uptake and the advancement of Information and Communication Technology (ICT), development of internet and a growing acknowledgement of the significance of ICT on teaching and learning, many educational institutions are migrating towards the use of internet and its technologies in order to foster knowledge dissemination in many developing countries [1–5]. A significant amount of studies have shown evidence about the positive impact of using ICT in teaching and learning conditions [3, 6–8]. This, therefore, has led to an increasing demand on the educational institutions to use ICT in teaching skills and knowledge relevant for their respective workplaces. Appreciating the outcomes of ICT on the workplace and everyday life, many educational institutions are trying to streamline their educational curricula, ICT facilities, classroom instructions, in order to bridge the

B. Asuman (✉)
Islamic University in Uganda, Mbale, Uganda

M. S. H. Khan · H. R. Mubarak
Islamic University of Technology, Gazipur, Bangladesh
e-mail: skha8285@iut-dhaka.edu; hamisiramadhan@iut-dhaka.edu



E. Eilu et al. (eds.), *Digital Literacy and Socio-Cultural Acceptance of ICT in Developing Countries*, https://doi.org/10.1007/978-3-030-61089-0_10

existing ICT (technology) gap in teaching and learning [9, 10]. Web-based learning (WBL) is generally considered as a platform of e-learning or online learning since it supports live instructions, videoconferencing, online discussion forums, and more advanced features available in the web [11–14].

A significant amount of studies have reported positive results of using WBL in providing and facilitating education opportunities in developed and less developed countries [10, 11, 15, 16]. More particularly, its usage as an educational platform has given learners and educators an extensive array of new and fascinating learning experiences and teaching environment that is not possible in traditional face-to-face atmosphere [6, 8, 17–19]. It provides students and teachers, in many cases, with a new and diverse range of teaching and learning opportunities such as accessing information at any time and at any place, sharing information and lecture presentations via internet, online group discussions, interactive learning activities, and long distance educational opportunities which is less possible in face-to-face classroom teaching contexts [16, 20, 21].

Due to a high demand of online learning and the potential benefits of WBL, many educational institutions in the world have already integrated it into their educational systems such as United States, Canada, Singapore, Australia, Taiwan, India, China, and so on [1, 3, 8, 22–25]. However, regardless of the huge growing demands of WBL and its benefits, investments on ICT infrastructure, and advancement in specialized teaching and learning opportunities, the adoption of WBL in many developing countries, including Uganda, has been limited due to many confronting constraints [1, 4, 5, 22].

This chapter is organized into 11 sections including Introduction. The rest of the chapter is grounded as follows: the next section presents the *background* of this chapter focusing on the government of Uganda and higher education institutions' emphasis on the use of WBL in the education system. The third section discusses the *Ugandan context* and rationale for incorporating WBL in HEIs. The fourth section explains *what WBL is* and how it is interpreted by different scholars. The next section presents the *literature review* on *usefulness* of web-based learning in Uganda. The sixth section discusses the *literature review* on *problems* related to Integrating WBL. The seventh section has a brief explanation of *methodologies* used in data collection and analysis. The eighth section discusses the *findings* on barriers of integrating WBL in Uganda. The next section discusses suggestions for improving the barriers to the integration of WBL which includes providing solutions and recommendations for eliminating barriers that are encountered in integrating WBL in Uganda and other developing countries. The tenth section gives the implications towards improving current practices and also a summary of how we can improve the problems being faced. Finally, the last section provides a brief conclusion that summarizes the overall ideas of this chapter.

## 2 Background of the Chapter

Uganda, like any other least developing countries, has always used education as a key factor to foster social-economic development and eradicating poverty. Different policies aimed at improving on school enrolment have long been formulated including Universal Primary Education (UPE) in late 1990s that has led to a dramatic increase in the enrolment in primary education countrywide [26]. With higher numbers of primary school graduates, the need for Universal Secondary Education (USE) to ensure free and equal access to secondary school was realized and constituted in 2007 [27]. Though these policies were formulated with good intention, they posed a big challenge to higher education institutions in accommodating the ever-growing number of applicants as a result of UPE and USE.

Many higher educational institutions adopted to the use of ICT, also known as web-based learning into their academic environment to facilitate learning and teaching and to accommodate the growing demands of higher education in the country [28]. Besides supporting sustainability of institutions' enrolments, web-based learning has the potential to enhance educational quality through increased motivation in learning, promoting inquiry-based learning, transforming traditional teaching and learning to beyond classroom environment, and most importantly to prepare individuals for the technology-driven world. Much as the implementation of web-based learning presented a massive potential to higher institutions, many higher institutions are still struggling to adjust with the new educational-technology environment. A number of HEIs have no enough funds to afford ICT technologies necessary for WBL. This is highly attributed to its economic nature and lack of financial support from the government and donor agencies. Other challenges are extensively addressed in the chapter with evidence from literature review, documentation, and the research findings.

In spite of the constraints and challenges that Uganda is facing, there is growing interest in integrating web-based education in teaching and learning process. Web based education or learning continues to be appreciated as an important novelty within higher education. This book chapter provides information about technologies, methods and tools of WBL process. We need an approach to ICT and WBL professional development with different ways to handle the many various situations teachers face at varied levels of teaching while integrating WBL. Thus, the main aim of this chapter is to identify the main challenges faced by teachers towards using WBL and to find out the possible remedial measures for the integration of WBL in educational institutions (HEIs) of Uganda and other less developed countries.

## 3 Ugandan Context and Rationale for Web-Based Learning

Uganda, a landlocked nation in East Africa, is bordered by Tanzania, Rwanda, DRC Congo, Kenya, and South Sudan. Uganda has approximately 34.6 million people distributed across a land area of 241,550.7 km$^2$ [29]. The country's population is expected to grow exponentially given its high population growth rate, attributed to its high fertility, mortality, and net migration rates, thus being more youthful. Youth, therefore, need to be equipped with necessary knowledge, skills, and attitude through proper and modern education for sustainable development.

Over the past few decades, donors have invested over $500 million into African institutions of higher education and over $3 billion to develop centers of excellence in science and technology [30]. The Government of Uganda (GoU), in its vision 2040, considers access to basic education a fundamental human right and a key strategic component for human resource development necessary to transform the country's economy towards a modern and prosperous society. To maximize these economic returns, Science, Technology, Engineering and Innovation (STEI) are considered the key fundamental strategies laid down towards achieving Vision 2040. Therefore, most African countries needs to embrace technology in their educational system in order to improve economic growth [22, 31].

Institutions that have adopted online learning in Sub-Saharan Africa have increased in the recent years. These online learning system, web-based learning, have helped in reducing the application fee and reduced travel cost to institution by allowing them to interact with the institutions remotely. Musiimenta [32] recommended web-based learning that provides flexible and unlimited access to educational resource, improve understanding of learning concepts and enhance engagement which in the long run could improve performance in STEI. Therefore, the GoU, and other stakeholders (donors, investors, and policymakers) need to realize the benefits of web-based learning and how it could significantly contribute towards attainment of nationals' vision and prosperity towards achieving economic growth.

## 4 What Is Web-Based Learning?

Conventional learning has faced a number of challenges that are especially attributed to its statically learning environment and limited learning resources [6]. To address these challenges, a number of institutions have started to transform towards digitalization of teaching and learning process. Web-based learning (WBL), which is commonly defined as learning via the web [5, 6], has played a significant role in reducing and addressing these challenges. Learning via web (WBL), enables the learner to learn at any time any place while utilizing different online learning materials to support and facilitate learning.

Web-based learning, in literature, is often interchangeably used to mean online learning or e-learning. Some authors define WBL as a self-paced learning podium or an educational opportunity which is grounded on the constructivism in which students can actively engage in their learning in a self-leading way and to construct their own knowledge in ways that are beneficial to them [1]. On the other hand, WBL is considered a process of providing online instructions as an instructional program using resources and attributes via the internet (World Wide Web) to create a purposeful and interactive learning experience [33, 34]. In this study we define: WBL with great learning potential that involves the use of web applications design with suitable software tools meant to run on web browsers by the users. It is an important learning platform, which deals with instructions through the internet. Here, students and teachers engaged through online. Students access learning resources and do learning activities via internet. Besides, in this study, when teachers use online platform in a blended context (face-to-face and online teaching), then the blended mode also belongs to WBL. The growth of WBL through internet has offered more opportunities than ever before. These include provision for users to communicate, learn, interact and access information, attending conventional seminars remotely and opening up possibilities that has influenced all public sector, economy and particularly the learning process [1, 8, 23].

The introduction and integration of WBL in teaching and learning of HEIs has been debated by researchers for over half a decade [1, 5, 23, 34]. WBL provides much benefits in teaching and learning, more particularly the use of the internet as an educational tool has provided instructors and learners with a range of interesting and novel learning experiences and teaching approaches, which is not possible in face-to-face classroom instructions [2, 22, 34, 35]. For example: offers independent and student centered learning to students, where they can learn and work at their own pace [3], when they want, influences their own learning by adapting different styles depending on their own needs [36], enables them to study more deeply, thereby motivated by their areas of interest [4, 37], encourages students to access their learning resources and necessary activities on their own while skipping over materials already mastered [35], enables students to join discussions at any time while encouraging those who may face difficulties to ask questions speak face-to-face classroom context to learn [8, 18, 38], accelerates learning through a variety of activities [1, 4, 17], remote access to courses materials enables students to reduce costs and travel time, and encourages interaction between instructors and students [4, 18, 39], students have opportunity to share their ideas with other students, which may support them understanding learning materials in a better way. WBL provides learning materials and learning resources in many different formats [33]. (Table 1)

**Table 1** Comparison between web-based learning and face-to-face learning

| No | Web-based learning | Face-to-face learning |
|---|---|---|
| 1 | Learning is situated via web (internet) (both teachers and students are connected via internet) | Learning is situated in person (both teachers and students are present in person) |
| 2 | This learning can occur anywhere, anytime as long as the learners have access to internet. | This learning occurs when the learners are physically present in a classroom (physical space) at a specific time and date. |
| 3 | The self-paced learning and/or self-regulated learning is the main teaching and learning strategy where the instructor mainly applies student-centered teaching approach | In this context, both teacher-centered and student-centered teaching approaches could be applied, with the learners interacting with the teachers in person |
| 4 | Technology is a condition for this learning. For example, internet, computer, mobile devices are required tools for learning to take place. | In this context, a physical learning space (classroom) is required where students need to present in person. Technology is not a condition for this learning. |
| 5 | Both synchronous and asynchronous communications could be generally used. For instance, students to instructors, and students to students involve mostly nonverbal communications through emails, instant messaging, audios, and videos. However, in few cases, live class could be organized via videoconferencing. | Face-to-face communication should be used where generally students and teachers involve in dialogue, group discussion, and other conversations, which may be done during the formal class. |

# 5 Literature Review on Usefulness of Web-Based Learning in Uganda

WBL is centrally aimed at helping student achieve better learning outcomes as compared to traditional system [40], other stakeholders also benefit in the long run (Fig. 1). In this section, opportunities that WBL brings to different stakeholders are discussed in details.

## *5.1 Policy Makers*

***Scalability*** With the number of higher secondary graduates growing exponentially over the years, institutions of higher education are facing challenges to accommodate all of them [41]. Matovu [41] highlighted that majority of institutions in Uganda are overwhelmed with an increasing number of higher secondary leavers. WBL thus provides policy makers with new possibilities to promote education by allowing qualified students attend classes remotely.

***Inclusiveness*** Policy makers face a great challenge of addressing discrimination of disabled students in a traditional learning environment. WBL systems provides

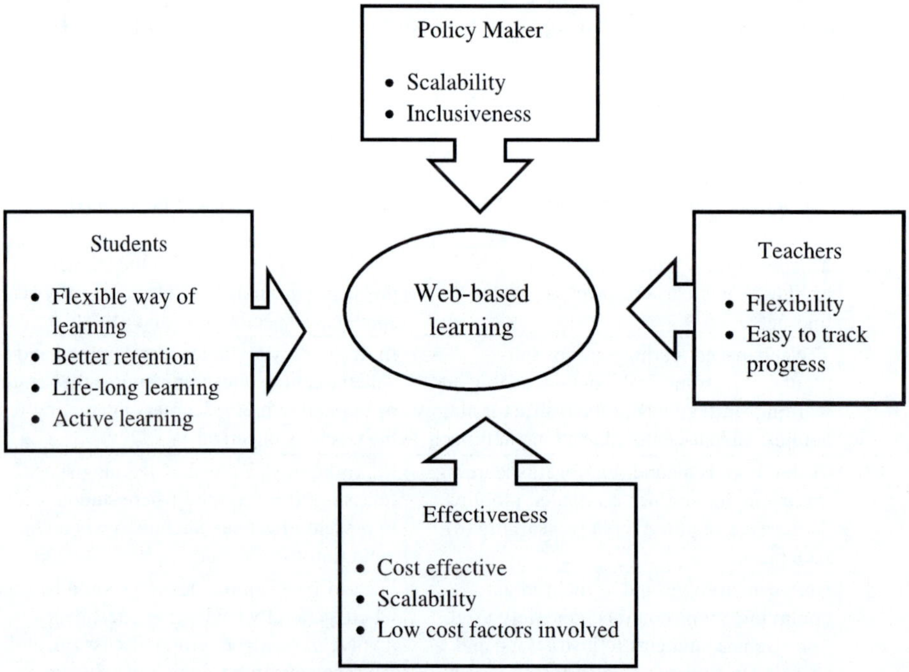


**Fig. 1** Opportunities of web-based learning to students, teachers, administrators, and policy makers

learning opportunities for misfortunate or disabled students who might fear to appear in public due to negative social attitudes or stigmatization [40].

## 5.2 Effectiveness

***Cost-Effectiveness*** As Uganda is among the least developed countries, WBL seems to be more cost-effective as compared to the traditional educational system (face-to-face) [41]. Many institutions lack enough facilities to accommodate students in a traditional education system unlike web-based learning system that grants students a provision to access learning remotely thereby cutting administration cost [42].

***Scalability*** Web-based learning has extended universities' students' level of enrolment to over 15% by allowing them attend classes remotely. With limited resources, institutions have gained confidence to handle an ever-increasing numbers of students who qualify for admission annually.

## 5.3 Learners

***Active Learning*** WBL encourages students to take charge of their learning by actively engaging themselves in building their own knowledge and thus achieving more learning outcomes [43].

***Better Retention*** As web-based learning offers different resources for students to interact with, for example, video, audio, and games, students tend to find learning more fun [42]. Students can interact with the learning material at any time, thus repetition makes information retention much easier.

***Flexibility in Learning*** Given its online nature, learning becomes more flexible to students to learn and complete their studies at their own convenience. Students naturally have different learning capabilities; some are slow learners others are fast learners. Thus, web-based learning approaches these dynamics by providing content with flexibility to suit the learners' needs [44].

***Life-Long Learning*** As learning can be accessed at any point in time, students have the opportunity to relearn and advance their learning thereby creating life-long learning or continuous learning [22].

## 5.4 Teachers

***Easy-to-Track Learning Progress*** As the platform is web-based, teachers can track students' learning progress whenever or wherever they want [42]. Teachers can easily isolate weak students and provide interventions to support and improve their learning.

***Flexibility*** Web-based learning system also provides instructors with increased flexibility in identifying their preferred learning tools that will best suit their learning needs and objectives [45].

## 6 Literature Review on Problems Related to Integrating WBL

A number of educationists have shed lights on the implementations of technology innovations in higher education. Few studies have highlighted on the barriers teachers encounter while integrating and implementing WBL in teaching and learning process. A number of literature studies have focused on the implementation of WBL innovations [1, 22, 23]. Meaningful technology incorporation depends on more than the device uses. It is important that educationists ensure that the integrated technology is in line with how teachers teach and what contents they are teaching [39].

Although, seen as easy and attractive learning means, there are some unresolved challenges which act as barriers in successfully adopting and implementing web-based learning in learning environments. These barriers to WBL can be defined as situations that hinder the integration of WBL in teaching practices. Studies highlighting the barriers encountered by teachers in diffusing WBL in education system have reported a slow pace of its acceptability and usage from different countries [2, 8, 35].

Furthermore, in Turkey, the barriers reported include lack of socialization of learners, poor teacher–learner communication, disregarding the individual differences, additional cost involved to ensure all learners acquire a computer and internet access, insufficient technical knowledge, and skills within learners in operating the system effectively [46]. Lack of technical ICT skills is among the barriers met by instructors while engaging with and implementation of online learning [47].

More challenges were reported in the United States; these included lack of feedback, inaccessible technical support (classroom, informally), unavailability of enough infrastructure (computer labs, software), lack of adequate administrative support, scarcity of resources, heavy workloads, negative attitudes and beliefs, strict policies on technological implementation (whether to administer online homework), and time for incorporating new technologies have been of a great challenge to teachers [48]. Studies have also shown that teachers with negative attitudes towards use of digital technologies are likely to fail to positively transform classes, support learning goals, and incorporate technology into curricular content [5, 19]. Technological stigma possessed among educationists has been observed as a potential barrier to the development and implementation of web-based learning. Educators have expressed the feeling of being overwhelmed with the entire process of adjusting and interacting with the new tools while showing little patience during navigating through minor technical issues [49].

Barriers like high connectivity costs, maintenance costs, ICT technical problems, language barriers, wars and political conflicts, poor education, financial hitches, and lack of qualified ICT instructors were reported in Arab speaking countries to be hindering the integration of WBL especially in pharmacy [5]. Alkoudmani & Elkalmi [5] further reported that countries such as Qatar, Jordan, Egypt, UAE, Saudi Arabia, and Kuwait have adopted the new attempts and approaches of instruction that provide interactive and computer-assisted learning, self-directed and case-based learning, and problem-based learning (PBL).

Challenges related to technology and pedagogy like financial limitations, funding restrictions, limited personal contact, weak internet connections, bandwidth limitations due to many users, software specific limitations were reported in Canada among the nurses doing the PhD [3]. Kurucz et al. [3] said that WBL is meant to bridge the gap between learners and instructors and meet the needs of Canadian nurses in doctoral education since most institutions have only traditional face-to-face programs at campus, affecting those residing in remote areas geographically distant from institutions or who might require more flexibility in their education.

Lack of social interaction, lack of training or knowledge, lack of time, lack of technical support, lack of appropriate content related to specific needs, lack of expertise made up the technical problems, lack of an adequate time-frame to implement online courses, resistant to change and lack of technological assistance were the most challenges reported in different countries affecting the integration of WBL [49]. The most critical barriers were individual's resistance to change, lack of support for the changing roles of students and teachers and difficulties in assessment of students, technological infrastructure of a college were identified as barriers related to the incorporating of WBL in the teaching and learning process in India [49]. Deshmukh & Deshmukh [49] further reported that faculty members who have not taught any web-based courses online listed time as a major barrier to its implementation as compared to faculties who have taught a web-based course.

More barriers to integration of WBL in education have been reported by many other previous studies, for example lack of time to use ICT technology in WBL and little time allocated to incorporate new technologies [22, 48], slow internet connectivity that could affect implementation and accessibility of WBL services [34, 50], lack of funds, getting or gaining access to the necessary equipment required to use during WBL costs good money, not only in terms of the capital cost or purchasing the equipment, but also in maintenance and servicing them, high costs of internet, the cost of hardware and software are high which may result in many institutions failing to accommodate, heavy teaching loads [51, 52], availability of infrastructure support, appropriate access to technical support (classroom, informally), and lack of technical support to incorporate new technologies are the major challenges for teachers and students [3, 5, 11, 50]. Teachers' negative attitudes towards the use of computers poses a potential barrier towards WBL implementation. A wide range of inquiries has established that if teachers don't have trust in using digital technologies (specially ICT and web), they will fail to transform classes, align with learning goals and incorporate technology into course content [15, 52]. Lack of web-based tools, updated hardware and software [25] and significant technology incorporation depends on more than device use. There are important steps like updating software and upgrading hardware to make sure integrating technology aligns with how you teach and what you are teaching [39]. In line with administration, a number of constraints faced were discussed by educationists, such as lack of administrative awareness and acceptability of need to integrate WBL in teaching and learning situation among administrators, academic policy makers, and other administrative offices [24, 35]. Teachers need ICT technologies for classroom implementation and to keep updated with continuous technical advancements.

A number of factors that affect the successful use of technology in teaching and learning, such as pedagogical skills of integrating technology into curriculum, a conducive environment of using technology which is both rewarding and encouraging, the availability of quality staff development opportunities, support from the community, supportive administration, collaboration among community, staff and students, and the availability of software and hardware [35, 46, 47]. Absence of either of these factors, however, will affect the smooth use of technology for teaching and learning. Users need to know about the system, its function, usability, policies, and regulation before embarking on the learning process. Hence, learning is required no matter how they have been previously been taught [48].

## 7 Methodology

Fifty teachers from five different universities in Uganda were purposively selected due to having prior experiences on using WBL in their teaching who were voluntarily participated. A mixed research method, that is both quantitative and qualitative, was used in this study. Specifically, a self-designed questionnaire was constructed, which contained both closed- and open-ended questions. Closed-ended questions were constructed from the validated questionnaire which was guided and created by getting comprehension from other previous studies [51, 53–55] focusing on barriers of using ICT and WBL in HEIs, whereas open-ended questions were designed for gaining much deeper understanding about the problems and to gain responses on possible solutions to overcome the barriers.

In order to analyze quantitative data from closed-ended questions, they were tabulated in the form of frequencies and percentages; the five-point scale items were analyzed using weighted average and chi square test through SPSS software. The qualitative data was analyzed using grounded theory. The main technique was reading the qualitative data several times for gaining central key ideas that draw the final outcomes. In this methodology section, we need to acknowledge that an article from the empirical data of this research was published in 2018 which is "Integration of Web-Based Learning into Higher Education Institutions in Uganda: Teachers' Perspectives." This book chapter is an extension of the published article.

# 8 Barriers of Integrating WBL in Uganda

WBL is faced with some challenges which slow the pace of its acceptability and usage. The barriers in the incorporation of WBL in the teaching and learning system in higher education in Uganda included but is not limited to slow internet speeds, insufficient web-based equipment and access to them, outdated computers, lack of enough monetary means to support WBL, limited technical support, lack of enough time, lack of government support and other organization, lack of support from administration, content relevance with web-based environment, lack of ICT technical skills to be used in WBL by teachers, and insufficient power supply. It is important to note that these factors are in line with other research findings [11, 38].

## *8.1 Internet Speed*

The slow internet speed is one of the major factors affecting the implementation of WBL according to research findings [56]. Most of the lectures in WBL take place through the internet connection and it therefore requires high-speed internet to have live streaming, video conferencing, and upload-download teaching and learning materials during the WBL teaching and learning process. In this research teachers reported that unreliable and slow internet speed hampered regular WBL activities. In many cases teachers were unable to conduct online classes due to on and off internet connection. This same challenge was reported by Wasim et al. [33], Avouris et al. [57], Grönlund [58], Kasse & Balunywa, [59], Liu [24] and Sife, Lwoga & Sanga [60]. Therefore the unreliability of the Internet and its slow speed connectivity is affecting WBL process.

## *8.2 Insufficient Web-Based Equipment*

Providing WBL in an institute in both developed and less developed countries requires ICT devices, such as: desktop computer for accessing online resources, projector with interactive whiteboard or television screen connectivity, a digital camera for showcasing work in film or photographs, USB or WIFI access for document sharing and device connectivity, laptops, cell phones, tablet devices for teacher and student, recording hardware/software for podcasting, e.g., headphone, microphone, and speakers. Most institutions have failed to provide most of these ICT devices to teachers to prepare teaching content and this is because they have no enough devices to support ICT usage in teaching and learning process. This barrier of not having enough equipment was found in the literature by other authors [2, 51, 54, 61–63]. Prior literature agreed with us that lack of ICT devices is the first-level barrier to the integration of ICT for web-based learning. Insufficient web-based equipment and access to them is another major blockade according to the findings and this was also reported in the literature by Chen et al. [64], Park & Ertmer [65]. Most of the HEIs in Uganda have limited web-based equipment that are mostly occupied by the students studying computers science, information technology (IT) programs. On the other hand, students pursuing other programs do not have the required level of opportunities to access WBL equipment. Thus, a teacher has limited access to WBL equipment when available equipment is mostly occupied and shared among students. This constraint is in line with other prior studies [2, 54, 63].

## 8.3 Outdated Computers

Most devices being used in Uganda's higher institutions are old machines and outdated which in many cases cannot support WBL. Teaching and learning using web-based requires ICT devices (computers, cameras, servers etc.) with high processing speed, memory, updated software, that can support the web-based platforms like blackboard, school keep, Moodle, articulate storyline which are required to facilitate classes regularly. Park & Ertmer [65] reported that using up-to-date hardware, software and other related resources is a key feature to uptake WBL at HEIs. It is important to keep devices and software of educational objects updated and integrated [8]. Thus, teachers from HEI's of Uganda perceived that due to having this constraint, many teachers face problems of using WBL in both blended and online mode.

## 8.4 Lack of Funds to Support WBL

Integration of WBL at any institutes and to provide service, require a substantial amount of funds [66]. Thus, to buy WBL devices and to use ICT equipment, to access the internet through an internet service provider (ISP) demand costs [5]. Additionally, training learners and teachers to acquire skills and also to get used to the web-based environment may be another expense to both the teachers, institutions and students. Unfortunately, Uganda as a developing country, is unable to locate required fund to support HEIs [67] and therefore, most institutions do struggle in managing funds for implementing WBL.

## 8.5 Lack of Time

Some teachers in Uganda have skills in using computers in the classroom, but they give little time to use WBL technologies because they are fixed up with other school activities and face-to-face lectures. As a result of their busy school schedule, teachers don't get enough time to prepare and update online content for their WBL and e-learning system. Teachers need to involve much of their time in adopting the skills of using the ICT devices in order to gain more experiences to present curricula content. Significant number of teachers identify time limitations as one of the difficulties in scheduling classes for WBL in their process of teaching and learning. In this research most of the teachers reported that time limitation is one of the difficulties in scheduling classes for WBL in their process of teaching and learning [11]. Most of the teachers in HEIs of Uganda do have many responsibilities, and therefore, they feel requiring additional time is an issue to integrate WBL in their teaching. Similar findings were also reported in the previous literature [47, 49, 60, 68]. This chapter, thus concludes that lack of time is another barrier in implementing WBL.

## *8.6 Lack of Technical Support*

This chapter indicates that lack of technical support during instructions is another challenge faced by teachers in incorporating WBL in HEIs in Uganda. A large number of teachers may not have adequate knowledge and skills in solving the technical problem dealing with WBL. As teachers are not given technical assistance with web-based learning technology, they tend to be reluctant and less motivated towards its usage in classroom environment. Teachers who do not implement WBL in the teaching and learning process, claim that *lack of technical support* is a constraining factor that prevents them from using WBL ICT devices. This was found in other countries struggling to integrate ICT in their education due to lack of technical support such as China [54], Oman [51], Nigeria [31], and Tanzania [22].

## *8.7 Lack of Funding by Government and Other Organizations*

Government and other organizations provide little support to institutions in implementing WBL in teaching and learning. These institutions find it very challenging to implement these activities because they mostly depend on institutes' fund, which heavily rely on students' tuition fee, to manage WBL initiatives. Government's support would reduce the cost to the institutions thereby granting institutions ease in managing and initiating WBL systems [69]. For examples if government could give web-based equipment, like for example computers to each HEIs then this can act as the starting point for them to implement WBL.

## *8.8 Content Relevance with Web-Based Environment*

Content relevance with web-based environment is also identified as a factor affecting incorporating WBL in the education system. For example, programs like medical, engineering, and other (applied disciplines) require hands on learning experiences. The content of these programs may not be handled through WBL environments because it requires face-to-face interactions and practical demonstration [35]. Therefore, teachers in Uganda face difficulties in teaching these contents through such platform [11]. While ICT literacy is generally on the rise, skills for the application of ICT to subject areas in engineering, medicine, and other applied disciplines have not been fully developed. However, few contents of the stated disciplines, and contents of other disciplines which may not have challenges to deliver via web could be suitable means to deliver via WBL. Moreover, WBL may offer many learning experiences that could have been difficult to organize in face-to-face contexts, especially with an interest of ensuring student-centered teaching [70].

### *8.9 Lack of Administrative Support*

In order to get successful implementation of WBL in HEIs of Uganda, the interest and positive initiatives of the top management and leaders at every level are considered to be a crucial factor. This is because without administrative support there will be no budget and therefore, only teachers' willingness would not be enough in incorporating WBL. The policy regarding incorporating WBL in HEIs of Uganda mostly depends on administrators (decision makers in Government, head of institution, and politicians, and the like). Some administrators in HEIs do not support such kind of learning environments due to having financial involvement in it. They claim that it would be a burden for institutions to handle in terms of purchase, maintenance, paying bills. In few cases, they have negative attitudes towards integrating ICT in educational system [2]. In contrast, a substantial amount of studies reported that the integration of ICT facilitates reducing the complexity and to accelerate performing administrative tasks of higher education [71]. This was also identified as constraint in other countries like Nigeria and South Korea [25, 31, 63].

### *8.10 Power Supply*

Most of the ICT and web-based equipment depends on the supply of electric power to operate. In Uganda large areas are still without reliable supply of electricity and the nearest power transmission lines are miles away. Electricity is available only in towns and in very few rural areas. Despite having electricity available in few areas, there is also frequent load shedding due to demand for electricity which is increasing day by day. Thus, the demand of power supply is more than the actual supply from the national grid. This was also indicated in Bangladesh where power affects implementing ICT in education [52]. Nonreliability of electricity supply negatively affects web-based learning because most students who want to enroll for web-based learning are the ones far from the good and high ranked institutions which are mostly in the urban areas and another category who is busy with their work.

## 8.11 Lack of ICT Skills

Online users require IT basic prerequisite skills. Absence of these skills may lead to difficulties in navigating through online system and thereby finding it not appealing nor user friendly. In Uganda teachers and most beginning students at High institutions lack of ICT skills which hinders the incorporation of WBL. There are two aspects: first, teachers should know how to conduct WB teaching (both blended and online) via ICT-supported infrastructure, such as upload learning materials, conducting online lectures, organizing group discussion, peer review, and many other learning experiences. Second, students should also require a certain level of digital skills for attending WBL. Therefore, both teachers and learners find it hard to navigate and interact with the WBL platforms hence affecting the teaching and learning.

Besides, WBL uses specific hardware and software, which in some cases are complicated to be used by some of the teachers/lecturers who have limited ICT skills. Lack of technical knowledge, skills, and competence in in creating interactive exercises on the Web was also reported by other studies [38, 54, 69]. There is a need to develop a strategic approach to initiate ICT professional development (PD) so as to prepare users to WBL acceptability (Table 2).

**Table 2** Brief summary of identified constraints

| No | Name of constraints | Brief descriptions |
|---|---|---|
| 1 | Internet speed | Unreliable and intermittent connectivity. |
| 2 | Insufficient web-based equipment | WBL equipment such as desktops, laptops, cameras, microphones, tablets, recording devices, and many others. |
| 3 | Outdated computers | Old and outdated hardware, software which have failed to support WBL activities. |
| 4 | Lack of funds to support WBL | HEIs of Uganda have lack of funds to manage and to operate WBL activities. |
| 5 | Lack of time | Teachers having little time to use WBL technologies because of their other priority tasks and heavy teaching responsibilities. |
| 6 | Lack of technical support | Lacks of technical support during WBL activities, many teachers feel uncomfortable of using WBL activities. |
| 7 | Lack of funding by government and other organizations | Lack of support from government and other organization of Uganda to support HEIs on the costs incurred in implementing WBL. |
| 8 | Content relevance with web-based environment | All contents from all disciplines are difficult to conduct teaching via WBL. |
| 9 | Lack of administrative support | Lack of commitment and interest of the top management affects the integration of WBL. |
| 10 | Power supply | WBL requires continuous power supply, which in many cases may not be possible in HEIs of Uganda. |
| 11 | Lack of ICT skills | Lack of ICT skills for both teachers and students is another factor that hinders WBL initiatives. |

## 9 Suggestions for Improving the Barriers to the Integration of WBL

In order to improve the contexts (problems facing integration of WBL) of HEIs of Uganda, first we should step forward to remove barriers by solving and reducing the reasons for the occurrence of these difficulties. Government, educators, teachers, rectors, university vice-chancellors and other stakeholders should work together for overcoming the above-mentioned barriers so that a meaningful integration of WBL into teaching-learning process could be done. The following strategies could be maintained in different stages, which were identified mostly open-ended questions asked for the participants (see the methodology section):

*First,* HEIs need to provide fast Internet to support WBL platform by increasing the bandwidth, upgrading devices, getting services from companies which provides fast speeds. Institutions need to increase the bandwidth so that both teachers and students can have easy access to academic instructional materials. Furthermore, in most cases, unused internet-connected devices need to be switched off because they keep interrupting marginal amounts of the bandwidth which may slow down the speed. Also students should be allowed to have few devices to avoid too much bandwidth usage which may result in slow internet speeds.

*Second,* HEIs and government needs to purchase ICT supported tools, including a teacher's laptop, desktops, projectors and storage devices to support the integration of WBL in teaching and learning process. Resource sharing techniques could be an alternative solution until HEIs of Uganda have enough fund for purchasing WBL equipment. For example, a well-equipped institute may allow another low-equipped institute for accessing their WBL equipment.

*Third,* outdated computers should be disposed of and thereafter, purchase new computers with high processing speed and memory, which can support WBL. New software needs to be installed and updated regularly on computers for smooth operation. HEIs and government can pattern with ICT companies to have computers serviced regularly and software provided and updated at a lower cost.

*Fourth,* in light of financial struggles that HEIs are experiencing, they therefore need to mobilize additional resources for investment in research and ICT from government, donors, public private partnerships (PPP) to improve innovation and scale up its use. Support for WBL at both the national and individual level is important, for example local government, parents, and other stakeholders should play a meaningful role in fundraising through different activities such as organizing fairs, events, football match, or the like for generating funds.

*Fifth,* in terms of lack of time for teachers, it is important for administrators to cooperate with them by providing sufficient time to implement new technologies in the WBL environment. For example, an institution can reduce the teacher's extra load for motivating them to engage in WBL. This will give them more time to concentrate on WBL activities.

*Sixth,* institutions should provide technical support to teachers for implementing WBL in HEIs of Uganda. The HEIs may introduce two policies to solve this issue:

first, each institute, which may not have ICT technicians, should recruit an ICT technician who will provide support to the students and teachers whenever technical problems arise. Second, HEIs may provide short training on improving basic ICT skills so that teachers and students have minimum skills for handling technical problems. Thus, the university stakeholders like teachers and administrators should be given support so that they can be able to teach and deliver without being interrupted with technical issues.

*Seventh,* the Government of Uganda should support HEIs towards implementing WBL by allocating the required funds, providing training to teachers, giving ICT equipment to support WBL, and encouraging the citizens to have further studies through web-based environment. In additional to the above, lack of funding by government and other organizations can be resolved by increasing the funds for Education sector, mobilizing resources for ICT and innovation to widen its use. Governments can allocate budget for research and development on innovative uses of ICT in HEIs, for using software and hardware more affordable and relevant for teachers and students. Government and other organization can also pattern with companies to provide services like maintenance of ICT equipment, supply of hardware on a cheaper price. Government should initiate a holistic approach towards the development of plans and policies in relation to WBL by involving different stakeholders on how to incorporate ICT skills in the curriculum of HEIs of Uganda.

*Eighth,* on content relevance, WBL development needs continuous update, change, and refinement of course content to match the contexts. Thus, the content has to be constantly evaluated against the course's learning objectives. For students with practical sessions which require their presence, face-to-face lectures could be organized and thereafter, the videos of these contents should be uploaded.

*Ninth,* in administrative support, administrators have a crucial role to adopt and invest in ICT in HEIs of Uganda. They should take initiatives to reduce the digital gap among teachers and administrators. They should allocate adequate money in developing the infrastructure of ICT and the educational system, such as initiatives of building and implementing WBL platforms. Additionally, they should formulate policies aimed at eradicating other problems which needs administrative supports, such as arranging technicians to support teachers for technical support, finding more cost-efficient options for implementing WBL.

*Tenth,* in order to ensure continuous power supply, government should take initiatives in building big power plants to supply and extend electricity in all rural areas. The construction of large hydropower plants through public and private investments can be one of the key strategic interventions to resolve the problem of power supply and load shedding in Uganda. Priority for continuous electric supply should be given to those areas where HEIs are located.

*Eleventh,* in terms of lack of skills, teachers should develop their profession by integrating the use of WBL and ICT into their teaching process and also have enough experience in such fields. In order to improve teachers' technical skills towards integrating WBL, two specific strategies could be implemented: first, preservice professional development training (before starting university teaching) focusing on ICT skills should be offered for those graduates who will be entering university teaching. Therefore, teachers should be equipped with basic ICT knowledge and skills before joining the teaching profession. Second, service training could be offered for those teachers who are offering WBL or who are interested to offer WBL by HEIs focusing on offering basic ICT skills. Uganda needs to adapt best practices and lessons learnt among other countries. In order to improve performance expectancy, proper training to students and lecturers on the benefits of using the WBL learning should be introduced. This knowledge, thereafter, can be used to inform the development of plans and policies to better support WBL in HEIs of Uganda.

## 10 Implications

This chapter has implications towards improving current practices. For instance, it will act as an input for teachers to realize the constraints of implementing WBL, to provide knowledge on how to overcome them. This will also act as an inspiration to use WBL in higher education of Uganda particularly and other developing countries in general. The results of this chapter will act as a means to understand various issues and increase public awareness in motivating institutions which have not integrated and implemented WBL in their teaching and learning process yet, hence reaping from the earlier mentioned benefits in widening educational opportunities for both the educators, teachers and students (Table 3).

**Table 3** Brief discussion on ways of improving this factor

| No | Name of constraints | Way of improving this factor |
|---|---|---|
| 1 | Internet speed | HEIs and government need to provide fast internet to support WBL platform through increasing the bandwidth, upgrading devices, getting services from companies which provides fast internet speed. |
| 2 | Insufficient web-based equipment | HEIs should purchase WBL equipment sets, including a teacher laptop, desk tops, projectors and storage devices to support the integration of WBL in teaching and learning contexts. |
| 3 | Outdated computers | HEIs should formulate policies to manage old computers and thereafter replace them with new and updated one with latest software installed. |
| 4 | Lack of funds | Sufficient funds should be allocated for integration of WBL in Uganda. |
| 5 | Lack of time | Teachers teaching loads in face-to-face might be reduced and replaced with WBL activities. |
| 6 | Lack of technical support | Technical support during WBL instructions should be ensured by each HEI for smooth operation of WBL. |
| 7 | Lack of funding by government and other organizations | Government and other organizations should ensure continuous support to HEIs by mobilizing funds, increasing sector money, providing training to teachers in WBL. Government needs to take a holistic approach towards the development of ICT in education plans and policies. |

(continued)

**Table 3** (continued)

| No | Name of constraints | Way of improving this factor |
|---|---|---|
| 8 | Content | WBL development needs continuous update, change, and refinement of course content to match the courses objectives. |
| 9 | Lack of administrative support | Policies should be developed for minimizing gap between administrations and HEIs of Uganda. |
| 10 | Power supply | Government should have big power plants producing required electricity and distribute in all rural areas. In short term, priority should be given those areas where HEIs are located. |
| 11 | Lack of ICT skills | In order to provide them basic ICT skills, PD program should be organized by each HEI in every year. |

This chapter will act as a tool for building knowledge and for facilitating learning because it contributes to the existing literature to fill the gap, since there was no much research that had been done concerning WBL in HEIs in Uganda and other developing countries so that they can support the implementation WBL in HEIs. This chapter provides useful information that will act as a guideline for those who are working on the implementing and designing WBL in Uganda and other developing countries. It will also provide guidelines to the policymakers working on new web-based policies in relation to implementing WBL in Uganda.

Furthermore, it will act as a way to identify lies and to support truths. Curriculum developers and course designers will use this chapter as source of content in understanding barriers and solutions to integrating web-based learning in teaching and learning in HEIs, which will be of much benefit and novelty in developing the syllabus for various programs that can be taught through web-based environments.

## 11 Conclusion

In this digital age, ICT is changing rapidly, therefore flexibility and compatibility to adapt different ICT tools and platforms are important issues in teaching at HEIs. Therefore, WBL has a new dimension, advancing classroom learning to a new level through the creation of internet-enabled virtual interactive communities of learners and teachers. Providing flexibility to both students and faculties, WBL will attract students outside a campus location to join institutions. This provides the institution an upper hand in a competitive environment since educational system is changing with IT advancement where interactive classroom, effective instructional materials, and technical skills improve the learning adaption of the learner.

Since slow internet speeds, lack of professional skills, competence, and accessibility to equipment have been found to be critical components for WBL technology integration in HEIs, WBL resources such as software and hardware, professional development programs for enhancing teachers' technical skills, arranging required time for teachers, technicians for technical support, all these should be provided to teachers for ensuring WBL in the HEIs of Uganda. Besides, continual support from Government of Uganda to institutions is necessary in order to have WBL run effectively and efficiently. In-depth research should be done in order to clarify each identified constraint of integrating WBL in Uganda. The future research will provide much concrete knowledge about understanding each barrier that is discussed in this book chapter.

## References

1. P. Chatwattana, P. Nilsook, A web-based learning system using project-based learning and Imagineering. Int. J. Emerg. Technol. Learn. **12**(5), 4–22 (2017). https://doi.org/10.3991/ijet.v12i05.6344
2. F. Mohammadi, A. Abrizah, M. Nazari, M. Attaran, What motivates high school teachers to use web-based learning resources for classroom instruction? An exploratory case study in an Iranian smart school. *Comput. Hum. Behav.* **51**(PA), 373–381 (2015). https://doi.org/10.1016/j.chb.2015.05.016
3. S. Kurucz, L. Rietze, A. Lim, M. Swamy, Web-based learning: A bridge to meet the needs of Canadian nurses for doctoral education. Can. J. High. Educ. **45**(1), 37–47 (2015)
4. C.S. Chai, L. Tan, F. Deng, J.H.L. Koh, Examining pre-service teachers' design capacities for web-based 21st century new culture of learning. Australas. J. Educ. Technol. **33**(2), 129–142 (2017). https://doi.org/10.14742/ajet.3013
5. R. Alkoudmani, R. Elkalmi, Challenges to web-based learning in pharmacy education in Arabic language speaking countries. Arch. Pharm. Pract. **6**(3), 41 (2015). https://doi.org/10.4103/2045-080x.160989
6. N. Hamzah, A. Ariffin, H. Hamid, Web-based learning environment based on students' needs. IOP Conf. Ser. Mater. Sci. Eng. **226**(1) (2017). https://doi.org/10.1088/1757-899X/226/1/012196
7. G. Burch, J.A. Heller, J.J. Burch, N.A. Heller, Web-based and face-to-face classes: Are there unintended outcomes? J. Manag. Dev. **35**(8), 1031–1044 (2016). https://doi.org/10.1108/JMD-06-2015-0088
8. V. Mkrttchian, I. Krevskiy, A. Bershadsky, T. Glotova, L. Gamidullaeva, S. Vasin, Web-based learning and development of university's electronic informational educational environment. Int. J. Web-Based Learn. Teach. Technol. **14**(1), 32–52 (2019). https://doi.org/10.4018/IJWLTT.2019010103
9. B.-A. Charles, Factors influencing teachers ' adoption and integration of information and communication technology into teaching: A review of the literature Charles Buabeng-Andoh. Int. J. Educ. Dev. Using Inf. Commun. Technol. **8**(1), 136–155 (2012)
10. M.S.H. Khan, L. Markauskaite, Technical and vocational teachers' conceptions of ICT in the workplace: Bridging the gap between teaching and professional practice. J. Educ. Comput. Res. **56**(7), 1099–1128 (2018). https://doi.org/10.1177/0735633117740396
11. B. Asuman, M.S.H. Khan, C.K. Clement, Integration of web-based learning into higher education institutions in Uganda: Teachers' perspectives. Int. J. Web-Based Learn. Teach. Technol. **13**(3) (2018). https://doi.org/10.4018/IJWLTT.2018070103
12. K. J. Coleman, The Presidential Nominating Process and the National Party Conventions, 2016: Frequently Asked Questions (2016)
13. P. Ifinedo, Acceptance and continuance intention of web-based learning technologies (WLT) use among University students in a Baltic country. Electron. J. Inf. Syst. Dev. Ctries. **25**(6), 1–20 (2006)

14. E.T. Lwoga, M. Komba, Antecedents of continued usage intentions of web-based learning management system in Tanzania. Educ. Train. **57**, 738 (2015)
15. H. Mahdizadeh, H. Biemans, M. Mulder, Determining factors of the use of e-learning environments by university teachers. Comput. Educ. **51**(1), 142–154 (2008). https://doi.org/10.1016/j.compedu.2007.04.004
16. C.S. Nam, T.L. Smith-jackson, Web-based learning environment: A theory-based design process for development and evaluation. J. Inf. Technol. Educ. **6**, 23–43 (2007)
17. R. Carrillo, E. Lavoué, Y. Prié, Towards qualitative insights for visualizing student engagement in web-based learning environments, pp. 893–898 (2016). https://doi.org/10.1145/2872518.2891076
18. M. Yağcı, D.A. Sırakaya, G. Özüdoğru, The investigation of attitude and readiness of information and communication technologies pre-service teachers toward web based learning. Procedia. Soc. Behav. Sci. **174**, 1099–1106 (2015). https://doi.org/10.1016/j.sbspro.2015.01.800
19. A. Cordier, F. Derbel, A. Mille, Observing a web based learning activity: a knowledge oriented approach, Research report, LIRIS UMR CNRS 5205 (2015)
20. A.B. Nordin, N. Alias, Web-based teaching and learning approach (WBTLA) usability in institutions of higher learning in. Malaysian Online J. Educ. Technol. **1**(2), 44–55 (2013)
21. M.S.H. Khan, M. Nambobi, M.S. Ali, MOOCs in TVET sectors of developing countries: Benefits, possibilities, and challenges, in *Emerging self-directed learning strategies in the digital age*, (IGI Global, Hershey, 2018), pp. 22–47
22. E.T. Lwoga, M. Komba, Antecedents of continued usage intentions of web-based learning management system in Tanzania. Educ. Train. **57**(7), 738–756 (2015). https://doi.org/10.1108/ET-02-2014-0014
23. C.W. Tsai, Applying web-based co-regulated learning to develop students' learning and involvement in a blended computing course. Interact. Learn. Environ. **23**(3), 344–355 (2015). https://doi.org/10.1080/10494820.2013.764323
24. S.F. Petra, J.H. Jaidin, J.S.H.Q. Perera, M. Linn, Supporting students to become autonomous learners: The role of web-based learning. Int. J. Inf. Learn. Technol. **33**(4), 263–275 (2016). https://doi.org/10.1108/IJILT-05-2016-0017
25. S.H.J. Liu, Y.J. Lan, Social constructivist approach to web-based EFL learning: Collaboration, motivation, and perception on the use of Google Docs. Educ. Technol. Soc. **19**(1), 171–186 (2016)
26. T. Weatherholt, R. Jordan, L. Crouch, E. Barnett, J. Pressley, Challenge and drivers of over-enrollment in the early years of primary school in Uganda. Int. J. Early Child. **51**(1), 23–40 (2019)
27. Y. Jafari, *Effects of Using Capitation Grants on Access and Academic Performance of Lower Secondary Schools in Uganda: A Case Study of Koboko Municipality* (Makerere University, 2019)
28. H. Wanga, F. Ngumbuke, J.O. Oroma, Challenges of e-learning in developing countries: The Ugandan experience. Retrieved **12**(12), 2015 (2012)
29. U. Statistics, National Population and Housing Census 2014 (2014)
30. C. for Africa, *Still our common interest: Commission for Africa Report 2010* (Commission for Africa, London, 2010)
31. S.S. Adebayo, The impact of application of information and communications technology (ICT) in the administration of polytechnics in Ogun state, Nigeria. Int. J. Students Res. Technol. Manag. **1**(5), 505–515 (2013)
32. A. Musiimenta, W. Tumuhimbise, M. Nankunda, E. Bangumya, J. Atuhaire, Electronic learning may improve the teaching and learning of mathematics and science in marginalized schools in Nakivale refugee settlement, Uganda: A baseline analysis. J. Edu. Develop. **3**(2), 63–72 (2019). https://doi.org/10.20849/jed.v3i2.611
33. J. Wasim, S.K. Sharma, I.A. Khan, S. Jamshed, Web-based learning innovations. Int. J. Comput. Sci. Inf. Technol. **5**(1), 5859–5864 (2014)

34. B. Hariadi, M.J.D. Sunarto, P. Sudarmaningtyas, Development of web-based learning application for Generation Z. Int. J. Eval. Res. Educ. **5**(1), 60 (2016). https://doi.org/10.11591/ijere.v5i1.4523
35. G. Marzano, L. Ochoa-Siguencia, Challenges of web-based participatory learning. Soc. Integr. Educ. Proc. Int. Sci. Conf. **2**, 458 (2017). https://doi.org/10.17770/sie2017vol2.2395
36. G. K, I. Tusubira, A framework for the integration of e-learning in higher education institutions in developing countries. Int. J. Educ. Dev. Using Inf. Commun. Technol. **9**(2), 19–36 (2013)
37. A. Sudha, S. Amutha, Higher secondary learners' effectiveness towards Web Based Instruction (WBI) on chemistry. Univ. J. Educ. Res. **3**(7), 463–466 (2015). https://doi.org/10.13189/ujer.2015.030706
38. B. Asuman, S.H. Khan, C.K. Clement, Constraints encountered by teachers to the implementation of web-based learning in Uganda. IUT J. Eng. Technol. **12**(2), 21–28 (2015)
39. Y. Ding, P. Zhang, Practice and effectiveness of web-based problem-based learning approach in a large class-size system: A comparative study. *Nurse Educ. Pract.* **31**(April), 161–164 (2018). https://doi.org/10.1016/j.nepr.2018.06.009
40. A.S. Weber, Web-based learning in Qatar and the GCC states. Weill Cornell Med. Coll. Qatar, Educ. City (2016)
41. M. Matovu, Distance education in Uganda: Issues, opportunities, and challenges. OIDA Int. J. Sustain. Dev. **4**(9), 63–70 (2012)
42. N. Guragain, E-learning benefits and applications. Bachelor Eng. Inf. Technol. Helsinki Metrop. Univ. Appl. Sci. (2016)
43. A. Hoerunnisa, E-learning for a better advancement of National Education. Aecon **267**, 5–8 (2018)
44. P. R. K. Raju, E-learning boon or bane, an HRD perspect. Retrieve. **24** (2019)
45. V. Chang, C. Gütl, M. Ebner, Trends and opportunities in online learning, MOOCs, and cloud-based tools. Second handbook of information technology in primary and secondary education, 935–953 (2018)
46. S. Suleyman, O.G. Nihat, I. Bayrak, A comparison of student views on web-based and face-to-face higher education. Turk. Online J. Dist. Educ. **15**(2), 209–227 (2014). https://doi.org/10.1177/0081246314555305
47. D.D.L.L. O'Doherty, Barriers and solutions to online learning in medical education – An integrative review. BMC Med. Educ. **18**, 130 (2018)
48. D.W. Surry, A.G. Grubb, D.C. Ensminger, J. Ouimette, Implementation of web-based learning in colleges of education: Barriers and enablers. Can. J. Learn. Technol. / La Rev. Can. l'apprentissage la Technol. **35**(3), 1–18 (2010). https://doi.org/10.21432/t22g6q
49. S.V. Deshmukh, V.P. Deshmukh, Analysis of various Barriers to implementation of Web based technology in education: Literature review. Int. J. Adv. Res. Comput. Eng. Technol. **6**(3), 214–217 (2017)
50. M. Shadreck, Integrating ICTs into the environmental science primary school classroom in Chegutu district, Zimbabwe: Problems and solutions. Eur. J. Sci. Math. Educ. **3**(1), 90–96 (2015)
51. S. Al-Senaidi, L. Lin, J. Poirot, Barriers to adopting technology for teaching and learning in Oman. Comput. Educ. **53**(3), 575–590 (2009). https://doi.org/10.1016/j.compedu.2009.03.015
52. S.H. Khan, M. Hasan, C.K. Clement, Barriers to the introduction of ICT into education in developing countries: The example of Bangladesh. Int. J. Instr. **5**(2), 61–80 (2012)
53. F.M. Ihmeideh, Barriers to the use of technology in Jordanian pre-school settings. Technol. Pedagog. Educ. **18**(3), 325–341 (2009). https://doi.org/10.1080/14759390903255619
54. X. Liu, J. Pange, Early childhood teachers' perceived barriers to ICT integration in teaching: A survey study in mainland China. J. Comput. Educ. **2**, 61–75 (2014). https://doi.org/10.1007/s40692-014-0025-7
55. K. Pajo, C. Wallace, Barriers to the uptake of web-based technology by university teachers. J. Dist. Educ. **16**(1), 70–84 (2001)

56. M. Sharples, Mike, *et al.* Innovating pedagogy: Exploring new forms of teaching, learning and assessment, to guide educators and policy makers. The Open University, (2014)
57. N. Avouris, C. Goutis, K. Markellos, P. Markellou, A. Tsakalidis, Implementing a Web-based e-learning environment for electrical and computer engineers. Int. Conf. Eng. Educ., 1–9 (2003)
58. Å. Grönlund, Y.M. Islam, A mobile e-learning environment for developing countries: The Bangladesh virtual interactive classroom. Inf. Technol. Dev. **16**(4), 244–259 (2010). https://doi.org/10.1080/02681101003746490
59. J.P. Kasse, W. Balunywa, An assessment of e-learning utilization by a section of Ugandan universities: Challenges, success factors and way forward. Int. Conf. ICT Africa **2013**, 15 (2013)
60. C. Sife, Lwoga & Sanga, New technologies for teaching and learning : Challenges for higher learning institutions in developing countries. Int. J. Educ. Dev. Using Inf. Commun. Technol. **3**(2), 57–67 (2007)
61. Becta, Barriers to the uptake of ICT by teachers. Br. Educ. Commun. Technol. Agency **1**(June) (2004)
62. S. Hadjerrouit, Developing web-based learning resources in school education: A user-centered approach. Interdiscip. J. E-Learning Learn. Objects **6**, 115–134 (2010)
63. S.-H. Liu, Factors related to pedagogical beliefs of teachers and technology integration. Comput. Educ. **56**(4), 1012–1022 (2011). https://doi.org/10.1016/j.compedu.2010.12.001
64. W. Chen, A. Tan, C. Lim, Extrinsic and intrinsic barriers in the use of ICT in teaching: A comparative case study in Singapore. **1999**, 191–196 (2009)
65. S.H. Park, P.A. Ertmer, Examining barriers in technology-enhanced problem-based learning: Using a performance support systems approach. Br. J. Educ. Technol. **39**(4), 631–643 (2008). https://doi.org/10.1111/j.1467-8535.2008.00858.x
66. E. D. Wagner, Minding the Gap: Sustaining eLearning innovation, in *Presented at E-Learn: World Conference on E-Learning 2008*, (2008)
67. W. Mugizi, The role of higher education in achieving Uganda vision 2040. Elixir Int. J. **115**(April 2013), 49831–49837 (2018)
68. M. Gosper, M. McNeill, R. Phillips, G. Preston, K. Woo, D. Green, Web-based lecture technologies and learning and teaching: A study of change in four Australian universities. ALT-J Res. Learn. Technol. **18**(3), 251–263 (2010). https://doi.org/10.1080/09687769.2010.529111
69. Y. Goktas, Main barriers and possible enablers of ICTs integration into pre-service teacher education programs. Educ. Technol. Soc. **12**(1), 193–204 (2009)
70. M.S.H. Khan, L. Markauskaite, Approaches to ICT-enhanced teaching in technical and vocational education: A phenomenographic perspective. High. Educ. **73**(5), 691–707 (2017). https://doi.org/10.1007/s10734-016-9990-2
71. D. Pohekar, Role of ICT on universities administrative services and management Deepak Pohekar. Int. Res. J. Eng. Technol. (IRJET) **5**(11), 266–271 (2018)

# Learner-Driven Education: The Use of Information Communication and Technology (ICT) in Social and Literacy Practices

**Mailane Mofana-Semoko and Julia Preece**

**Abstract** The study proposed to find out how literacies were practiced at grass-roots, in order to enhance post-literacy (PL) training. The study applied the New Literacy Studies (NLS) conceptual framework through ethnographic case study in selected communities to find out who participated in literacy activities, when they participated and the types of materials they used. The methods of data collection used included direct and participant observation, documentaries, face-to-face interviews, dialogue and group discussions. The data collected generated themes, one of which was information communication and technology (ICT). The data was analysed following LeCompt's (2000) approach. The findings were presented in three case studies of 'Tum's story in stonemasonry, piggery, and community councils' literacy practices. The cell phone use is common in literacy practices, which need enhancing. Thus, PL training should be an extension of existing literacy practices.

**Keywords** Literacy · Post-literacy · Information communication and technology · Literacy practices · Literacy events · Social practices · Home domain · Community domain · Life-long learning

## Glossary

| | |
|---|---|
| Literacy | is understood as social practices and not the 3 Rs, representing reading, writing and numeracy. Literacy refers to daily activities that need enhancing through improving IEMS' approach to literacy, as opposed to the autonomous model of literacy, that emphasises specific skills. It refers to the |

M. Mofana-Semoko (✉)
Department of Adult Education, National University of Lesotho, Roma, Lesotho

J. Preece
Durban University of Technology, Durban, South Africa



E. Eilu et al. (eds.), *Digital Literacy and Socio-Cultural Acceptance of ICT in Developing Countries*, https://doi.org/10.1007/978-3-030-61089-0_11

social uses of reading and writing. These are multi-literacies meant for different purposes and valued in different ways.

Post-literacy — involves the use of ICT in materials development for use in community development activities, supported by facilitators within the domains of literacy. It involves a notion of life-long learning that uses the principle of adult education, namely, that adults learn from their daily activities, in which their acquired skills are applied immediately. Post-literacy is an extension of existing activities, rather than something that is imposed from above.

Information Communication and Technology — refers to the gadgets in the form of cell phones used to communicate in the three case studies, by members of the selected community groups. Cell phones are the most common channels of communication, among the young and old people. ICT is used in the stonemasonry, piggery and community councils' activities.

Literacy practices — are the everyday uses and meanings or rationales that people give for the way they undertake actions or events. Maddox (2005:124) describes it as "how people value literacy and use it". Literacy practices differ according to different domains of literacy or different specific places, and according to different literacy events taking place. The literacy practice is the process of using literacy skills in daily activities.

Literacy events — take place within different domains and are actual activities or occasions in which literacy has a role. Heath (1983, 93) explains literacy event as written text involved in a social interaction. In an editorial in a special issue of *Language and Education*, Tusting 2012: 99) analysed that studies are yard sticks. Tusting argues that learning arises from everyday activities and that it is most effective when it addresses immediate needs.

Social practices — are the existing practices or activities people engage in at community level that need enhancing through PL training. They are based on the social practices approach to literacy, which emerged through evolution of the NLS movement. The approach is based on 'the social practice theory of literacy'. The theory regards literacy as social practice. Hence, literacy practices are embedded in social practices, based on everyday activities in different contexts. The social practice theory of literacy points to the multi-literacies that need to be enhanced through PL training. The theory emphasises starting where learners are in order to develop relevant curriculum for PL programmes. Recent studies indicate how literacy should be regarded. Parr and Campbell (2012: 562) assert that, according to the NLS movement, literacy is concerned with social practices and relationships, knowledge, language and culture and that they differ according to places. The use of the social practices approach to literacy is recommended for literacy researchers, so that they can find out about the many literacies that exist in different contexts.

**Home domain** constitutes any activity taking place in the family compound. Home domain may be subdivided into private or individual, and family work activities, which are for the purpose of personal and family benefit. An example of an activity in a family domain is provided by Mpoyiya and Prinsloo (1996: 184–186). They describe a South African township family member named Nowowo, who works in Khayelitsha. Nowowo uses sewing patterns, takes clients' measurements and records details, in addition to household literacy practices, such as shopping and financial transactions. Her children look after family cattle and goats, counting them to identify whether animals are missing.

**Community domain** constitutes any activity taking place in a social setting, in public, church or in the shops. These activities affect the lives of community members social, economic, political or developmental, for the purpose of communal benefit. For example, Malan (1996b: 145-149) says, "at the funeral service in the home, friends and neighbours said their condolences using oral or verbal communication; at the church the minister used oral or verbal communication and reading the Bible and narrating religious stories and at the graveside, the choir took over through singing, flower offerings and reading scriptures".

**Life-Long Learning** refers to a continuum of learning, which begins at birth and lasts until death. It continues throughout life within people's activities, in their different settings or contexts. In other words, it takes place at home, in family life, in working life and in social and civic life. Torres (2003: 36-37) regards lifelong learning to be organised for the 21stcentury. Torres explains that emphasis is on learning than on education. In other words 'learning' and 'lifelong learning' replace 'education' and 'lifelong education', respectfully.

## 1 Introduction

There are 774 million adults worldwide who have not mastered basic skills in reading, writing and numeracy (UNESCO 2014: 4, 11). The majority are in developing countries.

Lesotho regards literacy as reading and writing skills. The Lesotho appreciates the application of literacy skills in all the activities of life. The use of literacy skills in development activities is important if someone is to function effectively, as modern living requires reading and writing in almost every aspect of life.

For Lesotho literacy training has always been a component of non-formal education, and as a result, the government has always regarded non-formal education as literacy teaching; the concentration of literacy training has always been on adult literacy and functional literacy.

In African countries, the concept of PL came a long way. It has been developed over years through researches that took it beyond basic literacy. The term 'post-literacy' connoted step two in the process of literacy training, referred to as functional literacy, for the specified target audience. In the nineteenth century, it involved people's application of basic literacy skills, to prevent relapse into illiteracy as stated by, among others, Dave et al. (1989: 390–391), Srivastava (1989: 372) and NECC (1992: 23), so that they would not relapse into illiteracy, if literacy skill is not put into practice. The new development in researches over three decades now sees what used to be called PL, recognised as social practices (UNESCO 2014: 3).

In the olden days, the most commonly used media in developing PL activities included those that were suggested by Srivastava (1989: 374) and NECC (1992: 23). They include newspapers and magazines, pamphlets, posters, display charts, extension literature, radio, libraries and study groups. With the introduction of technology, the most common media used for learning were television and cellular phones in the two families studied by Lemphane and Prinsloo [7, p. 741]. Lemphane and Prinsloo mention other technological communication media used such as computers, Internet and e-mail which are used to communicate with people from long distances.

The globalised world has become a village, in which people communicate through technology, which is quicker than the post office. This era of computers created a need for education and training to incorporate ICT into curricula, so that everybody will be able to communicate via the Internet. ICT is a form of PL for all people, because of its relatively recent introduction as a communication medium. To enhance the use of ICT, the government developed an ICT policy for Lesotho in 2005 (GoL MoCST 2005: 29), which addresses the matter of training to produce ICT expertise among educators in the country, who, in turn, will promote the use of ICT in schools and in higher education institutions, as well as in lifelong learning.

Lesotho's definition of PL differs from those of Rogers (2005: 60) and Ghose [4, p. 125] which cover a broad spectrum or *multiliteracies*, because Lesotho is concerned with a few functional literacy skills that are related to income generation and specific community development areas. In Lesotho, PL is less well researched or defined than literacy, and its education initiatives are not well documented.

The specific problem that the study addressed is based on the argument that PL training needs to be premised on a better understanding of how to engage with communities in order to facilitate more relevant learning materials. In other words, knowledge on the existing social practices will lead to the development of relevant materials for PL training, for the purpose of development of literacy skills.

## 2 Literature Review

The literature review revealed the common use of technology in daily life and in teaching and learning. On the one hand, the findings of Snyder [15, p. 142] identify trends towards, as well as resistance to, developing digital literacies. On the other hand, some studies indicate a positive move towards the common use of technology worldwide. For instance, Barton [1, p. 50] indicates that the world engages in web-based literacy events, while Warschauer [21, p. 131] confirms an acceptance of the use of technology in social networks and in learning. In his research, Morgan [11, p. 32] also found that literacy practices embrace the use of technology and that where technology is the main channel of communication, other types of media, like print, diminish, losing its power to the variety of channels of communication. Research was carried out by Hull and Nelson [6, p. 201–214] over 5 years to explore new literacies, including media and morality derived from a young girl, Layla, from 'Kaap', a village between the Cape Winelands and the Klein Karoo, in South Africa. According to Hull and Nelson, in 2008, Layla composed an electronic story to introduce herself to her age group social network. Six years later, different types of technologies, including cellular phones, were used for different purposes by different people in different settings, for example, Lemphane and Prinsloo [7, p. 741] report that the Mahlale family, in their study at site C, Khayelitsha, a squatter settlement or slum outside Cape Town, had two cellular phones belonging to parents, who used both the phones. These parents used the phones exclusively for making and taking voice calls and for not texting or for any other purpose. One of the phones had features for receiving FM radio and two pre-installed animated games, which were played by the children when they got access to the phone. The parents used cellular phones to communicate with their family and friends.

Computer literacy is practiced or learnt in families, the home domain, that have access to technology. The second household in Lemphane and Prinsloo's study confirms this situation. Lemphane and Prinsloo present the Bolton family, who lived in Cape Town, as professionals, whose children manipulated computers at any time for school work and computer games and had PlayStation connected to the TV. They lived in a comfortable house with Internet connectivity, electricity and running water, and the children had ample space to study and play. They were free to use desktop computers at any time for school work purposes but were only allowed to play games on the computers on weekends. They watched TV and had their PlayStation connected to the TV. The situation is that extensive use of ICT is possible in families that have technologies available, as they are able to learn and practice without any hindrances.

Another perspective regarding the use of ICT is provided by education research, in which McDowall [10, p. 1] studied a group of teachers who used ICT for teaching English and the arts to promote the emergence of knowledge-building communities. These studies show the wide variety of domains, practices and meanings given to using ICT literacies. Rogers and Street (2012 : 141) provide a long list of researchers interested in this aspect, dating from 1999 (Rassool) to 2011 (Juffermans).

The most common channel of communication is short message service (SMS) on cellular phones. Verheijen [20, p. 583] says SMS became common among the young generation, who owned cellular phones and computers. Verheijen states that these devices function as shortcuts, reducing writing time and enabling a quick response. In short, they can save time, space and money. Verheijen [20, p. 584] explains that SMS texts did not necessarily mind grammar and all language rules, which were broken. What was important was getting the message through quickly.

All the issues discussed above are relevant to PL educational interventions. PL in the context of lifelong learning is part of the continuum of literacy learning, and PL activities will inevitably be multiple and context specific. It is important, therefore, to consider literacy as a social practice and in a cross-cultural perspective. Street [16, p. 2; 2003:91], Barton and Hamilton [2, p. 10] and Brandt and Clinton [3, p. 342] offer one framework that can be applied to map literacy practices in different settings and different situations. It is these different situations that need close scrutiny if PL interventions are to become more meaningful. By understanding what counts as literacy, and what forms of communication are already being used, the educator will be able to recognise the current and potential learning points suitable for PL interventions. Thus, PL will be an extension of existing literacy practices or activities described below, rather than something that is imposed from above.

## 3 Theoretical Framework

The investigation followed the ethnographic approach to research, encouraged by the New Literacy Studies' (NLS) researchers like Barton and Tusting (2005: 10–11) in the United Kingdom, Ghose [4, p. 10] in India, Shiohata (2009: 66) in urban Senegal, Openjuru (2011: 2) in Uganda and Ade-Ojo (2015: 3 PPT) in Europe. The study engaged an ethnographic approach to literacy to determine who participated in the already existing literacy practices, when and how the types of materials were used for such purposes. The research involved the participants in the data collection activity within the two selected areas of study, among the nine constituencies of Berea District Council.

Data was analysed on the bases of the NLS idea, and it revealed the domains, events and literacy practices. The recommendations were based on the direct and participant observations. The findings resulted in recommendations on how to develop post-literacy materials.

## 4 Methods Used in the Study

The data was collected with the use of multi-methods appropriate for ethnographic study, as follows: direct and participant observations, documentaries, face-to-face interviews, conversation and focus group discussions. The research used a variety of methods to achieve triangulation and to represent the participants' perceptions. It engaged people in conversations about their literacy practices, investigating how and why they did thing, as well as how literacy practices could be enhanced.

The direct observations were of the two community councils of Kana (KCC) and Seneke (SCC) constituencies of Berea district, which comprised 18 participants in SCC and 16 in KCC. Other activities that were observed were the engraving of sandstone by Tums, the owner of a private stonemasonry business and the construction of a keyhole plot by Mrs. Mots (pseudonym) and her training team, as existing literacy practices. My participant observations were conducted in the monthly meetings of the two councils, where I observed the literacy practices while I participated in the meetings. The documents collected and analysed included agendas, minutes, posters and letters. Activities taking place within the four case studies were photographed and integrated with data obtained using other tools. The in-depth interviews and conversations were carried out with the chairperson of the Marabi Piggery Association of Women Entrepreneurs (MPAWE), with the owner of the private stonemasonry industry, Tums and Mrs. Mots during my visits to her home. Interviews were also carried out with one PL provider at SCC and one staff member of KCC. The focus group discussions and some conversations were conducted with the four workers of the private stonemasonry business and the six members of MPAWE. The approach used multiple methods for triangulation purposes, an approach that aims to support the respondents' perceptions and gives a broad approach to the issue of literacy Barton and Hamilton [2, p. 57]. Ghose [4, p. 14] emphasises that this approach also includes engaging in conversation with relevant people about their literacy practices.

The study used the 'curriculum wheel' (Fig. 1) as a conceptual framework for analysing the relationships between domains of literacy practices (e.g. work, family, community and private life), the practices (the meanings and uses people give to using literacy) and literacy events (the actual literacy activity or activities). The wheel was adapted from the Scottish Executive's literacy policy framework (2001: 1). It was informed by NLS theory that addresses literacy in relation to issues of power, context and wider global influences.

In Scotland the curriculum wheel was used to develop a curriculum that covered all the domains of literacy practices for its government literacy programme. The curriculum was designed to cater for adult learning goals that develop a variety of capabilities that are relevant to the lifelong learning needs of modern life. In this study, the adapted and simplified curriculum wheel, presented as Fig. 1, is used as a framework for analysing empirical data about literacy uses in selected Lesotho settings that were relevant to lifelong learning needs of modern life and ongoing PL development. The framework is used to help learners work out their own individual learning plan and goals.

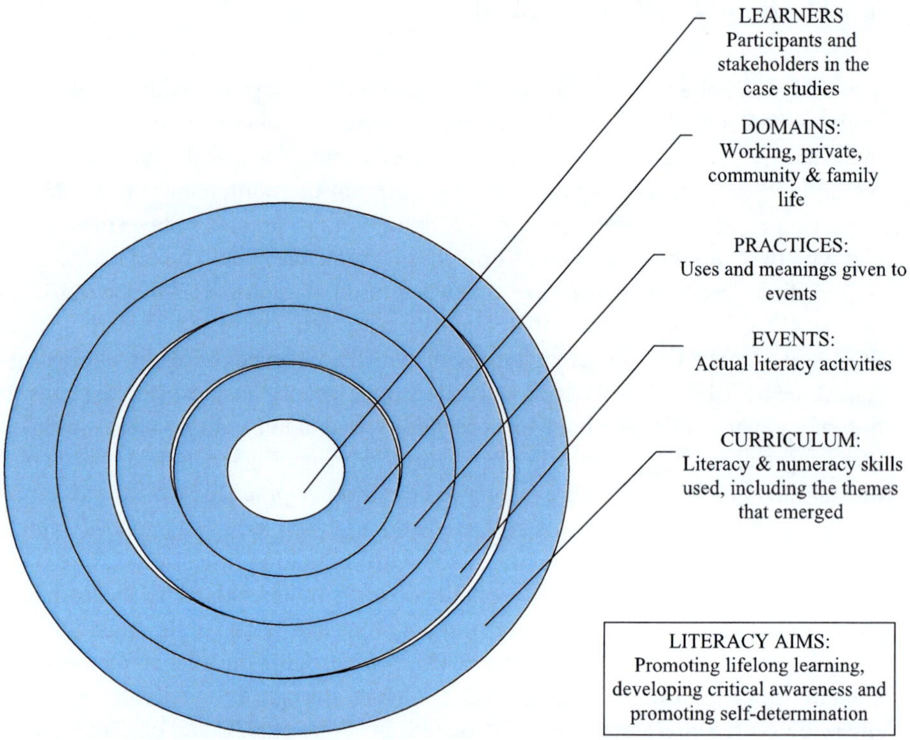


**Fig. 1** The curriculum wheel adapted from the adult literacy and numeracy curriculum framework for Scotland (2001: 1)

The NLS movement was adopted at a political level in Scotland (Scottish Executive 2001: 1). Scotland has adopted a literacy education approach that recognises literacy as involving local social practices that are specific to various domains of literacies. It is unusual for this approach to be adopted at a political level, but this precedent creates potential for wider consideration in other political contexts.

The results of the ethnographic study have implications for the way PL should be taught. The Institute of Extra-Mural Studies contributed to helping these communities within the ethnographic case study to develop relevant PL materials to enhance their literacy practices.

The data analysis was based on the approach developed by Le Compte (2000: 148–149), which goes through five stages, step by step, as follows: to tidy up data, as guided by Mouton (2001: 198); categorised and put together items that were similar; put together the items that belonged to a certain theme, so that I had categories of themes; put thematic patterns together according to relationships, such that when one thing happens, the other automatically follows; and, finally, linked the themes and patterns in serial order that helped to explain how and why things happened the way they did.

## 5 The Findings of the Study

The findings are presented in a table format and in a write-up. Table 1 below bears the results of the study on the use of ICT in the three case studies.

The study found out that using ICT could be very efficient for communication purposes. For example, cellular phone use has become common in the two sites selected for the study. The findings of the study included the use of technology in the form of cellular phones in stonemasonry; in piggery association, MPAWE; and in community councils' literacy practices. The stories of the three case studies in which ICT is used are given below:

### *5.1 The Use of Cell Phones in Stonemasonry Literacy Practices*

The use of ICT or cellular phones in the stonemasonry case study was explained by Tums as a stonemason producing tombstones in his private domain or site. In this case study, customers made use of smart cellular phones to take photos of tombstones, which they showed to their relatives, so that they could choose the design they wanted from the ready-made tombstones.

Tums himself did not communicate with his customers by cellular phone, nor did he use it to get information that would help him to expand his business. Cellular phones were not used extensively in Tums' business, even though his customers had cellular phones. This has an implication for training on the use of ICT even in businesses, like researchers such as Barton [1, p. 50] indicate that cellular phone use is becoming common worldwide and that people all over the world are engaged in web-based literacy events.

### *5.2 The Use of Cellular Phones in MPAWE Literacy Practices*

Besides verbal communication among the MPAWE members, another means of communication was technology. The secretary of MPAWE used a cellular phone to communicate with the members of the association. The members had no problem communicating by cellular phone, as they all had cellular phones, and it was convenient. It is situations like this that compare with what Warschauer [21, p. 131] found convenient for use in daily life social practices and in language learning, in four higher education contexts in Hawaii.

The secretary of the MPAWE association was responsible for convening meetings, as requested by the chairperson. The secretary used two modes of communication to invite members to meetings. During our conversations, she informed me that she invited the members by letter, and she also communicated by cellular phone to call meetings. She explained:

**Table 1** Categorisation of the use of ICT according to the curriculum wheel

| Literacy events (What) | Literacy practices (How, why, when) | Literacy domains (Where: home, family work, community settings) | Curriculum skills | Implications |
|---|---|---|---|---|
| 1. In stonemasonry business domain, Tums wrote messages on tombstones. | Tums clients used cellular phones to take photos of the tombstones to show their relatives, so that they could choose the one they liked. It might be assumed that they did the same with the products offered by other stonemasonry businesses, for the purpose of comparing products from different stonemasonry sites in the constituency<br>The tombstone chosen would be used on the special occasion, during the unveiling, for the purpose of marking the grave | The use of this cellular phone takes place in the two domains of literacy<br>Photo shooting takes place within the private, individual business site<br>The show activity takes place in the home or family domain | The use of cellular phones to keep record of the photos or pictures of tombstones | Tums had a cellular phone but did not make use of it to promote his business or even to copy designs from other stonemasons within the SADC region or internationally<br>He could also communicate with his clients electronically and make a lot of money<br>There is an implication for training of stonemasons on the wide use of cellular phones |
| 2. Cellular phones were used for the business of the MPAWE activities to invite members to meetings | The secretary of the association used SMS to communicate with members about business matters. She used two modes of communication to invite members to meetings for the purpose of oneself and community development.<br>ICT enabled her to write short messages<br>The MPAWE members also responded to the call through the use of cellular phones<br>The SMSs were send during the time of their seating | The secretary wrote SMSs to invite the members of the council to the meeting from her home (domain)<br>The members of MPAWE also read messages from their homes | The skills used in this case study are writing and reading of SMS messages<br>The secretary wrote SMS messages, and the members of the association read messages on the cellular phones and also responded in writing | This suggested that possession of different types of technology, including cellular phones, may be common to all classes of people.<br>Association members communicated among themselves through cellular phones, and communication was easy in this format<br>There is an implication for lifelong learning training for a wide community on the use of ICT for oneself and community development |

| | | | | |
|---|---|---|---|---|
| 3. Councillors were invited to attend meetings either through letters or through SMSs | Reading and writing was used to send and receive messages on cellular phones<br>The secretary wrote SMS messages to invite members of the council to attend council meetings<br>These text messages were also sent to members of the council if there was going to be an emergency meeting<br>Such invitations and the usual agenda could be sent quickly, even though there was some resistance towards the use of SMS for communication purposes in one of the councils, Kana Community Council<br>The councillors wanted SMS communication but not for official purposes – when dealing with developmental issues of the council | The secretary wrote SMS messages from the workplace, Kana Community Council offices, and sent them to individual members of the council<br>The individual members read the messages in their homes.<br>Literacy learning continues throughout one's life | Writing and reading skills were used on the cellular phones both ways | The necessity to enhance the use of cellular phones was critical for smooth operations of the councils.<br>Training on the use of ICT is necessary for oneself and community development |

***Hape ke sebelisa 'cellular phone' ho ba memela liphuthehong*** (I also use a cellular phone to invite members to attend meetings).

'Na (I): ***Litho li na le li 'cellular phone' kaofela?'*** (Do all members have cellular phones?)

Secretary: ***E, litho li na le li 'cellular phone' kaofela*** (Yes, all members have cellular phones).

The members' cellular phone numbers were written against their names (pseudonyms) on the attendance register in Fig. 2 below.

The use of technology, particularly cellular phones, had become the order of the day within the association. That meant that all the members of the association had cellular phones, and they used them to write and read messages. The use of cellular phones was becoming common – even very old people could be seen with strings around their necks with cellular phones hanging against their chests, either outside or inside their clothes, and when they answered the call, they just pulled up the hanging cellular phone to hold it against the ear. This meant the members of the association communicated through the cellular phones among themselves. Communication was also easy if they needed to consult Mrs. Hlasi, their volunteering trainer, during emergencies regarding the health of pigs.

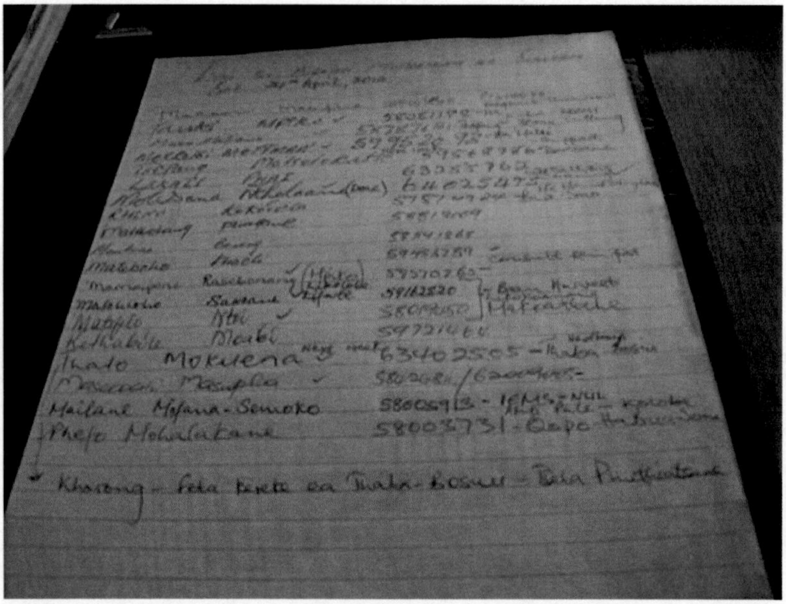

**Fig. 2** Photo of members' cellular phone numbers

## 5.3 *The Use of Cellular Phones in Community Councils' Literacy Practices*

The councillors were invited to attend meetings either by letter or SMS. Council members were also informed about emergency meetings by SMS. Reading and writing was used to send and receive cellular phone messages. Cellular phone messaging made it easy for secretaries to communicate with councillors. Invitations and agendas could be distributed quickly and easily.

Figure 3 below shows an example of an SMS communicating information relating to a council meeting. The SMS message in the photo is not complete.

The complete SMS reads as follows:

> Masek.
>
> *TULO EA LEKHOTLA E KA LABORARO LA LA 30.01.13, HA- SENEKE, KA HORA EA LESHOME HOSENG* (The sitting of the council will take place on Wednesday, 30 January 2013, at Ha Seneke, at 10 o'clock in the morning). 26/01/2013, 19: 56.

The SCC secretary used all capital letters with correct Sesotho language. She wrote a full sentence, while the SMS sent by KCC secretary to invite me to the council meeting was written in a mixture of upper and lower case and used a mix of languages. The English phrase or expression, 'evaluation team', was used together with Sesotho, and it was understandable. The message from Maleso, KCC secretary, reads as follows:

> *'M'e oa ka! Le ha ke thotse ke ntse ke u rata hle. Tulo e labobeli lehoja ke tla be ke le siko hle. Ke karolo ea evaluation team ea setereke* (Madam! Even though I am quiet, I still love you. The sitting is on Tuesday, though I will not be there. I am part of the evaluation team for the district councils). 10/06/2013, 05:27:00.

The two text messages are represented word for word as they appeared on the screen for the purpose of this paper. The Sesotho language was grammatically correct. The KCC secretary is a university graduate and her writing had developed. She did not use shorthand when communicating by SMS but wrote the message out in full. The second SMS was specifically directed to me, as an invitation to that sitting.

During its sitting of 11 June 2013, the council of Kana supported a complaint by one of the councillors, the Lady Chief, representing chiefs in the council, that the office had a messenger who could deliver letters of invitation to meetings, instead of councillors being invited by SMS.

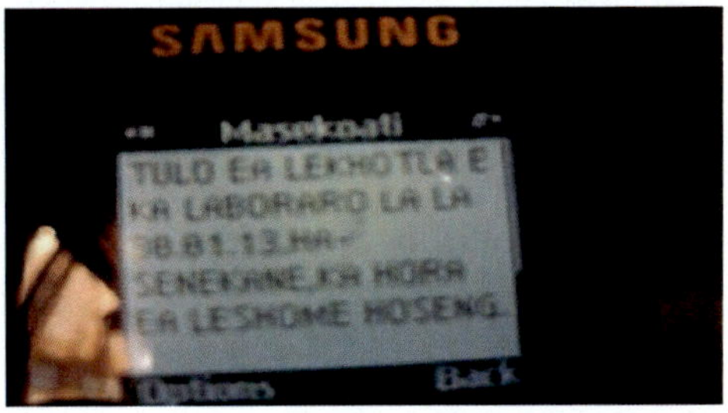


**Fig. 3** SMS inviting councillors to a meeting

> The *Lady Chief: Office ena e na le Messengera, ea ntsa lefuoa a sa sebelisoe ho delivera mangolo. Re mengoa ka li sms, empa ha ho le ka mohla ho keng ho be le ts^alo morao ea hore na re fumane li sms kapa che. Re tla li phuthehong tsa 'council' mona ho se bopaki li officing tsa rona ba hore na re ile kae. Re kopa hore kamoso ho mpe ho be le ts^alo morao, ha re hopotsoa ka phutheho* (The council office has a messenger who is underutilised, yet he is being paid. We were invited to meetings through SMS, and there had never been a follow-up whether we received the SMS or not. We come to meetings without leaving any evidence, as chiefs, of our whereabouts for our offices. We are pleading that, next time, there should be a follow-up on the invitation.

The councillors wanted the council to use tangible letters to supplement the SMS invitations and cellular phone calls used to invite councillors to meetings. On the one hand, the three channels of communication, used together, formalised the invitations. In other words, invitations should not be extended casually but be written down and formal. On the other hand, the complaint was based on the fact that airtime for the chairperson and secretary was covered by the council budget, while the councillors ever received calls to follow up on the SMS to confirm whether the members had received the messages or not. The Lady Chief was supported by one of the councillors, who raised her hand and was given permission to speak:

> *Kea leboha Molula-Setulo, le 'na taba ena e nkhathalitse ka ho ts^oana haholo hobane 'na sms ha ke ea e fumana. Ka boikokobetso re kopa hore re 'ne re mengoe ka mokhoa o formal oa lengolo* (Thank you, Chairperson, I was also bothered by the same issue, especially because I did not get the SMS. With due respect, we requested that we be invited in a formal way, through letters as well).

> Another councillor: *Molula-Setulo ha u ntumella, ke tlatsa taba eo* (Chairperson, if you permit me, I second the move).

The kind of literacy the councillors valued was a formal letter, even though the kind of literacy commonly used was SMS. The councillors wanted different communication channels to be used for different purposes. They regarded a formal letter as an important document. They wanted to receive letters, as it had always been a way that the Basotho communicated – they were accustomed to receiving letters from the post office from, for example, relatives in Gauteng, South Africa. I observed that councillors had no access to minutes, an important document. Letters were the only documents they received from the council that they could keep in their possession. Councillors wanted SMS to be used for casual communication and not for official purposes. The reaction of the councillors to the use of ICT in their activities implies the training gap that needs to be closed to enhance their computer literacy skills.

Much as some councillors had their own preference with regard to official communication, in the community development councils' activities, the use of technology is used to communicate, in their language, the issues are related to the development of themselves as individuals and the community at large. The same way as indicated by Warschauer [21, p. 131] that the use of technology in language learning strengthens learners' appreciation of the values of their community and culture.

## 6 Conclusions

This ethnographic study shows that the use of technology is becoming a common feature of literacy practice. The use of technology, particularly cellular phones, had become the order of the day within the stonemasonry, the MPAWE activities and community councils' literacy practices in Lesotho, but it is also used in entrepreneurship worldwide.

Technology is already regarded as important in teaching and learning. However, the opportunity to learn new literacies that are associated with rapid technological change has to be built into policy for literacy programmes. Thus, new literacies should be researched in order to ensure that appropriate new literacies' learning is done in a way that is relevant to context.

The use of technology in the form of cellular phones is mentioned in the three case studies. Tums customers used their cellular phones to take photos of the tombstones to show their relatives, so that they could choose the one they liked. It might be assumed that they did the same with all the products offered besides stonemasonry businesses, for the purpose of comparing products from different sites of the stonemasonry in the constituency. Tums had a cellular phone but did not necessarily benefit from it. He did not use it to enhance his business or even to benchmark his designs with those from other stonemasons within the SADC region or internationally.

Cellular phones were used to invite members of MPAWE to meetings. The secretary used SMS to communicate with members about business matters. She used two modes of communication to invite members to meetings for the purpose of oneself and community development. ICT enabled her to write short messages.

The cellular phone numbers of the association's members were readily available to the secretary. That meant all the members of the association had cellular phones and could be contacted through them and they used them to write and read messages. This proved that possession of different types of technology, including cellular phones, is common to all classes of people. Association members communicated among themselves through cellular phones, and communication was easy in this case.

Technology, for example, cellular phones, was also used for communication among councillors and between secretaries and councillors. It was the quickest channel or means of communication for the two councils, in the two constituencies. The use of cellular phones by the KCC secretary meant she could communicate with the councillors quickly and easily, even though KCC members indicated that they preferred to be invited to attend council meetings formally by letter.

Councillors were invited to attend meetings either through letters or through SMS. These text messages were also sent to members of the councils if there was going to be an emergency meeting. Reading and writing was used to send and receive messages on cellular phones. Such invitations and the usual agenda could be sent quickly, even though there was some resistance towards the use of SMS for communication purposes in one of the councils. The councillors wanted SMS to be used for less formal communication and not for official purposes. In this case, the necessity to enhance the use of cellular phones was critical for smooth operations of the councils. This means there is a need for PL training among the councillors.

The need for post-literacy (PL) education cannot be overemphasised, especially on the use of cellular phones within the existing literacy practices, in the three domains of literacy. It is critical for developing a nation. It helps people to develop further the skills they are already using in their different contexts. It includes everybody because literacy starts with individuals in the family and is used in the community and then the workplace. It is necessary for improving people's daily activities. It also has an influence on the economy of the country, as ongoing literacy updating is related to lifelong learning. Individuals who have access to computers and use them, need to keep learning due to technological changes, as new software and computer programmes are introduced from time to time. So there is a need to develop a more literate environment, in which people will be able to participate fully and be able to use different literacies in different contexts. The notion of PL has broadened in the 1980s and 1990s, to include the challenges of the influence of new technologies, electronic media (Internet) and other ways of knowing (UNESCO 2005: 15).

It is the role of the tertiary institutions to address literacy as lifelong learning through the training of facilitators in literacy. The aim of post-literacy training is to ensure that training addresses the principle of adult learning that requires that adult learners' education is immediately applicable in their everyday activities.

## 7 Recommendations

In three of the selected case studies in the three domains of literacy, the use of technology, in the form of cellular phones, is limited to communication, such as making calls and sending SMS, as described in the above paragraphs. Cellular phone has become a common gadget within the communities studied for this study. Cellular phones have become the easiest and the fastest way of communication in rural areas. There is a potential PL opportunity for training on using cellular phones and how to access information on the Internet for the purposes of improving oneself and the community as a whole. The recommendations listed are for future training interventions by the Institute of Extra-Mural Studies and for further research:

- Tums could enhance his use of ICT by sourcing new designs from the Internet and even marketing his products online. If he used ICT in a greater variety of ways, he would motivate his customers to do the same in order to access his products. This would result in an organic process of increasing his own and the community's ICT skills.
- The secretary of the MPAWE should be encouraged to enhance the use of ICT by using it to obtain information on piggery activities, as well as to establish a new poultry activity as a new initiative by the association. Using ICT for these pur-

poses would motivate members to use their ICT skills to improve their production of pork and poultry products, thereby improving members' and the community's ICT skills.

- The secretaries of the community councils should be encouraged to use ICT to communicate with the councillors on daily basis; doing so will give the councillors the opportunity to enhance their ICT skills and encourage them to use these skills to obtain information. Such information would become PL materials, which can motivate other staff and councillors to use ICT extensively.

Further research is recommended, in particular further ethnographic studies of different contexts. These studies could investigate further how people or community facilitators use ICT in literacy practices.

## References


1. D. Barton, Understanding textual practices in a changing world, in *The Future of Literacy Studies*, ed. by M. Baynham, M. Prinsloo, (Palgrave Macmillan, New York, 2009)
2. D. Barton, M. Hamilton, *Local literacies. Reading and writing in one community* (Routledge, London/New York, 1998)
3. D. Brandt, K. Clinton, Limits of the local: expanding perspectives on literacy as a social practice. Journal of Literacy Research **34**(3), 337–356 (2002). https://doi.org/10.1207/s15548430jir3403_4. Downloaded on 14 Jan 2009
4. M. Ghose, *Exploring the everyday. Ethnographic approaches to literacy and numeracy* (Nirantar and ASPBAE Publication, New Delhi, 2007)
5. S.B. Heath, *Ways with words. Language, life and work in communities and classrooms* (Cambridge University Press, New York, 1983)
6. G.A. Hull, M.E. Nelson, Literacy, media and morality. Making the case for an aesthetic turn, in *The future of literacy studies*, ed. by M. Baynham, M. Prinsloo, (Palgrave Macmillan, New York, 2009)
7. P. Lemphane, M. Prinsloo, Children's digital literacy practices in unequal South African settings. J. Multiling. Multicult. Dev. **35**(7), 738–753. http://www.tandfonlie.com/loi/rmmm20 Downloaded on 13 Sept 2014 (2014). https://doi.org/10.1080/01434632.2014.908894
8. B. Maddox, Assessing the impact of women's literacies in Bangladesh. An ethnographic inquiry. Int. J. Educ. Depart. **25**, 123–132 (2005)
9. L. Malan, Literacy learning and local practice in Bellville, in *The social uses of literacy. Theory and practice in contemporary South Africa*, ed. by M. Prinsloo, M. Breier, (SACHED BOOKS and John Benjamins Publishing Company, Cape Town, 1996)
10. S. McDowall, Using ICT to develop knowledge-building communities in subject English and the arts. *SET* **1** (2013) http://www.nzcer.org.nz/nzcerpress/set/articles/using-ict-develop-knowledge-building-com Downloaded on 03 Oct 2015
11. W. Morgan, Computers for literacy. Making the difference. Asian-Pacific J. Teach. Educ. **29**(1), 31–47 (2010)
12. P. Mpoyiya, M. Prinsloo, Literacy, migrancy and disrupted domesticity. Khayelitshan ways of knowing, in *The Social Uses of Literacy. Theory and Practice in Contemporary South Africa*, ed. by M. Prinsloo, M. Breier, (SACHED BOOKS and John Benjamins Publishing Company, Cape Town, 1996)
13. M. Parr, T.A. Campbell, Understanding literacy as our WORLD inheritance. Re-visioning literacy discourse and its implications for teaching practice. Int. Rev. Educ. **58**, 557–574 (2012). https://doi.org/10.1007/s11159-012-9297-1

14. Rogers, A., Street, B.: Adult literacy and development. Stories from the field. National Institute of Adult and Continuing Education (NIANCE) England and Wales (2012)
15. I. Snyder, Shuffling towards the future: The enduring dominance of book culture in literacy education, in *The future of literacy studies*, ed. by M. Baynham, M. Prinsloo, (Palgrave Macmillan, New York, 2009)
16. B.V. Street, *Social literacies. Critical approaches to literacy in development, ethnography and education* (Longman, London/New York, 1995)
17. B.V. Street, What's "new" in new literacy studies? Critical approaches to literacy in theory and practice. Curr. Issue. Compar. Edu. **5**(2), 77–91 (2003)
18. Torres, R. M.: Lifelong learning. A new momentum and a new opportunity for Adult Basic Learning and Education (ABLE) in the South. Adult Education and Development, Supplement to Vol. 60 (2003)
19. K. Tusting, Literacies and sites of learning. Lang. Educ. **26**(20), 99–104 (2012). https://doi.org/10.1080/09500782.2011.642877
20. L. Verheijen, The effect of text messaging an instant messaging on literacy. English Stud. **94**(5), 582–602 (2013). https://doi.org/10.1080/0013838x.2013.795737. Downloaded on 09 Sept 2014
21. M. Warschauer, Digital literacy studies. Progress and prospects, in *The future of literacy studies*, ed. by M. Baynham, M. Prinsloo, (Palgrave Macmillan, New York, 2009)

# Logged In or Locked Out of the Twenty-First Century? Implications for Adult Learners with Special Needs

**Selloane Pitikoe and Peles Biswalo**

**Abstract** The global rapid growth of technology comes with tools that address learner diversity and inclusivity. However, research on special needs education is at an infancy stage. Set against the backdrop of Eswatini, this chapter explores alternatives for reshaping traditional education in order to accommodate the learners with special needs. The authors' concern is on the adult learners with special needs who join the learning environment with multiple and diverse learning needs. However, the traditional education is ill-prepared to fully cater for learners with special education needs. The authors applaud the intense aggression of technology in re-imaging the traditional education provision. Assistive technologies such as Envision Artificial Intelligence (Envision AI) compliments traditional teaching for visually impaired learners. Paucity of literature on Envision AI misinforms education policy and decision-making. The findings harvested from this chapter could benefit education policy makers in addressing the diversity and inclusion of special needs education.

**Keywords** ICT · Assistive technology · Learners with special needs · Inclusive education · Adult learners

## 1 Introduction

This chapter argues from the context of the Kingdom of Eswatini (formerly Swaziland), the only remaining absolute monarchy in Southern Africa other than Morocco and Lesotho. Literature estimates a total area of 17,000 km$^2$ and a population of one million people [1]. The landlocked nature where 75% of the Kingdom lies within South Africa while the remaining 25% is surrounded by Mozambique limits the country's access to the sea. Henceforth, the economy is heavily reliant on the shocks and stresses that affect South Africa [2].

S. Pitikoe (✉) · P. Biswalo
University of Eswatini, Kwaluseni, Eswatini
e-mail: spitikoe@uniswa.sz; pbiswalo@uniswa.sz



E. Eilu et al. (eds.), *Digital Literacy and Socio-Cultural Acceptance of ICT in Developing Countries*, https://doi.org/10.1007/978-3-030-61089-0_12

Relative to most African countries, most of "her" population resides in the impoverished and inaccessible rural areas. Studies on disability indicate that 80% of people with disability are found in developing countries and live in extreme poverty where the elderly, the disabled and child-headed households are the mostly affected [3, 4]. The Kingdom of Eswatini is not immune to these conditions as evidenced by an estimated 3% of the total Emaswati – (natives of Eswatini) who are disabled. People with disability constitute 86% of the rural dwellers with limited access to basic amenities such as primary health care and education. Seemingly, rural dwelling also presents an infrastructural challenge such as inaccessible roads, no electricity, limited access to technology and sporadic network connections [5]. This is indicative that despite the Kingdom of Eswatini's high per capita income, the majority of the population lives below the poverty line [11].

### *1.1 Education and Disability Discourses*

[6] defines disability as "a physical or mental condition" that means an anomaly in the bodily functioning. Sharing similar sentiment, [3] laments that disability comes with impairments which could be physiological and psychological incapacitation that could also be temporary or permanent [3], limiting one's full capability or functionality in carrying out activities as it were in a "normal" being. The Disabilities Education Act Amendment of 1997 article c (1) views disability as a human experience that changes the normal bodily functioning. The Act further cautions that any form of disability should be viewed as an impediment to education access. This implies a need for the mainstream education to be flexible and accessible for such learners while also engaging the people with disability to voice out their diverse education needs [3].

We argue that despite being a smaller fraction of the entire population, the need for such learners to be equally represented in decision-making forums is underlined. Henceforth, the popularity of Special Education Needs Disability (SEND) which refers to an approach tailored specifically to address the uniqueness of the disabled learner [7]. In other words, SEND is designed with the disabled learner in mind and aims to maximize education access to be as close as possible to that which is accessed by "normal" learners for effective social and economic functionality [8].

Scholars argue that in Africa education is compulsory and yet there is a mismatch between the conventional provision and the diverse needs of the learners with disability. Henceforth, higher numbers of children with disabilities who are out of school yet are of a schoolgoing age are alarming [9]. [10] estimates that out of a total school enrolment in Africa, learners with disability constitute 10% of the formal education provision. On the contrary, South Africa has reported a remarkable 70% of South African learners with special needs aged 6 years and above who are out for school [7]. These arguments indicate the unevenness of the conventional system and how it perpetually relegates learners with disability to the periphery [7, 8].

The notion was well captured by [11] through a study that was conducted in 2011 on the living conditions of people with disability in the Kingdom of Eswatini. The findings affirmed the challenges of access to basic services, poor infrastructure and sporadic network connectivity. Evidently, some information gaps on the "indicators on the level of living" (p. 6) for people with disability were also identified. It can be argued that these gaps misinform policy and decision-making on how to support people with disability in the Kingdom of Eswatini.

In light of these arguments, we focus on the University of Eswatini (UNESWA) as the only national university acclaimed an equal opportunity institution that offers both full-time and part-time courses in different disciplines using the dual-mode approach. The study will focus on the Adult Education department of UNESWA with the hypothesis that UNESWA does not fully cater for the learning of the learners with disabilities.

For the purposes of this chapter, the focus is on the part-time programmes that are offered by the Department of Adult Education in the four main academic domains of Certificate, Diploma, Bachelors and Masters. Previously, enrolments into the Department of Adult Education attracted "normal" learners – those who did not need any special accommodation. [10] estimates that out of the total school enrolment in Africa, learners with disability constitute only 10% of the formal education provision. On the contrary, South Africa has reported a remarkable 70% of South African children with special needs aged 6 years and above who are out for school [7]. While UNESWA's context is in congruent with [10] on the fewer enrolment of learners with disability, such learners though are mainly concentrated within the part-time/non-formal education provision. For instance, the period 2015–2020 saw a drizzle of the visually impaired learners who enrolled into the Department of Adult Education to the dismay of both the institution and the Department.

The infrastructure remains unfriendly; teaching and learning is largely traditional and non-accommodative; the lecturers have not been trained on how to handle these learners. In other words these learners are logged in UNESWA's education system and yet locked out of content delivery – hence the title of the chapter. Paucity of literature on national disability discourses and how the gaps continue to marginalize disabled men and women is noted [3, 4, 11]. A question may be correctly asked: How inclusive is the current UNESWA's Adult Education provision in addressing the needs of the visually impaired learners?

## 1.2 Education Provision in Eswatini

Education provision in Eswatini is under the auspices of the Ministry of Education and Training (MoET). MoET among others strives for equitable education that is accessible to all members of society. [13] found out that the use of Internet in Swaziland (before the name change to Eswatini) was on the rise although not at the same level with the rest of the international communities. This discovery is in spite of the commitment made under Vision 2022 of taking Eswatini into the first world countries category in 2022. In spite of the numerous efforts taken by the Kingdom to develop the national ICT policy, [13] reports that the ICT policy remains in a draft form since 2006. [36] identifies some of the challenges in the implementation process to include (a) non ratification of the ICT policy to support the implementation process, (b) lack of political will and financial resources and (c) limited capacity to drive the process forward. UNESWA is the only national university which has been having its fair share of the challenges concerning ICT access in the country.

Empirical evidence from Eswatini that examined the views of the learners with disability revealed the struggles that UNESWA's adult learners with visual impairment faced [14]. Firstly, upon enrolment/registration, there were no efforts made to find out the medical condition that the learners with disabilities presented with. Secondly, in as much as the lecturers were made aware of how to deal with the learners with disability, the learners felt a need to be accommodated during assessment through extra time, user-friendly buildings and computer software that addresses their incapability [14]. One of the key challenges was the incompatibility of the software with the multiple needs of such learners. One of the learners lamented as follows:

> As a student in the institution you are expected to do research on your own but for me it is difficult because I cannot go beyond the lecturers' notes since the library does not have books with braille. Again, it is difficult to surf the internet because there are no computers with JAWS software, which enables the computers to talk so that I can also be able to use the information for my research. [22, p. 5]

The implication from this study is that more effort is needed from UNESWA to employ a fully blown inclusive education and appropriate infrastructure that caters for the diverse needs of the learners. Looking at this excerpt, the current technology and resources still lag behind to accommodate the visually impaired learners to fully attain equitable and quality education provision. The grey literature indicates that there is a room assigned for learners with visual impairment that has been equipped with some braille machines. However, access to this important resource is a challenge because the room is located on the first floor where learners require additional support to climb the stairs. The question then is: How inclusive and accessible is UNESWA education provision?

## 2 Human-Computer Interaction (HCI)

Simply put, human-computer interaction refers to communication process between man and the computer [15]. [16, p. 5] defines human-computer interaction (HCI) as:

> ...a discipline concerned with the design, evaluation and implementation of interactive computing systems for human use and with the study of major phenomena surrounding them.

In other words, computers are designed such that they simulate the functionality of man. The transition from the third industrial revolution into the fourth one is characterized by use of smart machines that emulate "human-like abilities" to perform simpler to more complex human tasks [17]. For such machines to be effective, collaboration and interaction between man and the computer are significant [12]. An introduction of HCI sought to transform the computer workstations into "computational machines" [16, p. 6] where jobs become more task oriented. [10, p. 174] asserts that effective integration of the "human-computer interaction requires a deeper understanding of the dynamism that humans face to let go the traditional approaches while migrating into the global network of computer-mediated interactions" (ibid).

Echoing similar sentiments, [12] sums up HCI through two approaches, namely, (a) the emulation approach and (b) the complementary approach where the former highlights the need to program the machine with simulated human activities that can be easily retrieved when such machines are presented with a similar task. On the other hand, computers and humans are not the same, which implies a need to create a synergy that will allow "collaboration and interaction" [p. 66] between the computer and humans. [17] identifies three ways through which the HCI can be used. Firstly, they can make predictions of human tasks; secondly, the devices can play an assistive role in task performance; and thirdly, these models can substitute the user and perform the task independent of the user but dependent on computer networking. The arguments presented above are mainly concerned with the cognitive processes negating physical and the physiological context that may surround the user. In this study, the authors explore the effectiveness of assistive technologies in teaching and learning of learners with disability especially the visually impaired learners.

## *2.1 Assistive Technologies in Education*

Scholars define assistive technology (AT) differently. For instance, [18] define AT as "a wide range of devices, services, strategies, and practices that are developed and used to assist individuals with disabilities to deal with their problems" (p. 18). [19] defines AT from the cognitive point of view as "intelligent machines" [20, p.106] that mimic the cognitive functions of humans. A comprehensive definition is provided by [8] who refers to AT as:

> …any item, piece of equipment, or product system, whether acquired commercially off the shelf, modified, or customized… to increase, maintain, or improve functional capabilities of a child with a disability…. that directly assists a child with a disability in the selection, acquisition, or use of an assistive technology device

The definition reiterates the assistive role of AI being to complement the inadequacies that people with disabilities may encounter in their daily life routine. One such AT is the artificial intelligence (AI), a "machine-learning" which is trending in institutions of higher education and its transformative role in teaching and learning activities [21]. For this duo, AI refers to "computing systems, which engage human-like processes such as learning, adapting, synthesizing, self-correction and the use of data for complex processing tasks" (p. 107). What remains unknown from this literature is how the visually impaired learners of UNESWA can benefit from these devices.

The effectiveness of the computing systems relies heavily on machine learning, a software that helps the machine to predict and "identify patterns, and apply" [21, p. 107] them to reality. It can be argued that such computing systems may help assist the visually impaired to make sense of their environment through the prediction and pattern identification function. The history of AI dates back to 1950 through the works of Alan Turing, who argued for one to affirm that a computer is artificially intelligent; such computer must be fully capable to imitate human behaviour without leaving any room for doubt [21]. The need for such technologies to transform learning and teaching into a flexible offering that caters for the learners' convenience is underscored. [22] view this modified computers as telerobotics that provide learners with an opportunity to learn even in times of emergencies and/or hospitalization. Notably information and communication technologies (ICT) do not only modify learning into a flexible and affordable activity, but come with assistive applications – a ray of hope for people with visual impairment [21].

## 2.2 *Envision Artificial Intelligence (Envision AI)*

While AI has become a global topical research phenomenon, in Africa the phenomenon is under-researched. The current conventional education lacks the African context, a gap which the chapter intends to fill. One such assistive technology that compensates the learners with visual impairment is Envision Artificial Intelligence (Envision AI).

Envision Artificial Intelligence as the name indicates comes with technological/machinery intelligence that "pushed technical boundaries" [23, p. 81] for people with visual impairment. For instance, if the application is successfully installed to the mobile phone's camera and speaker, then the application can familiarize the visually impaired person about the immediate surroundings [24]. According to [25], Envision AI was developed to serve three objectives, namely, (a) to replace the specific hardware with a user-friendly smartphone application, (b) to improve the usability of smartphones by all people and (c) to develop "a robust, high performance obstacle detection method" (p. 129).

Some of the functions of application include voice recognition, navigating directions, detection of obstacles and merging phase. Although other devices can also scan the environment, the advancement of Envision AI is the new innovative video application that detects obstructions in real time [26]. By so saying, the application plays an assistive role to the visually impaired person. While these modified applications portray a potential resource for learners with visual impairment [27], the gap on the technological need of people of the visually impaired people remains unknown.

The discussions on Envision AI have to first take into cognizance the diverse nature of the clientele – in this case being the visually impaired people. For instance, they are diverse in terms of age, gender and their various levels of visual impairment [27]. WHO [12] laments that globally the statistics of visually impaired people is estimated at 36 million out of which 220 million are either moderately or severely affected. Such differences imply a need for different approaches and specifications to resonate "a sighted companion" (p. 7) that guides the mobility and safety of the visually impaired people. [29] echoes with emphasis that the visually impaired populace is invisible in decision-making forums, which misinforms strategies that affect their lives.

Based on these gaps, we argue that with proper research, applications such as the Envision AI could be explored further to find out the value that they could bring to the UNESWA's visually impaired learners. The findings from this paper could benefit the Ministry of Education and Training in Eswatini to further develop policies that facilitate an effective inclusion of the diverse needs of the learners with special needs in Eswatini.

There were three objectives that the chapter sought to address. First is to identify the accessibility of mobile devices to learners with visual impairment in UNESWA. Second is to determine how the learners with visual impairment use their mobile devices in teaching and learning. Third is to explore ways of adapting Envision AI as a learning resource for UNESWA's visually impaired learners. In order to achieve these objectives, the chapter is outlined as follows: the methods, results and discussion followed by the conclusions.

## 3 Methods

The discussions in this chapter were informed by an interpretivist study that was conducted at UNESWA on learners with visual impairment. Interpretivists assume that knowledge is subjectively constructed based on the individual perception of the world. This approach was used in order to give the participants an opportunity to give their subjective views and opinions based on their learning experiences and the challenges that such learning presents to learners with disability. This study was conducted on two learners: one who has graduated and anticipating to enroll in future programmes in Diploma in Adult Education and the other who is currently enrolled in the university. Data were collected using semi-structured interview guides and analysed thematically.

## 4 Findings and Discussions

The study revealed that the institution had made some strides in accommodating the learners with disabilities which took place in the first half of 2019. For instance, the learners were now accommodated during assessment whereby they received preferential treatment such as extra time as well as being allocated a different building from the rest of the learners. This approach according to the learners enabled them to work at their own without feeling under pressure to compete with the rest. However, it appears teaching and learning remains unfriendly to such learners. The findings were categorized under the following three themes.

## *4.1 Access to Mobile Devices*

When asked if any of them owned a mobile device, the participants acknowledged that they had cellular phones which help them to communicate with their friends and families. For instance, Dalida (not a real name) had this to say:

> I have a phone you see (waving the phone in the air); but is a *skorokoro* (meaning an old model). I use this phone mainly to receive and make calls to my work colleagues as well as family. I can also log onto WhatsApp with my phone.

Similar sentiments were echoed by Dudu (not a real name) who stated that:

> My phone is still new but it is a cheap model so much that sometimes the screen locks itself; in the process, I cannot make or receive any calls. Sometimes the screen can be locked for some days while I am still looking for some help to fix the phone. In the meantime people are getting worried whether I am ok or not.

Dalida added then that whenever he needed to engage in intensive research, he relies on his workplace:

> If there is some work that needs an intensive research then I go to my office and work from there. The problem is even though there are computer labs and the library around they are not accessible for visually impaired people because they do not have the JAWS (Job Access With Speech) software.

The participants stated the challenges they experienced with their mobile devices and how the challenges impede the communication process. Equally important was the fact that even the institutional resources were also not conducive for learners with visual impairment as they lacked the much desired JAWS software which reads the on-screen writing for the visually impaired learners. These stories affirm the literature which identified one of the roles of AT as being assistive to the visually impaired learner.

When asked how come they are stuck with old devices that are already proving to be problematic, Dudu had this to say:

> I am unemployed and live on handouts from the family. I usually safe the little that they offer to cater for my education needs. Although I graduated from this university, I still intend to pursue my studies with UNESWA may be up to a Masters level. So a new mobile device is the least of my priorities.

On the other hand, Dalida responded that:

> I am currently intending to enroll for a diploma level which I am not sure of its feasibility now that we have COVID-19 in the way; I am scared because I feel more vulnerable with

> my visual impairment. However, back to the phone issue, I have been a self-sponsored for the past two years which will still be the case if I further my studies. That makes buying a new mobile device impossible; unless a miracle happens.

It is clear from these responses that the economic challenges are rife. Nonetheless, their stories also indicate their determination to pursue their education against all odds. Given the nature of their medical condition, the institution could take these two cases as part of the social responsibility to support their further education intentions.

## *4.2 Application of Mobile Devices in Education*

When asked how they used mobile devices for learning activities, not much was done really based on the inadequacies of their gadgets. For instance, Dalida had this to say:

> I can read PDF files only using my laptop. So one of the lecturers usually converts the word documents such as the lecture notes, the course outline as well as the power point presentation into PDFs and sends them to me via e-mail. That way I am able to access and read the material using the JAW software.

As a follow-up question, the researcher asked his experience with other lecturers in terms of teaching and learning, and Dalida responded thus:

> It is not easy because in some classes we are taught using a one-size-fits-all approach with no consideration for our diversity. For instance, as a class we will be asked to read a passage. Thereafter, we identify the key messages in that passage. My disability is assisted by braille; meaning I read with both hands. Thereafter, I must go back to do the second part of the assignment. Obviously I will not complete the assignment at the same pace as those who can see; hence poor performance.

On the contrary, Dudu indicated that there has been some improvement though in the way the assessments are conducted. For instance:

> At first we used to sit together with the rest of the class for assessment activities. Of late I am told that during examination time, the visually impaired learners are allocated a separate room, with an invigilator/s and flexible time to work without competing with the rest.

Echoed by Dalida:

> When I sit for a test I go to my departmental office for learners with disabilities where I can access Braille. I am also allowed an extra hour to complete the test.

Based on these stories, mobile devices are minimally used in teaching and learning. However, laptops have become an AT that helps the learners to access learning resources. The two stories further indicate that other lecturers were more accommodative than others in teaching and learning. However, the participants did indicate that there has been some flexibility where special arrangements were made during assessment to visual impairment.

### *4.3 Envision AI as a Learning Resource*

When asked how they would like UNESWA to assist them in terms of equitable access to quality education, the participants alluded to the important role that technology can play in education by making education flexible and accessible by all learners regardless of their medical conditions. Dalida stated:

> I wish we could be given a special preference. For instance now following the current COVID-19 pandemic UNESWA has embarked on education transformation where all teaching and learning will be conducted online. This calls for a need to have smartphones which we do not have in order to access the e-learning platforms easier. The question is; was there any thought on how to help us as a minority group?

What do you wish UNESWA could do in order to accommodate your educational needs?

> We know times are tough and the pandemic caught the world off guard. But we also want to learn online like the rest of the learners through UNESWA's support. The current computers need to be installed with a user friendly software to help us learn independently. At the moment there is just talk about a wholesale transition into online learning but not much has been said on how we will be accommodated in the whole process (Dalida).

Echoing similar sentiments Dudu stated that:

> I also think that effective online learning must include the needs of the learners with disabilities by equipping facilities such as the library and the computer labs with disability friendly corners well equipped with technology and human capital to accommodate visual impairment learning needs. We want to acquire education that is at *par* with the need of the 21st century. Other institutions provide preloaded tablets to their visually impaired students which make learning alone enjoyable. Sadly, our institution still lags behind (Dudu).

The consistent concern that has been raised by the student is the inaccessibility of the institutional resources to the visually impaired learners. Emphasis was on the need to provide the software that would aid the visual impairment towards independent learning.

## 5 Conclusions

The study sought to explore how learners with disabilities accessed and utilized technology in the teaching and learning environment. While these learners do own the technology, it does not meet the requirements of online learning. The eminent transition from traditional teaching to online requires a paradigm shift that encapsulates transformation of software. We therefore recommend that UNESWA explores AT such as Envision AI for learners with visual impairment. Given the smaller percentage of such learners, a pilot could be implemented using the portable devices of the current learners to test the adaptability of the software to the poor infrastructural conditions of Eswatini and amend as necessary.

**Acknowledgments** The authors would like to acknowledge UNESWA for the support provided in carrying out the study.

# References

1. C. Magagula, *Implementing educational policies in Swaziland. World Bank Discussion Papers Africa Technical Department Series 88* (World Bank, Washington DC, 1990)
2. M.N.K. Boulos, D.M. Giustini, S. Steve Wheeler, Instagram and WhatsApp in health and healthcare: An overview. Fut. Int. **8**(37), 2–14 (2016)
3. J. O'Brien, *Swaziland National Disability Plan of Action (NDPA) 2015–2020, 2nd Draft* (UNDP, Mbabane, 2015)
4. A.H. Eide, B. Jele, *Living conditions among people with disabilities in Swaziland: A national representative study* (SINTEF Technology and Society, Oslo, 2011)
5. S. Khenkara, H. Alsulaimana, S. Ismaila, A. Fairaqa, S.K. Salma Kammoun Jarrayaa, H. Ben-Abdallah, ENVISION: Assisted navigation of visually impaired smartphone users. Prog. Comput. Sci. **100**, 128–135 (2016)
6. A.S. Hornby, *Oxford Advanced Learners' Dictionary*, 7th edn. (Oxford University Press, New York, 2005)
7. Barbotte, Guillemin, Chau, & the Lorhandicap Group, Prevalence of impairments, disabilities, handicaps and quality of life in the general population: A review of recent literature. Bull. World Health Organ. **79**(11), 1047–1055 (2001)
8. O.C. Abosi, Trends and issues in special education in Botswana. J. Spec. Ed **23**(3), 11–16 (2001)
9. A.S. Awoniyi Samuel Adebayo, K. Ngwenya, Challenges in the implementation of inclusive education at Elulakeni cluster primary schools in Shiselweni district of Swaziland. Eu. Sci. J **11**(13), 246–251 (2015)
10. V. Devedzic, Education and the semantic web. Int. J. Art Int. Ed. **14**, 39–65 (2004)
11. G. Awada, Effect of WhatsApp on critique writing proficiency and perceptions toward learning. Cog. Ed. **3**, 1–25 (2016)
12. World Health Organization [WHO], *World Report on Disability*. Geneva: WHO Press, (2011)
13. J. Hollan, E. Hutchins, D. Kirsh, Distributed cognition: Toward a new foundation for human-computer interaction research. ACM Transac. Comput-Hum Interact. **7**(2), 174–196 (2000)
14. S. Isaacs, *Survey of ICT and Education in Africa Swaziland Country Report* (World Bank, Washington, DC, 2007)
15. Maseko, N. Feasibility study on the ICT industry in Swaziland, (2011). https://www.academia.edu/8898658/Information_Communication_Technology_in_Swaziland
16. N. Ngozwana, S. Maduna, D. Jele, L. Ngcobo, P. Biswalo, Openness and access of learners with disabilities in the department of adult education: Challenges and prospects. Ar. and Hum. Op. Acc. J. **2**(5), 292–298 (2018)
17. H. Sayan, Affecting higher students learning activity by using WhatsApp. Eur. J. Res. and Refl. in Ed. Sci. **4**(3), 88–93 (2016)
18. F.E. Ritter, G.D. Baxter, G. Jones, R.M. Young, Supporting cognitive models as users. ACM Transac. Comput-Hum Interact. **7**(2), 141–173 (2000)
19. G. Fischer, User modeling in human–computer interaction. User Model. User-Adap. Inter. **11**, 65–86 (2001)
20. L. Chmiliar, B. Cheung, Assistive technology training for teachers – Innovation and accessibility online. Dev. Dis. Bul. **35**(1), 18–28 (2007)
21. D. Maor, J. Currie, R. Drewry, The effectiveness of assistive technologies for children with special needs: A review of research-based studies. Eu. J. Spec. Ed. **26**(3), 283–298 (2011)
22. R. Manduchi, J. Coughplan, (Computer) Vision without sight. Commun. ACM. **55**(1), 96–104 (2012)

23. R.A. Koweru, C.M. Omoke, J.A. Orodho, The role of assistive technologies on quality educational outcomes of student with visual impairment in Kisumu County, Kenya. *IOSR J. Hum. Soc. Sci. (IOSR-JHSS)* **20**(3), 39–50 (2015)
24. E. Cutrel, A. Dhareshwar, Imagining artificial intelligence applications with people with visual disabilities using tactile ideation. ASSETS'17 **1**, 81–92 (2017)
25. S. Panchanathan, S. Chakraborty, T. McDaniels, Social interaction assistant: A person-centered approach to enrich social interactions for individuals with visual impairments. IEEE J. Sel. Top. Sig. Proc. **10**(5), 1–1 (2016)
26. UNDP, *National Disability Policy: Empowerment of people with disability in the 21st Century*, Government of Swaziland: Mbabane, (2013)
27. S.K. Thwala, K. Ntinda, B. Hlanze, Lived experiences of parents' of children with disabilities in Swaziland. Journal of Ed. and Tra. Stu. **3**(4), 206–215 (2015)

# Mismatch Between Persuasive System Design and Learner's Perspective: An Interdisciplinary Approach

**Nhlanhla Andrew Sibanyoni**

**Abstract** There are many learners who are still unable to develop a learning habit, particularly on a mobile device, despite of the introduction of persuasive technology such as the LevelUp app. Due to the lack of support and formation of other competing habits that satisfy their personal interests and their social and information needs but not necessarily to study. Therefore, this study applies persuasive system design to systemically analyse the LevelUp app as a persuasive mobile app and the activity theory as a lens. This exploratory case study gathered data from the focus group as well as from observations. The results show that there are contradictions between theory and practice. The study draws both from information systems and social science to contribute to interdisciplinary knowledge. More case studies with different combination of theories should be conducted by other researchers using a multidisciplinary approach, particularly mobile learning after school time.

**Keywords** Persuasive technology · Persuasive system design · Interdisciplinary · Activity theory

## 1 Introduction

We live in a mobile society, where many learners spend a significant part of their time on the move, tech-savvy and acquainted with the latest mobile technologies, and are continuously exposed and stimulated by numerous forms of entertainment such as social media apps like Facebook, Instagram, WhatsApp and Twitter. Social habits are formed where learners expect similar stimulation in mobile learning, which result in competing habits to mobile learning, as formal education cannot provide learners with all the knowledge and skills they require throughout a lifetime [24]. The interest in youth's interaction with mobile games using ubiquitous devices has expanded the traditional classroom learning to encompass mobile learning

N. A. Sibanyoni (✉)
University of Johannesburg, Johannesburg, South Africa
e-mail: nhlanhlasi@gmail.com



E. Eilu et al. (eds.), *Digital Literacy and Socio-Cultural Acceptance of ICT in Developing Countries*, https://doi.org/10.1007/978-3-030-61089-0_13

(m-learning) leading educators and researchers to conduct investigation into the influence of m-learning in education. Research suggests that m-leaning can create more active learning experiences and to help learners to increase their motivation to engage in learning activities. Technology can be used to provide information to satisfy learner's curiosity and encouragement and thereby amplify motivation to solve problems. Also, the difficulties in supporting, guiding and motivating learners in m-learning environments that combine real-world and digital-world learning have been reported. Hence, models to help understand the motivation features of using mobile devices and to enhance learner's motivation have been proposed. Other researchers experimented with a game-based learning activity designed for secondary school learners in order to test their engagement, motivation and learning [23, 30]. Gamification is a concept of applying game design elements to engage and motivate learners in a learning environment. It has been exploited since its discovery in the 1990s to take advantage of today's adolescents between ages 12 and 17 who are (nearly all of them) social gamers in house or online environments [13]. Gamification has shown to engage and enhance, in real or virtual worlds, the learner's motivation and academic performance in complex and critical thinking for learners to complete tasks. For example, gamification makes education experience both challenging and fun where learners' engagement is increased when a teacher gamifies an activity or teaching a concept by including mechanics such as achievement badges, levels and XPs for mastering a given level [10]. Gamification is mostly used to change behaviour, to educate and to motivate using game elements such as points, badges and leader boards [22]. Changing the learners' behaviour to study maths, particularly after school, where there is no adult supervision can be a daunting task. The challenge is how to design mobile learning environments that can be accepted and positively change the learner's behaviour to study after school using a mobile device. It has always been the researchers' aim to change the human mind to behave positively. Persuasive technology refers to the design that works towards changing attitude and influencing human minds, which falls under the human-computer interaction (HCI) discipline. HCI deals with techniques that leverage both the insights and capabilities of importance of environment and human psychology in driving human behaviour and the ubiquity of technology [6]. Learning also takes place at times when there is a break in the formal education routine, with its pre-specified times and place. The focus should therefore be placed on learner's mobility and learning on a mobile device. The LevelUp app is an educational mobile app that uses real rewards such as airtime and data bundles to persuade and motivate learners to study maths after school time [25]. Developers of educational apps, such as LevelUp, continuously grabble with reinventing themselves in being innovative, particularly in developing educational app, in an attempt to keep learners motivated and engaged in the learning process, in particular when they are mobile after school and using their mobile advices [26]. Despite the use of real rewards such as airtime and data bundles to persuade and motivate learners to study maths after school time, many learners are still unable to form and maintain good study habits which usually lead to improved academic achievement. The current study aims to answer this question: Why are the learners not sufficiently motivated to use the LevelUp app after school to study mathematics?

An interdisciplinary approach (Fig. 1) is required when attempting to understand why learners are less interested in maths, particularly when using a mobile device outside school time, as this activity is a complex phenomenon [32]. The interdisciplinary approach (see Fig. 1), provides a more complete understanding of the phenomenon being studied by integrating and synthesising methods and knowledge from different disciplines into a coordinated and coherent whole [29, 31]. The study of persuasive technology is described as being a subdiscipline within human-computer interaction [15], and critics of [3] claim that it is not a new discipline but has rather brought aspects of other existing disciplines together [6] as cited by [15]. This supports the argument that this field should be considered as interdisciplinary. The current study contributes knowledge information system research in interdisciplinary approach (see the highlighted box of Fig. 1) by using the combination of two theories that intersect at the human behaviour, that is, persuasive technology refers to the design that works towards changing attitude and influencing human minds at the system level, while activity theory deals with human activity and the environment.

HCI can be both social and collective [27]. In social cognitive theory, involvement in a collective activity is important in assisting the participants to develop their understanding of concepts by making these explicit and sharing meaning, particularly in the context of education [1]. The use of computers, such as persuasive technology, in education might develop the ability to learn using tools, such as a mobile device, that might afford or constrain their actions while also allowing computer-supported collaboration. Activity theory (AT) assists researchers to understand the

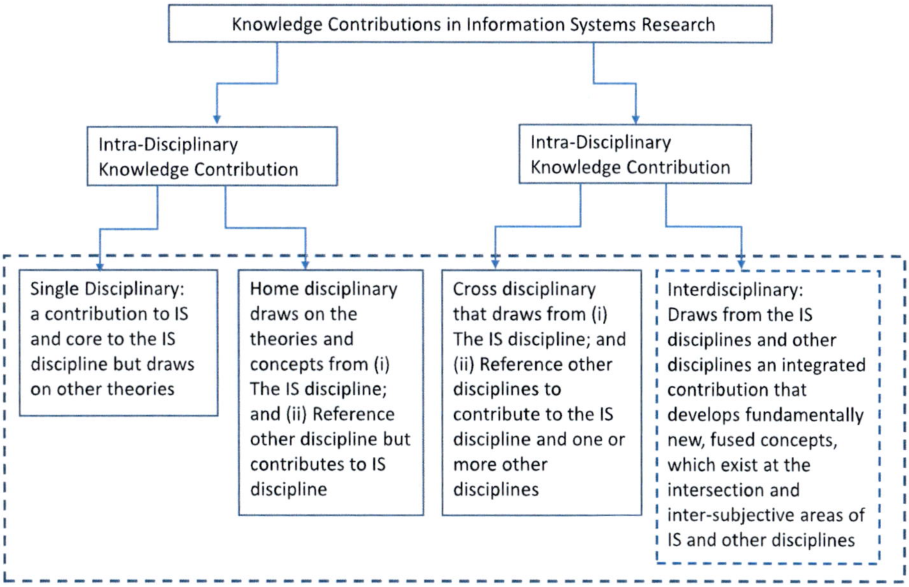


**Fig. 1** Interdisciplinary and intra-disciplinary research in information systems. (Adopted from [31])

ways in which people interact with each other, using tools over time, and has also been used to extensively study human-computer interactions [5, 8, 9, 16, 21]. AT draws on methodologies from the humanities, psychology and social studies to make sense of the complex activity of human learning [27].

This study therefore explores the learners' perceptions and experiences about the effectiveness of the LevelUp app as a persuasive mobile app to foster motivation and engagement for learners to study maths after school time. The study was guided by activity theory which is used as a lens in order to better understand studying maths on a mobile device as it is a complex social activity. Persuasive system design is then used to analyse the persuasion design aspects of the LevelUp app. To capture the learner's perceptions and experiences, the triangulation methodology was employed through observations, focus groups and semi-structured questionnaire. The results show that there is a possible mismatch between learner perception and the LevelUp design to foster the learners to study maths on their devices after school time. This chapter is organised as follows: it explores the related work in activity theory and persuasive technology design followed by the methodology, discussion and conclusion.

## 2 Related Work

### 2.1 Activity Theory

Activity system is applied in activity theory to understand a mobile individual activity which gives an activity a meaningful context (see Fig. 2); a conceptualisation by [2] is a systemic model of an activity system which contains relationships between the nodes, that is, tools, object, community, rules and the division of labour. All the

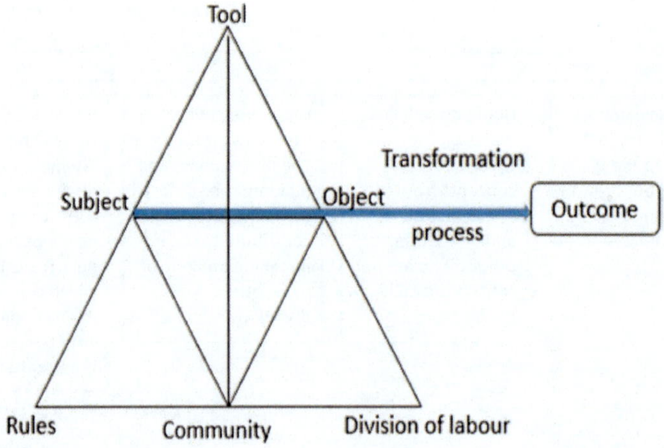


**Fig. 2** [2] Basic activity system

elements in the activity system are linked together which makes an activity a system whole. The object often changes during the activity which might eventually present itself in a different way. Hence, in activity theory, the activity system is the basic unit of analysis where a common object is shared by a community or a group of people, who can transform that object into an outcome through the use of tools, such as computers and mobile devices [5].

An object can be a shared material, tangible (e.g. a plan) or completely intangible (like a common idea) for manipulation and transformation by the participants of the activity.

The relationships between the subject and the object within the activity are mediated by tools [11]. The rules in the system control the relationships (primarily the relationships between the subject and the community) which might support or constrain the behaviour. The relationship between the community and the object is facilitated through the division of labour. During the division of labour, some contradictions may occur, manifesting themselves as problems, ruptures, breakdowns, new ideas and so on. In other words, different people or groups, who are assigned different parts of the work (be these operations or actions), may not understand their role in the same way that others do, may not carry out their tasks as expected by others, etc. But this is not the only source of contradictions as will be seen in the discussion of this case study. Activities work through contradictions resulting in a change leading to improvements [33]. Figure 2 explains [2] basic activity system.

## 2.2 *Persuasive System Design*

Persuasive system design is a framework developed to designing and evaluating systems that influence behaviour of users [12]. PSD principles provide a systemic way to investigate the effectiveness of a persuasive technology design [34]. The model describes 7 propositions which are mapped to users of persuasive systems, persuasion strategies and system features and 28 design guidelines that must be addressed when designing or evaluating persuasive systems. There are four system feature categories: primary task support, dialogue support, system credibility support and social support [20].

### 2.2.1 Primary Task

It describes seven ways in which computers might persuade as interactive tools: reduction, channelling or tunnelling, tailoring, suggestion, self-monitoring, surveillance and conditioning [15].

*Reduction technology* reduces a complex activity by giving a few instructions or simple steps, making the target behaviour easy. This may increase a person's motivation to participate as the perceived benefit/cost ratio of performing the activity is favourable.

*Tunnelling technology* is predetermined sequences of actions or events that lead users through a series of steps. For example, a website that guides the user through the registration process may motivate the user to continue with the registration as it is easy and quick to complete.

*Tailoring technology* persuades the user through customising. It provides relevant information tailored to the user's needs in an attempt to change behaviour or provide empowering information.

*Suggestion technology* offers suggestions at opportune moments. For example, a pop-up message reminds the user to save their work before exiting.

*Self-monitoring* encourages individual users to achieve their predetermined goal or outcome by enabling them to monitor themselves through a computer technology in order to adjust or correct their behaviour, for example, a wristwatch that monitors heart rate and alerts the users when the heartbeat is out of the normal range [4].

*Surveillance technology* enables individuals to monitor the behaviour of other individual users using the same technology and performing the same target behaviour. This is to encourage individual users to learn from each other and improve theirs in turn.

*Conditioning technology* uses positive reinforcement, such as reward, to increase the likelihood that an individual will perform or change a particular attitude or behaviour.

### 2.2.2 Dialogue Support

*Praise*: Based on user behaviour, the application should provide the user with relevant praise through feedback.

*Rewards*: The application should encourage users to perform a target behaviour through virtual rewards.

*Reminders*: The application should encourage users through reminders to perform their target behaviour.

*Suggestion*: The application should encourage users to carry out target behaviour while using the application.

*Liking*: The application should have a look and feel design that appeals to the users.

*Social role*: The application should assume a social role

### 2.2.3 System Credibility Support

*Trustworthiness*: The application should ensure that the information provided to the user is truthful, fair and unbiased.

*Expertise*: The application should show that information provided is obtained from knowledgeable, experienced and competent experts.

*Surface credibility*: The system should provide content that reflects competence and have a look and feel that make the user feel it is credible (e.g. a professional appearance).

*Real-world feel*: The application should provide background information of the organisation, content and services and the experience of the actual people behind it.

*Authority*: The application should refer to people in their role of authority.

*Third-party endorsement verifiability*: The application should provide outside sources that can verify accuracy of website content.

### 2.2.4 Social Support

Techniques that are effective in facilitating social interactions are provided and are described. These are normative influence, social comparison, competition and social recognition that enable individuals to enhance their physical activity behaviour [36]. These are explained below:

*Social learning*: A person is able to use a mobile device application to observe other people or peers performing a behaviour. This will possibly be motivating to learners to persuade them to perform that same target behaviour.

*Social comparison*: A person is able to use a mobile device application to compare his/her performance with the performance of other; the chances become greater that they will perform the target behaviour.

*Normative influence*: A mobile device application can increase the possibility of a person to perform a target behaviour by leveraging a normative influence, for example, trying again if you fail or practicing every day.

*Social facilitation*: A mobile device application can make users aware that there are others performing the same behaviour along with them; these users are more likely to be motivated to perform the target behaviour.

*Cooperation:* A mobile device application can leverage human beings' natural drive to co-operate, to motivate users to adopt a target behaviour.

*Competition*: A mobile device application can leverage human beings' drive to compete, to motivate users to adopt a target behaviour.

*Recognition*: A mobile device application can increase the likelihood of a person or group to adopt a target behaviour by offering public recognition for an individual or group.

## 3 Methodology

Information systems (IS) is a social phenomenon which has triggered an interest in the application of interpretive research paradigm [18]. The goal of the current study is, therefore, to use interpretive research in a form of a case study, in order to analyse and understand the social and cultural contexts of learners' lives in their natural setting when using the LevelUp app. The use of a persuasive technology, namely, LevelUp app, by secondary school learners from a single school is examined within the context of their socio-economic environment. The current study is interested in understanding why the learners are not sufficiently motivated to use the LevelUp application to study mathematics. This study adopted focus groups as a technique to interview the participants. Learners of a Grade 11 class were interviewed in small groups of three to five for 1 h each.

## 3.1 Data Gathering

***Case Study*** The research reported on in this study used a case study to examine LevelUp as a persuasive technology that attempted to persuade and motivate study behaviour after school for learners at one school. Case study research is widely used in the social sciences and can use any of a variety of research methods such as experiments, surveys, histories and modelling [35]. Research case studies can be used in exploratory research to discover new knowledge and in explanatory research to test, to explain or to compare theories [13].

***Ethical Consideration*** According to [17], technology might influence human lives either incidentally or accidentally through affecting attitudes and behaviours. Therefore, this section looks at research processes and how they might impact on the participants. [13] says that when conducting research that involves real people, the researcher should consider ethical practices. It is a requirement for all researchers at a university to obtain ethical clearance to show the authorising body that the research will not harm the participants and measures will be taken to protect them from undesirable consequences. Parents were requested to provide consent for their children to be interviewed. In order not to be biased, to minimise errors when conducting the interviews and to validate the discussion with the interviewee's permission, the researcher used a recording device, a mobile phone, and also took notes [19]. Approval to conduct the study was obtained from the school and participants voluntarily participated in the study.

***Focus Group*** Group discussions may uncover unexpected information as group members encourage one another, share information, jointly follow lines of thought, become more open and may in fact start analysing data contributed by individuals. So they might uncover a "big picture" that they were unaware of. Groups may stimulate the discussion and generate consensus views [19].

**Questions for the Learners**

- What makes you keep on using a tablet or cell phone?
- Do you think you could be encouraged by an App designed like that to study?
- What playing ideas for games, learning, chatting, innovation, pastime, etc. do you have for using tablets or cell phones?
- How do you feel about studying after school using LevelUp?
- What other Apps do you use to study after school?

- What would motivate you to study using LevelUp after school?
- Do you think a person or people are better than technology in motivating you to study? If so, why?

***Observation*** Observation is a data generation method that involves closely watching and paying attention to the actions and interactions of participants. In this way, the researcher can obtain first-hand information about participants, that is, information revealing what they do rather than reports of those actions relying on memory which might be selective. The overt research was applied to inform the participants that they are being observed. This kind of research is considered to be more ethical because the researcher would have obtained consent from the people being observed [19].

### *3.2 Persuasive System Design*

The LevelUp app is systematically analysed using PSD in order to understand the effectiveness of its persuasive features to persuade the learners to study maths after school.

### *3.3 Activity Theory*

Activity theory attempts to understand human activity particularly what people do and how they use artefacts. It is always important to look at the factors that influence the activity such as the environment and other factors that shape the activity. Activity theory is applied in this study as it has proven to assist researchers to meaningfully examine the individual's interactions with the technology in a social context that represents the historical and cultural perspective [28].

## 4 Discussion

### *4.1 Focus Group Result Analysis*

Tables 1, 2, 3, 4, 5 and 6 discuss and analyse the learners' perspective collected from the focus group discussion and follow the outline in Sect. 3.1.

Tables 1 and 2 describe additional factors that encourage or motivate learners to study after school. Firstly, daily homework from school is priority. Secondly, there is an expectation from parents that their children should study; hence they play an important part in motivating their children to study. This might be because parents pay school fees and exert pressure and fear. This is unclear and does not form part

**Table 1** Do you think you could be encouraged by an App designed like that to study?

| Learners' comments | Researcher's analysis |
|---|---|
| … I think with any type of app that has to do with helping you study or do your homework is a great convenience, especially with maths given it's a subject that requires a lot of practicing, so I think it does help essentially to help give you a new broader question base | Learner appreciates any type of support after school to study. Here LevelUp application is in competition with other methods of studying such as textbooks and school homework The learners acknowledge the convenience of studying anytime or anywhere with a mobile device |
| … I think that any kind of afterschool practice or study is good, so whether it's in the form of a hardcover book or a soft cover book, it's practice at the end of the day, and it helps | Learners are looking for assistance after school |
| … I'm kind of old school so I don't like using technology to study; I would try it but I'll probably just get bored afterwards and probably just go back to books | This response contradicts with earlier responses by the developer in Table 4.9 that a lot of people prefer technology to study. Also, rewards earned are used for calling friends and possibly surfing the internet and using other applications |
| … Yes it will motivate you because there are rewards involved, and as a teenager, I would want free airtime to call my girlfriend | Learners recognise that rewards are a short-term motivation. Parents and teachers reward learners with positive feedback |
| … I think the reason why LevelUp would work is the rewards system; it is awesome to study and also get short-term rewards such as airtime or data bundles in the process | Combining this with concrete rewards could possibly enhance learner's motivation. Learners do not recognise positive feedback from parents as a reward |
| … Well because our parents don't reward us, our teachers don't reward us for studying we just get a well done, I would like to get a reward for studying | School homework is number one priority for learners because it is part of curriculum and counts towards their school mark |
| … Our parents would not reward us because studying to them is something that we must do; however, to a lot of us as youth, we are not motivated enough to study | Studying with LevelUp is regarded as a "nice to have" not a must use, possibly because it is not part of school curriculum |
| … Homework is important, so we obviously do homework first because we get it every day and it is compulsory | LevelUp daily challenges introduced after school in addition to school homework which might be unfair or too much for the learners |
| … For me I think studying with an app would be nice only because I can have a calendar there instead of always making one and then it can tell you today we studied this for this long and that. I'd use it for that, but to study on my phone I'll always get distracted by the messages that come in or going to play games. … I think that knowing that you are working for something like getting the points I think that it can be some sort of a motivation because you know you are doing this to get something although you are helping yourself at the same time | Learners recognise that they cannot rely completely on rewards to motivate them to study; they still need intrinsic motivation (self-sustained study habit) |
| … We all want rewards for something, but sometimes you also have to look in the long term so if you mess up it's about you | |

**Table 2** Do you think a person or people are better than technology in motivating you to study? If so, why?

| Learners' comments | Researcher's analysis |
|---|---|
| … My opinion is that when you study with a partner, you'll achieve more mentally with the support of that person compared to technology or doing it alone<br>… I think people are better than computers because people can understand and have feelings; in fact, they know what we go through; they understand the stress<br>… I agree with her opinion, and I also think that people are better than computers. You'll have a better experience with a person than with a tablet or a phone; they can't tell you what you are doing is wrong and whether you understand it or not. The tablet or computer does not care, so a person is better<br>… I think a person can do much better, because if you do not understand something, a person can try and explain it in a different way, whereas a tablet won't be able to. With that you will have a better understanding with a person explaining, because they can think of a way you can relate to and understand the topic that's being tested<br>… I think a person is better because if you like that person, then you start to have fun with maths because not only are you spending time with that person, but you are learning in a fun way<br>… I think I agree with everyone that interaction with people is better; you immediately get either a positive or negative feel, for example, emojis can't really tell a person's feelings, so I think people are better. I don't know about the app. If the app can give feedback as though we've done something, I think it can be better<br>… I feel like when it comes to comparing people to computers, people are honestly better. I also feel like an app is kind of programmed and rigid to tell you; it isn't on a one-on-one discussion where you can share your frustrations; an app is more programmed; it doesn't seem realistic to me | Learners expect LevelUp application to assist them to study mathematics in a human manner; in other words, the app should leverage the power of social influence to motivate and persuade<br>Sympathy and empathy are an important factor<br>The developers of LevelUp should take advantage of the social support principles, namely, social learning, social comparison, normative influence, social facilitation, cooperation, competition and recognition, to better mimic a person |

**Table 3** What makes you keep on using a tablet or cell phone??

| Learners' comments | Researcher's analysis |
|---|---|
| ... It's convenient<br>... For convenience, for the fact of having everything at the tip of your hand and being able to do what's needed<br>... The convenience it serves; you run multitask<br>... It is a convenient way to connect with family and friends. It is also convenient when it comes to social media and getting work done<br>... It's fast and convenient, and I can use it anywhere<br>... Versatility<br>... It is efficient and convenient<br>... It is easy to use, fast, efficient and versatile<br>... It allows us to do different things at once | The learners use a mobile device for convenience, to multitask and to connect with family and friends and because it is fast and can be used anywhere and anytime |

**Table 4** What ideas do you have for playing games, learning, chatting, innovation, pastime, etc. for using tablets or cell phones?

| Learners' comments | Researcher's analysis |
|---|---|
| ... The apps should be designed in such a way that it prevents distractions ... The apps should be interrelated<br>... Create an app that stops you from using other apps while studying<br>... Have classes through videos to help me study for maths. Have an app that locks my other apps when I want to study<br>... The bigger the screen, the better<br>... Social networking<br>... I use my phone for social networks and developing a good profile for myself<br>... Create an app to stop distraction | The learners seem to be concern about being distracted when studying with a mobile device. Some suggestions were made by the learners, for example, an app that prevents distractions from others such as WhatsApp and Facebook and mobile devices with bigger screens |

of Table 1. The learners' comments in Table 2 show that people are honestly better than computers; it illustrates the importance of human interaction and social influence. Peer pressure, social comparison and group polarisation are the types of influence that arise from a social environment. When designed to act as social actors, mobile applications can advantage to leverage the social influence principle, motivate and persuade [3]. In the current study, the LevelUp application is viewed as a tool to support and facilitate learner's studying activities after school.

**Table 5** How do you feel about studying after school using LevelUp?

| Learners' comments | Researcher's analysis |
|---|---|
| ... I think it is a good app, but the level of difficulty is not high, and it's fun to use<br>... It's a good app for primary school learners because it has their level of standard<br>... If the app provided more of an option in terms of the testing standard, like giving a Grade 11 section for the proper testing standard that will motivate you to study<br>... It is very beneficial looking at the fact that I spend so much time on my phone<br>... If feel comfortable<br>... I still feel the same about studying. I'm not motivated to use the app<br>... It helps, but I wish it had sections where I can just go search for trigonometry<br>... Optimistic, I feel this app could really help<br>...N/A. Not available on IOS devices<br>... It's a good app, but not for Grade 11 standard<br>... I prefer to study with books and paper so it would not be as beneficial | Learners are looking for an application that provides options to personalise and have choice<br>LevelUp does not seem to provide the Grade 11 CAPS aligned content; this may be due to not involving the school or teachers<br>Contradicting statements<br>Some learners still see studying on LevelUp as traditional studying and therefore still not motivated to do so<br>CAPS aligned content need here<br>This learner makes it clear that studying on a mobile device is not his/her preference |

**Table 6** What other Apps do you use to study after school?

| Learners' comments | Researcher's analysis |
|---|---|
| ... Mi.Ebooks, YouTube for educational videos, internet, WhatsApp, and Gmail | Learners already have tools that they use to study. LevelUp is competing with other applications |

**Table 7** What would motivate you to study using LevelUp after school?

| Learners' comments | Researcher's analysis |
|---|---|
| ... The rewards and the pressure of exam coming up<br>... Grade 11 standard exercises and questions<br>...More challenging questions from the App; it can be very effective if it helped you see your problem areas<br>... The rewards LevelUp offer<br>... If the rewards were reasonable<br>... Rewards for finishing sections<br>... Positive results from other surveys and tangible rewards | The learners are looking for more challenging, relevant exercises and questions to prepare for exams from LevelUp. They also suggest that rewards would motivate them |

From the learners' comments from Table 3 through to Table 7, it is evident that the learners are looking for personalisation options, CAPS-aligned content, and are easily distracted when studying on a mobile device. It is also evident that the learners use a mobile device for convenience and social networking; these are clearly competing activities which LevelUp developers might not have anticipated.

## 4.2 Observations

In the afternoon, after school time, learners are observed walking or waiting and socialising in groups (see Fig. 3). Observation method is used in this case to closely watch and pay attention to the actions and interactions of the learners with the LevelUp in their natural setting or environment after school.

Social pressure seems to lead learners to perform activities like group formation with peers and use personal mobile devices to access applications that are restricted by the school such as YouTube, listening to music and so on (see Fig. 4).

Learners are socialising while waiting for their transport in groups at their usual spot next to the lapa and others on the benches (see Figures 3 and 4 (right corner of the pictures)). The learners use different tools such as paper, textbook, tablets, cell phones and earphones to mediate their activities.

**Fig. 3** Learners waiting in groups and interacting with their mediating tools

**Fig. 4** Learners studying in groups and individually using papers

## 4.3 Analysis Through Persuasive System Design

In Table 8, the LevelUp app is systematically analysed using PSD in order to understand the effectiveness of its persuasive features to persuade the learners to study maths after school.

**Table 8** Analysis of the LevelUp application using persuasive system design

| Category | Persuasive feature | Mapping persuasive feature to learner's responses |
|---|---|---|
| Primary task support | Personalisation | Learners suggested some personalisation:<br>... The apps should be designed in such a way that it prevents distractions ... The apps should be interrelated<br>... Create an app that stops you from using other apps while studying<br>... Have classes through videos to help me study for maths. Have an app that locks my other apps when I want to study<br>... The bigger the screen the better |
| | Self-monitoring | LevelUp does not provide options to choose the level of difficulty for Grade 11 learners (Table 4.1.5)<br>Self-monitoring allows learners to compare their progress to other learners<br>"...Positive results from other surveys and tangible rewards" |
| Dialogue support | Rewards | The tangible rewards are not enough to motivate, but the exams: "...The rewards and the pressure of exam coming up" |
| | Social role | LevelUp can take advantage of using a social role feature to attempt and motivate the learners to study: "...my opinion is that when you have a study partner you can achieve more mentally with that person compared to doing it alone or with a computer compared to technology" |
| System credibility support | Expertise | Expertise: the application should show that information provided is obtained from knowledgeable, experienced and competent experts, for example, CAPS content: "...If the app provided more of an option in terms of the testing standard, like giving a Grade 11 section for the proper testing standard that will motivate you to study" |
| Social support | | LevelUp should take advantage of the social support principles such as social learning, social comparison, normative influence, social facilitation, cooperation, competition and recognition to mimic a person<br>"... I think people are better than computers because people are flexible and they have feelings, and also understand what you are going through as they have been there or done it before; in other words, they understand your feelings. ... I also agree with other opinions, and I think that people are better than computers, because you'll have a better experience with a person than with a tablet or a phone. The table or a phone cannot understand your emotions of what you are going through and whether you understand it or not; the tablet doesn't really care, so a person is better" |

## 4.4 Analysis Using Activity Theory

This section examines what learners do, the support available, rules and the division of labour when using artefacts and other factors that influence the outcome of using LevelUp activity to study maths after school.

Table 9 is an analysis of Figure 5. Table 9 analyses the learners' activity system: studying mathematics using LevelUp. This is done to meaningfully examine the individual learner's interactions with the LevelUp in a social context that represents the historical and cultural perspective.

**Table 9** Analysis of activity system and its elements

| Activity system and its elements | Analysis |
|---|---|
| Activity system | Taking studying mathematics as an activity system, as depicted in Fig. 5, the learner (the subject) interacts with the LevelUp application daily challenges on a mobile device. The application can be accessed with the following supporting tools: to either sign up and log in using Facebook or Google login details or directly on the internet called a mobi site. All these tools require data to function (as mediating tools), in order to act on them and to solve mathematics problems (the object) and in doing so to achieve a motivated study behaviour and indirectly develop a learning habit (the outcome). The learner is governed by the rules of the LevelUp application. The division of labour ensures that different roles are identified while sharing an object. In this case, the learners are expected to engage with the LevelUp application without other people being involved |
| Subject | The learners have been identified as individual subjects who are motived through a reward system, which uses real rewards, such as airtime, in order to use a mobile device to act on LevelUp daily challenges, out of their free will after school |
| Rules | The learner should register and can only redeem coupons when they have enough points. For the learners to obtain the rewards, they first need to register and log in to complete a few math exercises. A certain number of points are required before they can be converted to rewards, for example, 15MB Vodacom data requires 2500 points. The learners do not determine their own question category or level of difficulty; instead, the system automatically moves to the next level based on the learners' weekly challenge progress |
| Object | In an activity system, an object is transformed into an outcome when the subject acts on it, which is a reason for the existence of an activity. In the current study, the object is for the learners to perform daily challenges by solving mathematics problems and in return obtain real rewards once completed successfully. At the same time, motivation is what makes us do or act on something. It is usually difficult to understand what motivates learners to act or not to act on daily challenges. Additionally, learners stated that their purpose for using a mobile device overall is for convenience as it enables them to communicate and socialise with friends and for information, news and so on |

(continued)

**Table 9** (continued)

| Activity system and its elements | Analysis |
|---|---|
| Division of labour | The learners who use LevelUp are implicitly expected to tell their peers about their experiences of using the tool and hence to increase the use of the app by word-of-mouth recommendation. Additionally, the learners are advised to send their queries via the Report a Problem tool within LevelUp to a dedicated support team (self-service help) |
| Community | According to the developers, the daily challenge is something which the learners can do or perform on their own without the assistance of the community (school administration, teachers, parents or peers). A community is a group of people who are involved and share the same interest with the subject towards the object of the same activity system. The community and developers share interests with the learners, that is, to improve math results by practising maths using LevelUp; however, the community is not involved in this activity. The developers are therefore seen as the only community supporting or assisting with the studying mathematics activity system using LevelUp after school (Fig. 5) |
| Contradiction | (a) Between the designer activity object and the learner activity object and (b) between learner activity system and the more general using-a-mobile-device activity (not shown). These seem to be design mismatches. There is a "use value" contradiction, for example. Figure 5 is relevant for studying mathematics using LevelUp; hence, the objective of mobile phone activity, where it is used by the learners for communicating, calling, internet surfing and so on, is not reflected well in the learners' LevelUp activity; hence, this points to a contradiction between the activity that the App supports (Fig. 5) and the object of the more general activity of using a mobile device |

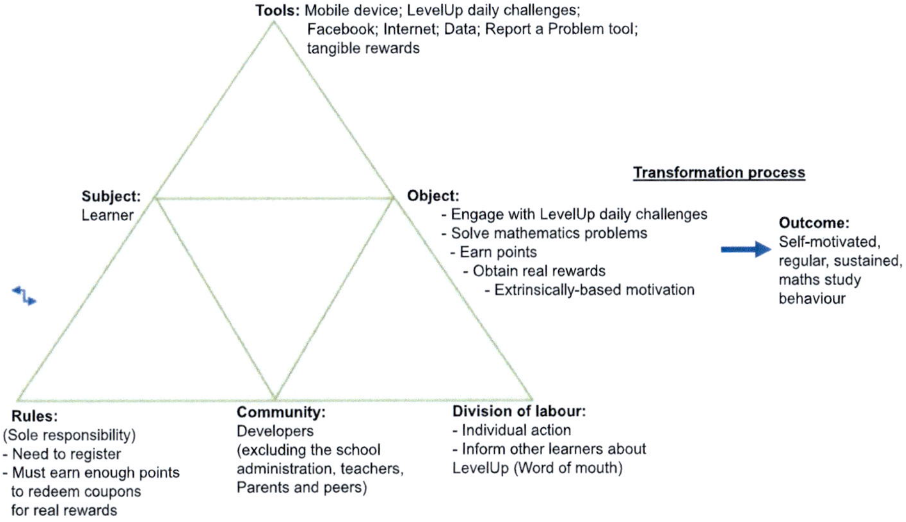


**Fig. 5** Learners' activity system: studying mathematics using LevelUp [26]

## 5 Conclusion

This study has used persuasive system design (information system) and activity theory (social sciences) to understand the complexity of studying mathematics after school using a mobile device. The aim of combining the two theories was to understand the learner's activity in a social context while using an information systems tool and to also contribute to interdisciplinary knowledge. The PSD used alone cannot show the contradictions; however, when supplemented with AT, this study was able to highlight the contradictions (a) between the designer activity object and the learner activity object and (b) between learner's activity system and the more general using-a-mobile-device activity. In addition, this study has used a practical real-life example of a persuasive technology for South African learners studying mathematics outside school. The study, therefore, suggests that further empirical research is needed, using different combinations of theories in a case study, in order to thoroughly understand how persuasive technology that uses real rewards can be effective in persuading and motivating learners to develop new study behaviours after school hours.

**Acknowledgements** This chapter is based on my unpublished thesis titled "Towards Assessing Persuasive Mobile Technology for South African Learners Studying Mathematics" supervised by Prof. PM Alexander at the University of South Africa.

## References

1. A. Bandura, Social cognitive theory, in *Annals of Child Development*, Six Theories of Child Development, ed. by R. Vasta, vol. 6, (JAI Press, Greenwich, 1989), pp. 1–60
2. Y. Engeström, Learning by expanding, in *Learning by Expanding: An Activity-Theoretical Approach to Developmental Research*, ed. by Helsinki, (Orienta-Konsultit, Finland, 1987)
3. B.J. Fogg, Persuasion in the digital age, in *Persuasive Technology: Using Computers to Change What We Think and Do (Interactive Technologies)*, ed. by Drew, Jeannine, Bluehler, 1st edn., (Morgan Kaufmann, San Francisco, 2002), p. 312
4. T. Fritz, E.M. Huang, G. Murphy, T. Zimmermann, *Persuasive Technology in the Real World: A Study of Long-Term Use of Activity Sensing Devices for Fitness*. CHI 2014, One of a CHInd, Toronto, ON, Canada (2014)
5. J. Hardman, Activity theory as a potential framework for technology research in an unequal terrain. S. Afr. J. Educ. **19**(2), 378–392 (2005)
6. A. Harris, S.U. Islam, J. Qadir, U.A. Khan, *Persuasive Technology for Human Development: Review and Case Study*. EAI Endorsed Transactions on Serious Games. (2017) https://doi.org/10.4108/eai.8-12-2017.153401
7. R.B. Johnson, Mixed methods research: A research paradigm whose time has come. Educ. Res. **33**(7), 14–26 (2004)
8. S. Karanasios, Framing ICT4D research using Activity Theory: A match between the ICT4D field and theory? Inf. Technol. Int. Dev. **10**(2), 1–17 (2014a)
9. S. Karanasios, Toward a unified view of technology and activity: The contribution of activity theory to information systems research. Inf. Technol. People **31**(1), 134–155 (2018)

10. T.L. Kingsley, M.M. Grabner-Hagen, Gamification: Questing to integrate content knowledge, literacy, and 21st-century learning. J. Adolesc. Adult Lit., 1–11 (2015)
11. K. Kuutti, Activity Theory as a potential framework for human- computer interaction research, in *Context and Consciousness: Activity Theory and Human Computer Interaction*, ed. by B. Nardi, (MIS Press, Cambridge, 1996), pp. 17–44
12. J. Matthew, K.T. Win, H. Oinas-Kukkonen, M. Freeman, Persuasive technology in mobile applications promoting physical activity: A systematic review. J. Med. Syst. **40**, 72 (2016)
13. S.N.M.M. Mohamad, N.S.S. Sazali, M.A.M. Salleh, Gamification approach in education to increase learning engagement. Int. J. Humanit. Arts Soc. Sci. **4**(1), 22–32 (2018)
14. M. Myers, *Qualitative Research in Business & Management*, 2nd edn. (Sage, London, 2013)
15. J. Mintz, M. Aagaard, The application of persuasive technology to educational settings. Educ. Technol. Res. Dev. **60**, 483–499 (2012)
16. A.B. Nardi, Activity theory and human-computer interaction, in *Context and Consciousness: Activity Theory and Human-Computer Interaction*, ed. by N. Bonnie, (MIT Press, London, 1996), pp. 7–16
17. C. Neuschwander, *Sir Cumference and the Dragon of Pi: A Math Adventure* (Charlesbridge, Watertown, 1999)
18. F. Niederman, S. March, Design science and the accumulation of knowledge in the information systems discipline. ACM Trans. Manag. Inf. Syst. **3**(1), 1–12 (2012)
19. B. Oates, *Researching Information Systems and Computing* (Sage, London, 2006)
20. H. Oinas-kukkonen, M. Harjumaa, Communications of the association for information systems persuasive systems design: Key issues. Process Model Syst. Features **24**(28), 485–500 (2009)
21. D. Russel, The RoutledgeFalmer reader in psychology of education, in , ed. by H. Daniels, A. Edwards, (RoutledgeFalmer, London/New York, 2004)
22. M. Sanmugam, N.M. Zaid, H. Mohamed, Z. Abdullah, B. Aris, S.M. Suhadi, Gamification as an educational technology tool in engaging and motivating students; An analyses review. Adv. Sci. Lett. (2015)
23. G. Schwabe, C. Goth, Mobile learning with a mobile game design and motivational effects. J. Comput. Assist. Learn. **21**(3), 204–216 (2005)
24. M. Sharples, D. Corlett, O. Westmancott, The design and implementation of a mobile learning resource. Pers.Ubiquit. Comput. **6**, 220–234 (2002)
25. N.A. Sibanyoni, P.M. Alexander, *Assessing the potential of levelup as a persuasive technology for South African learners*, in (IADIS) International Conference on Mobile Learning, 13th, Budapest, Hungary, April 10–12, 2017, (2017)
26. N.A. Sibanyoni, P.M. Alexander, *Insights into persuasive technology for M-learning using activity theory*, in IADIS International conference cognition and exploratory learning in digital age 2018 (2018)
27. R.E. Slavin, When and why does cooperative learning increase achievement? Theoretical and empirical perspectives, in *The RoutledgeFalmer Reader in Psychology of Education*, ed. by H. Daniels, A. Edwards, 1st edn., (RoutledgeFalmer, London, 2004), pp. 271–325
28. F. Ssozi-Mugarura, U. Rivett, E. Blake, *Using activity theory to understand technology use and perception among rural users in Uganda*. ICTD'16, June 3–16, Ann Arbor, Michigan, USA (2016)
29. M. Stember, Advancing the social sciences through the interdisciplinary enterprise. Soc. Sci. J. **28**(1), 1–14 (1991)
30. C.H. Su, C.H. Cheng, A mobile gamification learning system for improving the learning motivation and achievements. J. Comput. Assist. Learn. **31**, 268–286 (2014)
31. M. Tarafdar, R.M. Davison, Research in information systems: Intra-disciplinary and interdisciplinary approaches. J. Assoc. Inf. Syst. **19**(6), 523–551 (2018)
32. M. Turpin, M. Matthee, A. Kruger, The teaching of creativity in information systems programmes at South African higher education institutions. Afr. J. Res. Math. Sci. Technol. Educ. **19**(3), 278–288 (2015)

33. L. Uden, Activity theory for designing mobile learning. Int. J. Mob. Learn. Organ. **1**(1), 81 (2007)
34. Y. Wang, L. Wu, J.P. Lange, A. Fadhil, H. Reiterer, Persuasive technology in reducing prolonged sedentary behavior at work: A systemic review. Smart Health **7–8**, 19–30 (2018)
35. R. Yin, in *Case Study Research: Design and Methods*, ed. by G. Knight, V. Young, J. Koscielak, K. Bauhaus, B. Dickens, 5th edn., (Sage, London, 2013)
36. D. Yoganathan, S. Kajanan, *Persuasive technology for smartphone fitness apps*, in PACIS 2013 proceedings (2013), p. 185. https://aisel.aisnet.org/pacis2013/185

# Monitoring Tuberculosis (TB) Treatment Adherence Using eCompliance Digital Tool by Community Healthcare Workers

**Georgies Mgode, Dickens Bwana, Christophe Cox, Shelly Batra, Charles Richter, and Lena Fiebig**

**Abstract** Tuberculosis (TB) caused by *Mycobacterium tuberculosis* kills over one million people yearly. TB diagnosis and treatment are complicated, and hence innovative diagnostic and treatment adherence monitoring tools are important. In this chapter, the application of innovative model of TB detection, linkage to care, and monitoring the adherence to treatment regimen using the India-made digital technology created by Operation ASHA (eCompliance) in Dar es Salaam, Tanzania, is described. The holistic approach showed to be successful in increasing TB case detection yield by 36%, strengthened linkage to care, and led to 98% treatment adherence in terms of doses taken among enrolled patients who stayed throughout the study. The use of digital technology proved feasible in the setting with little training, and support has been well accepted by patients and healthcare providers including community healthcare workers with no prior knowledge of digital adherence tools.

**Keywords** Directly observed therapy · Treatment adherence · Patient tracking · Trained TB detection rats · Community healthcare workers · Digital literacy · Treatment adherence technology · e-health · Health monitoring · Treatment success · Tuberculosis diagnosis

---

G. Mgode (✉)
Pest Management Centre, Sokoine University of Agriculture, Morogoro, Tanzania

Africa Centre of Excellence for Innovative Rodent Pest Management and Biosensor Technology Development, Pest Management Centre, Sokoine University of Agriculture, Morogoro, Tanzania

APOPO Project, Sokoine University of Agriculture, Morogoro, Tanzania
e-mail: gmgode@sua.ac.tz

D. Bwana
MUKIKUTE - National Patient Organization, Dar es Salaam, Tanzania

C. Cox · C. Richter · L. Fiebig
APOPO Project, Sokoine University of Agriculture, Morogoro, Tanzania

S. Batra
Operation ASHA, New Delhi, India



E. Eilu et al. (eds.), *Digital Literacy and Socio-Cultural Acceptance of ICT in Developing Countries*, https://doi.org/10.1007/978-3-030-61089-0_14

## 1 Introduction

Tuberculosis (TB), caused by bacteria of the *Mycobacterium tuberculosis* complex, is the leading cause of death from an infectious disease. TB is curable and preventable, but its diagnosis is challenging, and it takes at least 6-month treatment with daily doses to achieve cure and prevent the emergence of drug resistance [1]. Efforts are needed to improve case detection, yet to unfold health benefit, also holistic and patient-centered models of care are required, which assure that TB will be effectively treated.

In this chapter, we describe the innovative approach and findings from our ongoing project "Test, Treat, and Track" in Dar es Salaam, the port city in mainland Tanzania that contributes to 20% of TB burden in the country [2].

The innovative model comprised TB detection services that include TB case detection using trained African giant pouched rats (Cricetomys spp.) that are for research use, the extension of TB diagnostic testing capacity and service in Dar es Salaam with same-day TB testing and reduced result delivery turnaround time (within 24 h); and a scale-up of linkage to TB care services provided by community health workers and volunteers (CHW); directly observed therapy (DOT) and tracking of hard-to-reach patients in poor urban areas. The DOT and patient tracking are supported by the digital health technology named eCompliance, which aims to reduce the numbers of necessary clinic visits and improve treatment adherence and cure rates.

The chapter describes these interventions, the role of digital adherence technologies, and eCompliance in particular and shows how the community members were trained and helped to acquire the digital literacy.

## 2 Detection of Missed Tuberculosis Patients with the Help of Trained Rats

In Tanzania, TB is the third major cause of morbidity and mortality after malaria and HIV/AIDS [3]. The case detection rate is low mainly due to the lack of adequate diagnostic methods. Like in most developing countries, Tanzania has been largely relying on smear microscopy, which is known to have good specificity, usually above 95%, but a lower sensitivity that varies between 20 and 60% [4]. Rapid molecular tests, mainly the GeneXpert MTB/RIF platform, which detects deoxyribonucleic acid (DNA) of the *Mycobacterium* causing TB and resistance to rifampicin drug [5], are about to be rolled out in Tanzania although the rollout of the Xpert machines is not yet full to cover all TB diagnostic facilities [2], yet there are remaining challenges such as the ability to provide this expensive test to all presumptive TB patients [6]. New simple, rapid, accurate, and affordable tests are still much needed [7]. This goal is pursued by training giant African pouched rats, as an innovative scent-based detection technology being developed by APOPO and Sokoine University of Agriculture, Morogoro, Tanzania. Trained rats target the smell of the volatile organic compounds (VOCs) specific for TB-causing mycobacterium [8]. TB detection rats have shown to stand out through low cost and high speed, throughput, and sensitivity; the refinement of the tool to further boost accuracy and expand the range of sample materials is ongoing. The research use of rats in the field has been well established as described elsewhere [9–12], briefly, sputum samples from presumptive TB patients indicated by trained rats as TB positive is subjected to bacteriological confirmation using endorsed test before patients are linked to care. The diagnostic innovation has been the starting point of the innovative intervention involving model comprising also eCompliance technology. Since efforts to increase TB case detection should be always paired with efforts to effectively treat and reliably cure patients, this work consisted these elements. The long duration required to complete TB treatment involving taking antituberculosis doses every day for at least 6 months for drug-susceptible TB requires an effective method of monitoring the adherence to treatment regimen to ensure treatment success. Paper-based documentation of TB treatment record has been in place over years, and recently digital tools have been introduced to strengthen the recording and monitoring the treatment. These efforts aim to prevent the emergence of drug-resistant TB [13].

## 3 Digital TB Treatment Adherence Monitoring Options

Adherence to TB treatment regimen is certainly difficult; hence it requires supervision involving someone dedicating time and availability to ensure a TB patient under supervision takes daily doses of medication accordingly. Without direct supervision, it becomes a challenge and difficult to derive reliable treatment adherence data. Monitoring treatment adherence can thus be an integral part of TB control especially due to extraordinarily long period required to complete the first-line drug treatment regimen that takes a minimum of 6 months for drug-susceptible TB [1]. Treatment of the drug-resistant TB takes much longer period which can be over a year [14]. A stringent digital monitoring tool is thus ideal for recording and making relevant data easily accessible for assessment of adherence. More powerful mobile devices, new health apps, greater network coverage, and better Internet access open up interesting new opportunities to improve the healthcare of people worldwide including low-resource settings. There is also a growing variety of digital tools for TB care and control [15]. Particular attention is paid to monitoring therapy [16]. The reasons for this are that the TB treatment lasts for at least 6 months and that patients have to take medication every day so that the therapy works and the development of antibiotic resistance is prevented [14]. In many countries, directly observed therapy (DOT) is a standard procedure (as part of the so-called "DOTS" strategy). DOT consists of regular personal contact between the patient and the healthcare worker for controlled medication delivery and directly monitoring medication intake. Patients without DOT, on the other hand, receive medication for a longer period in advance and take the medication independently without supervision (self-administered therapy – SAT) [17]. DOT proves to be practical in many places but does not necessarily improve the TB treatment results and can significantly restrict patient autonomy and pose logistic challenges as well as imply high transport costs to the patients, although the treatment is supposed to be for free.

This is how the idea of digital adherence technologies (DAT) came up to electronically monitor the doses taken (eDOT). A further distinction is made between SMS reminders [18–20], video-monitored therapy (VOT or vDOT), [21] and the use of the medical event monitoring systems (MEMS), e.g., intelligent pill boxes that remind patients of the medication they are taking and send signals to their case care staff when opened [22]. MEMS also include blister packs with numerical codes, which are only visible after the tablet has been removed and which the patient must send to the case manager by SMS. If there is no signal, the case worker will personally discuss the reasons with the patient. Other approaches such as oral biomedical sensor technologies (ingestible sensors) are currently being tested [23]. DAT can also be used in addition to face-to-face meetings with the case worker (instead of replacing the face-to-face meeting). For example, case workers can carry a digital tool (such as a mobile device with compliance software) that electronically records every meeting.

Expectations on how DAT can improve TB treatment adherence and outcomes are high, yet evidence, especially from low-resource settings, is still limited. The finding from the Test, Treat, and Track project in Dar es Salaam provides insights on successful implementation of digital adherence tool in low-income settings involving community healthcare workers with no prior digital literacy.

## 4 Equipping the Community with Digital Tool for Monitoring Treatment Adherence

Operation ASHA (OpASHA) an Indian non-governmental organization came up with an innovative digital technology of registering TB patients and monitoring their adherence to TB treatment regimen [24–26]. The consenting TB patients did provide biometric information including name, gender, age, HIV status, and contact information including telephone number and physical address. These are stored in the eCompliance biometric system (Fig. 1) and are identified by unique fingerprint of the particular patient also stored in the eCompliance gadget consisting of a tablet and fingerprint reader (Fig. 1) maintained by the trained healthcare workers and the TB coordinators in the specific area. It is also important to mention that providers have to scan their fingerprint, too, at each visit and are thus also accountable for making the patient visit happen.

The eCompliance component (Fig. 1) adds up to examples of human-computer interactions (HCI) in which it provides a complete linkage of the TB patient treatment adherence record stored in the digital form and the community healthcare workers, the project supervisors, and TB coordinators with real-time information while at different geographic locations. It is similar to the renowned HCI involving

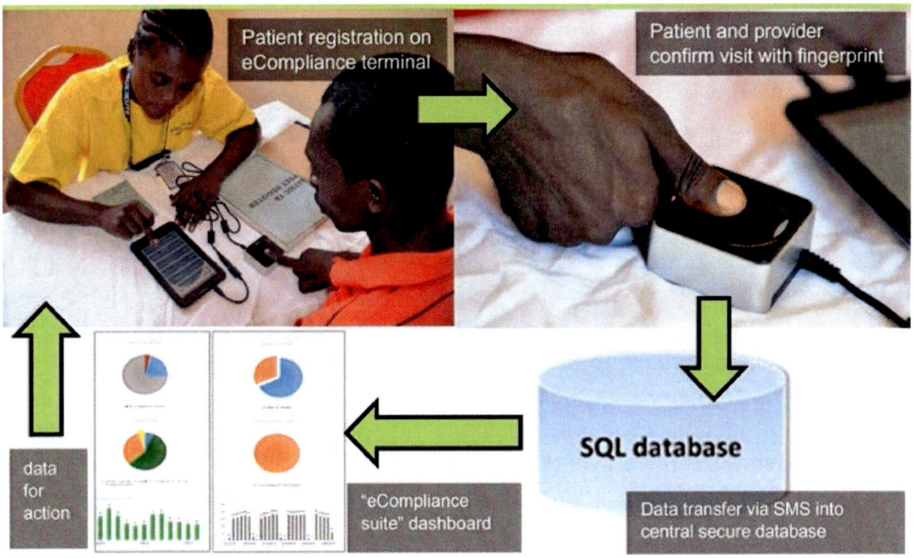


**Fig. 1** The components of eCompliance biometric system gadgets (tablet and fingerprint reader) and data storage

multi-touch technology which enables graphic interaction through computer touch panel [27]. The eCompliance mode of operation has been simplified using color indicators (orange, green, and red) like the "traffic lights" to enable quick identification of individuals taking their doses appropriately (green) and not taking doses (red). This color enables quick intervention and finding out the causes. With eCompliance the TB coordinators can monitor the treatment adherence in reality although they are away. This innovative patient tracking may be related to HCI virtual reality [28] in which through the three-dimensional display, users see the real world indistinguishable from virtual argumentation [28].

## 5 Training Community Healthcare Workers

The local community group known in Swahili language as Mapambano ya Kifua Kikuu na Ukimwi Tanzania (MKUTA), which is a national patient organization grouping the former TB patients as community volunteers and health workers, with its branch known as Mapambano ya Ukimwi na Kifua Kikuu Temeke (MUKIKUTE) based in Temeke District, Dar es Salaam, Tanzania, the area where the reported experience was gathered, is pivotal in demonstrating the massive role that the community has in achieving the sustainable development goal number 3 (SDG 3, i.e., health). MKUTA is a 2018 winner of the global award for Community Leadership and Mobilization given by the Stop TB Partnership [29]. The community healthcare workers became conversant with the digital technology and acquired the digital literacy from a 3-day practical training offered by IT expert from OpASHA in 2016 and 2017 in Dar es Salaam, Tanzania. Following successful training of 15 community healthcare workers from MUKIKUTE and 9 hospital nurses and local community volunteers working under MUKIKUTE, some of the trainees became supervisors to oversee the application of the digital technology in respective areas. This technology is applied on tablets that are similar to smartphones that are now widely used at community levels including remote rural areas. This makes possible for community members to adapt and efficiently monitor the patients as instructed by the supervisor. The community healthcare workers continued receiving further training through a WhatsApp group formed to enable quick communication and sharing technical challenges (but no confidential patient information) that need solutions from supervisors and/or IT experts from OpASHA based in India. This is achieved through the question and answer support given by OpASHA. For many of the community healthcare workers and nurses, this was the first exposure to digital health technology. English language of the tool posed an initial challenge but was later resolved as they got used to the system. The fact that these community healthcare workers were already working on community-based directly observed treatment (DOTS), involving visiting TB patients homes to assess whether they have taken medication, means that they were interested on practicing the eCompliance. Periodic 1-day training provided by MUKIKUTE including onsite training to address challenges that users of the eCompliance face made the trained community healthcare worker more conversant and led to increased digital literacy in the community. This is demonstrated by consistent trend of increasing adherence to treatment regimen measured by dosage completed.

Each community volunteers delivered treatment to three to five patients' households daily. Normally, a patient has to return to TB clinic to take medicines weekly, each of these weekly visits needs about 30 min for paperwork documentation apart from patients travel time from home to hospital. With eCompliance patients stay at home and are visited by the community healthcare. Nurses equipped with eCompliance tool at hospitals monitored the patients during the initial phase of the treatment (commonly the first 2 weeks) and recorded the patients' visit at clinic.

## 6 ECompliance Value on Treatment Adherence

An increase of treatment adherence (in terms of % doses taken) up to 98% has been observed from July 2017 to December 2019, enrolling over 3000 TB patients. A challenge is that no baseline data are available from traditional community-based DOT, since the records on doses administered and taken are not systematically captured and analyzed. A community healthcare workers supervise 10–20 patients until they are fully cured. The community healthcare workers therefore helps to reduce the number of TB patients defaulting TB medication, which increases transmission of this disease from untreated patients to the community. The eCompliance technology potentially increases treatment adherence and reduces the risk of treatment

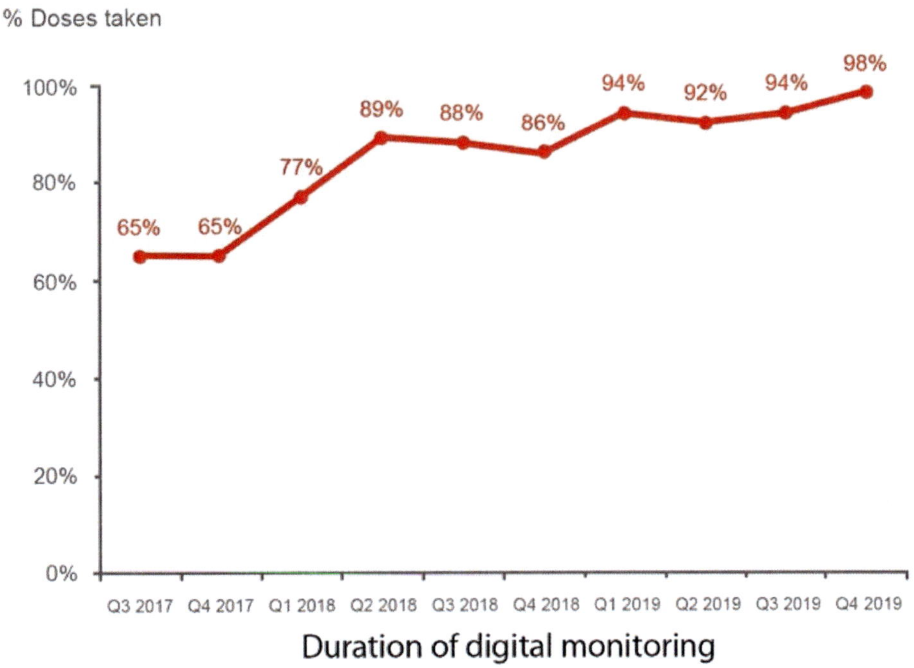


**Fig. 2** Treatment adherence trend generated from eCompliance tool operated by community healthcare workers

interruptions that may lead to drug resistance or clinical complications. It provides robust data collection and accurate monitoring of treatment adherence (Fig. 2).

## 7 Common Challenges

Two categories of challenges can be encountered when engaging the community to implement digital technology in health, namely, technical and logistical or geographical.

### *7.1 Technical Literacy*

It takes tangible efforts to ensure that the community is aware of the new digital technology for application in health system. Some trainees might be slow learners and hence require frequent training and supervision including onsite spot visits. Monthly team meetings are vital for resolving common challenges and sharing expert solution. Unavailability of spare parts and infrastructure for effective connectivity to the eCompliance system and synchronization may be a challenge although this is common at initial stages of implementation.

### 7.2 *Logistical and Geographical Challenges*

Applying a digital technology for monitoring TB treatment adherence in a small geographic area may suffer when patients enrolled move out of the area to another area without eCompliance digital tool. Transfer-out patients who were initially enrolled in the study can affect the overall success expected from the monitoring of treatment adherence because the eCompliance system considers the patients moving out as incomplete doses. However, this can be modified to accord them a different status upon availability of proof that dosage was completed accordingly elsewhere.

## 8 Can Digital Literacy Installation in the Community Yield Expected Outcomes?

The answer to the question on whether digital literacy can be instilled in the local communities and yield the expected result such as improvement in health system is yes. When a new innovation is taken to the community and the community realizes its benefits that it can help them to address their challenges such as a problem of ensuring that TB patients complete dosages according to the existing treatment guidelines from the National Tuberculosis and Leprosy Programme and/or the World Health Organization (WHO), the perception toward such digital tool (eCompliance) becomes positive, and they develop interest to learn. This is observed in this chapter which shows an increasing proportion of treatment adherence determined by dosage taken. It is intriguing to observe that the community healthcare workers adapted and successfully implemented the technology regardless of their education level.

How may the future of DAT for TB care look like? It could be an interesting focus for tool developers to focus more on education and support than predominantly the monitoring and supervision aspect: for example, with particular attention to functions that could make everyday life easier for patients and case workers, promote communication, and encourage patients to complete their therapy. Such support DATs could be used regardless of whether the therapy is being monitored. Digital tools cannot cure tuberculosis and not (yet) replace traditional medical care. The more carefully the digital tool design takes into account the realities of life for people with TB, the more they can improve TB care. The implementation of eCompliance technology on monitoring TB treatment adherence and the successful contribution to increasing appropriate dose taking depict further the importance of human-computer interactions (HCI). The TB patients feel comfortable and cared when they realize that their treatment is well cared. Similarly the community healthcare workers linking their patients to a digital platform that enables secured storage of real-time patient treatment adherence information that is visible to their supervisors and data coordinators located away from the field demonstrated the actual human-computer interaction usefulness [27, 28].

The "Test, Treat, and Track" monitoring intervention involving a group of 15 healthcare workers trained together with five (5) qualified nurses achieved to successfully introduce in the community the eCompliance digital tool which improved significantly the adherence to TB treatment. The use of this technology enhanced timely visits and provided more time for public awareness of the disease since patients are no longer needed to go to clinic after initial treatment. Timely visits to patients' home enabled instant data capturing and rapid health intervention evaluation. Accountability of both the health provider and the patient becomes vivid as this is incorruptible.

## 9 Conclusions

Digital health monitoring tools are much needed for enhancing treatment adherence especially for disease with prolonged dosages such as TB. Carefully designed and user-friendly digital tool can penetrate local communities with no prior knowledge or background or digital skills. Engagement of former TB patients serving as community healthcare workers in the implementation of digital monitoring of treatment adherence can yield expected outcomes in tracking TB treatment due to their familiarity with the problem. Providing digital treatment adherence monitoring to community and subsequent technical support as well as localization to meet local needs may enhance successful uptake of the technology by the end users and increase digital literacy in the community. This can contribute to improved public health and socio-economic livelihood.

**Acknowledgments** We highly appreciate the contribution from Mr. Daniel Thobias Opalla for providing incredible supervision to community healthcare workers and day-to-day site visits and Joseph Soka and Fidelis John for supervising the detection rats' component and the entire team from MUKIKUTE, OpASHA, and APOPO TB, Tanzania. We thank the enrolling patients, partnering health centers, the National Tuberculosis and Leprosy Programme, the National Institute for Medical Research, and the Sokoine University of Agriculture. The UKAID through the Human Development Innovation Fund (HDIF), Tanzania, is highly acknowledged for financial and technical support.

## References

1. World Health Organization, Stop TB Initiative (World Health Organization). *Treatment of tuberculosis: Guidelines* (World Health Organization, 2010)
2. National Tuberculosis and Leprosy Programme, Annual report 2018. https://www.ntlp.go.tz/site/assets/files/1046/annual_report_2018_final.pdf. Visited 16 March 2020
3. Global tuberculosis report 2019. Geneva: World Health Organization; 2019. Licence: CC BY-NC-SA 3.0 IGO

4. G.S. Mfinanga, E. Ngadaya, R. Mtandu, B. Mutayoba, B. Doulla, G. Kimaro, T.M. Chonde, P. Ngowi, S. Mfaume, A.M. Kilale, S. Egwaga, The quality of sputum smear microscopy diagnosis of pulmonary tuberculosis in Dar es Salaam, Tanzania. Tanzan. J. Health Res. **9**(3), 164–168 (2007)
5. C.C. Boehme, P. Nabeta, D. Hillemann, M.P. Nicol, S. Shenai, F. Krapp, J. Allen, R. Tahirli, R. Blakemore, R. Rustomjee, A. Milovic, Rapid molecular detection of tuberculosis and rifampin resistance. N. Engl. J. Med. **363**(11), 1005–1015 (2010)
6. C.A. Evans, GeneXpert – A game-changer for tuberculosis control? PLoS Med. **8**(7), e1001064 (2011). https://doi.org/10.1371/journal.pmed.1001064
7. World Health Organization, *Pathways to Better Diagnostics for Tuberculosis: A Blueprint for the Development of TB Diagnostics by the New Diagnostics Working Group of the Stop TB Partnership* (WHO Press, 2009)
8. G.F. Mgode, B.J. Weetjens, T. Nawrath, C. Cox, M. Jubitana, et al., Diagnosis of tuberculosis by trained African giant pouched rats and confounding impact of pathogens and microflora of the respiratory tract. J. Clin. Microbiol. **50**, 274–280 (2012)
9. B. Weetjens, G.F. Mgode, R.S. Machang'u, R. Kazwala, G. Mfinanga, et al., African pouched rats for the detection of pulmonary tuberculosis in sputum samples. Int. J. Tuberc. Lung Dis. **13**, 737–743 (2009)
10. A. Poling, B. Weetjens, C. Cox, N. Beyene, A. Durgin, A. Mahoney, Tuberculosis detection by giant African pouched rats (*Cricetomys gambianus*). Behav. Analyst. **34**, 47–54 (2011)
11. C. Mulder, G. Mgode, S.E. Reid, Tuberculosis diagnostic technology: An African solution… think rats. Afr. J. Lab. Med. **6**(2), 1–4 (2017)
12. G.F. Mgode, C.L. Cox, S. Mwimanzi, C. Mulder, Pediatric tuberculosis detection using trained African giant pouched rats. Pediatr. Res. **84**(1), 99–103 (2018 Jul)
13. World Health Organization, *Tuberculosis fact sheet N 104*. Retrieved February 15, 2017. http://www.who.int/mediacentre/factsheets/fs104/en/
14. World Health Organization, *WHO Treatment Guidelines for Drug-Resistant Tuberculosis* (World Health Organization, 2016)
15. D. Falzon, H. Timimi, P. Kurosinski, G.B. Migliori, W. Van Gemert, C. Denkinger, C. Isaacs, A. Story, R.S. Garfein, L.G. do Valle Bastos, M.A. Yassin, Digital health for the end TB strategy: Developing priority products and making them work. Eur. Respir. J. **48**(1), 29–45 (2016)
16. World Health Organization, *Digital health for the end TB strategy: An agenda for action* (World Health Organization, 2015)
17. R. Belknap, D. Holland, P.J. Feng, J.P. Millet, J.A. Caylà, N.A. Martinson, et al., Self-administered versus directly observed once-weekly isoniazid and Rifapentine treatment of latent tuberculosis infection: A randomized trial. Ann. Intern. Med. **167**(10), 689–697 (2017)
18. D.J. Finitsis, J.A. Pellowski, B.T. Johnson, Text message intervention designs to promote adherence to antiretroviral therapy (ART): A meta-analysis of randomized controlled trials. PLoS One **9**, e88166 (2014)
19. S. Mohammed, R. Glennerster, A.J. Khan, Impact of a daily SMS medication reminder system on tuberculosis treatment outcomes: A randomized controlled trial. PLoS One **11**(11), e0162944 (2016)
20. X. Liu, J.J. Lewis, H. Zhang, et al., Effectiveness of electronic reminders to improve medication adherence in tuberculosis patients: A cluster-randomised trial. PLoS Med. **12**, e1001876 (2015)
21. R.S. Garfein, R.P. Doshi, Synchronous and asynchronous video observed therapy (VOT) for tuberculosis treatment adherence monitoring and support. J. Clin. Tuberc. Other Mycobact. Dis. **17**, 100098 (2019)
22. J.R. Wu, D.K. Moser, M.L. Chung, T.A. Lennie, Predictors of medication adherence using a multidimensional adherence model in patients with heart failure. J. Card. Fail. **14**(7), 603–614 (2008)
23. R. Belknap, S. Weis, A. Brookens, et al., Feasibility of an ingestible sensor-based system for monitoring adherence to tuberculosis therapy. PLoS One **8**, e53373 (2013)

24. S. Batra, S. Ahuja, A. Sinha, N. Gordon, *eCompliance: Enhancing tuberculosis treatment with biometric and mobile technology*, in Proceedings of M4D 2012 28-29 February 2012 New Delhi, India, 28(29), (2012), p. 36
25. S.J. Snidal, G. Barnard, E. Atuhairwe, Y.B. Amor, Use of eCompliance, an innovative biometric system for monitoring of tuberculosis treatment in rural Uganda. Am. J. Trop. Med. Hyg. **92**(6), 1271–1279 (2015)
26. N. Agapitova, C.N. Moreno, *OpASHA – Improving Tuberculosis Treatment and Outcomes* (The World Bank, 2017)
27. The PLATO-IV system for computer aided instruction. https://ub.fnwi.uva.nl/computermuseum/PLATO.html. Accessed 3 June 2020
28. I.E. Sutherland, *A head-mounted three dimensional display*, in Proceedings of the Joint Computer Conference, Part I 1968 Dec 9, pp. 757–764
29. Stop TB Partnership. http://www.stoptb.org/news/stories/2018/ns18_073.asp. Accessed 16 March 2020

# Proposed Expanded TAM in the Sub-Saharan African Context: Theoretical Underpinnings Towards the Acceptance of Technological Tools for Supporting Co-teaching, Co-researching, and Co-learning

**Martin Mabeifam Ujakpa and Delene Heukelman**

**Abstract** An enormous amount of research has been carried out in examining the technology acceptance model (TAM) as propounded by Davis, and as a result, TAM has been applied in several technological acceptance settings and enhanced with additional constructs, hence leading to the availability of TAM2 and TAM3. TAM has widely been used to evaluate various technological tools globally and has been successful in predicting technological tools' adoption behavior in some international settings. However, using the existing versions of TAM (1, 2, and 3) concepts to choose technological tools for co-teaching, co-researching, and co-learning for the sub-Saharan African environment may not give accurate results. As a result, there is a need to expand the existing TAM for the sub-Saharan African environment. Expansion of TAM for the said environment, firstly, requires examination of the concepts (constructs and variables) of the existing TAMs to understand if they may be applicable in the said environment and thereafter developing an expanded TAM for the said environment. To examine the concepts of TAM that may apply to the sub-Saharan African environment, this chapter discussed the findings of literature review and theoretical underpinnings towards features of technological tools for co-teaching, co-researching, and co-learning for the sub-Saharan African environment and thereafter performed confirmatory factor analysis to confirm the applicability of the identified constructs and variables in the said environment. Based on the findings, this chapter concludes that four (4) main constructs of TAM (perceived ease of use, perceived usefulness, perceived performance, and perceived benefit) are applicable to the sub-Saharan African environment for choosing technological tools for co-teaching, co-researching, and co-learning. Within the 4 constructs, 13 variables were identified as follows: ease of use, usefulness, frequency of use, system quality, culture, satisfaction, experience, student or user centeredness, training,

M. M. Ujakpa (✉) · D. Heukelman
Durban University of Technology, Durban, South Africa
e-mail: deleneh@dut.ac.za



E. Eilu et al. (eds.), *Digital Literacy and Socio-Cultural Acceptance of ICT in Developing Countries*, https://doi.org/10.1007/978-3-030-61089-0_15

information quality, use mode, external factors, influence by peers, and influence by important persons. Hence, when choosing technological tools to support co-teaching, co-researching, and co-learning in sub-Saharan Africa, the 4 constructs and 13 variables should be considered as they could have implications on the adoption and use of technological tools for co-teaching, co-researching, and co-learning. Further study is needed to deeply understand the 4 constructs and 13 variables and how they influence technological tools' acceptance for co-teaching, co-researching, and co-learning; hence, it is recommended that future studies consider developing an expanded TAM for the sub-Saharan African environment.

**Keywords** Expanded TAM · Technological tools · Co-teach · Co-research · Co-learn · Sub-Saharan African environment · Sub-Saharan African context · Technology acceptance

## 1 Introduction

An enormous amount of research has been carried out in examining the technology acceptance model (TAM) as propounded by Davis, and based on the results, TAM has been applied in several technological acceptance settings and enhanced with additional constructs, hence leading to the availability of TAM2 and TAM3 [26]. Additionally, TAM has been widely used to evaluate various technological tools globally and has been successful in predicting technological tools' adoption behavior in some international settings [1, 3, 11]. Among these technological tools include audio for instruction such as Vocaroo for making audio recordings; video for instruction such as WeVideo for making videos with narrations; safe searches such as YouTube or Safe YouTube or SafeShare.tv, Flipgrid, TinyTap, Padlet, Labster, VideoNot.es, Evernote, and Symbaloo; and learning management systems (LMS), just to mention a few. These tools among others facilitate co-teaching, co-learning, and co-researching.

Application of the TAM concepts to diverse cultural environments (apart from the United States where it originated) requires laborious examination. McCoy et al. [20] confirm that caution needs to be taken when applying TAM in at least 20 identified countries as a result of cultural differences. In addition to this cultural limitation, Woraporn et al. [31] and Lee et al. [17] point out additional limitations of the existing TAM versions. As such, using the existing versions of TAM (1, 2, and 3) concepts to choose technological tools for co-teaching, co-researching, and co-learning within the sub-Saharan African context may not give accurate results [20]. According to [33], an extensive comprehension of the concepts influencing the constructs of perceived ease of use and perceived usefulness is needed. Furthermore, the assessment of diverse information systems for diverse contexts is neglected in the current TAM versions. The sub-Saharan African culture and social factors are no exempts to the neglected cultural and social factors.

According to [31], the existing versions of TAM presume that technological tools are used individually and hence group, team, or organizational perspective (which may be influenced by culture) is missing. Additionally, TAM assumes that technological tools' use is voluntary; hence, use as a result of compulsion is missing. Use by voluntary act or compulsion may be influenced by culture (which varies from society to society). As a result of TAM's limitation, [13] suggested that an expanded TAM be reflected as an obligatory part of "any pre-implementation study of information and communications technology among health workers" in Africa. This indicates that the choice of technological tools based on the existing TAM to support co-teaching, co-researching, and co-learning in sub-Saharan Africa may not give the best results. As a result, this research suggests that the choice of technological tools to support co-teaching, co-researching, and co-learning in sub-Saharan Africa should be based on an expanded TAM for the sub-Saharan African environment.

Furthermore, applying the frequently used TAM external concepts in e-learning adoption, [1] proposed a "general extended technology acceptance model for e-learning" (GETAMEL). Environmental factors, including cultural and social factors, were not considered in [1] work. Also, system quality, information quality, and service quality were not considered in developing GETAMEL [1]. Since [1] research had not focused on the sub-Saharan African environment, applying GETAMEL's concepts to choose tools for co-teaching, co-researching, and co-learning in sub-Saharan Africa may also not give the best results. There is therefore the need to expand the existing TAM in order to use it to choose technological tools for co-teaching, co-researching, and co-learning in sub-Saharan Africa.

As a first step, expansion of TAM for the sub-Saharan African environment requires examination of the concepts (constructs and variables) of the existing TAM to understand if they may apply in the said environment and, after, developing an expanded TAM in the said environment. To achieve this, this chapter discussed literature and theoretical underpinnings towards identifying the existing TAM constructs and variables that may apply in the sub-Saharan African context and hence be geared at achieving the first step in expanding the existing TAM in the sub-Saharan African context. Discussion of the literature and theoretical underpinnings was presented in five parts: While part 1 discussed studies that applied the original TAM in elucidating technology adoption, part 2 discussed studies that applied the modified TAM in clarifying technology adoption. Parts 3 and 4 discussed studies that applied the information systems success model (ISSM) and technology-organization-environment (TOE) framework in explaining technology acceptance, respectively. Finally, part 5 identified the potential constructs and variables from the existing TAM that could be included in the proposed expanded TAM for the sub-Saharan African environment. After identifying constructs and variables from literature, factor analysis was performed to confirm the constructs and variables.

## 2 Literature Review

### 2.1 Research That Applied Original TAM

The original TAM, as in Fig. 1, was found in very recent studies to still be an acceptable tool for predicting technological acceptance [2, 3, 11, 18]. While [2, 11, 18] conducted research in sub-Saharan Africa, the study conducted by [3] was conducted in the Netherlands.

### 2.2 Existing TAM Constructs to Be Included in an Expanded TAM in the Sub-Saharan African Context

In Helia et al. (2018) study, [11] examined the technology adoption level among university academics. With 355 academics drawn from different Nigerian universities, [11] analyzed collected data and concluded that TAM is an ideal theory that enables one to comprehend user technology adoption [11]. Further, it was found that the following influence changes in behavioral intention (BI) towards technology adoption:

- Self-efficacy (SE)
- Social influence (SI)
- System accessibility (SA)
- Perceived usefulness (PU)
- Perceived ease of use (PEOU)

The researcher [11] also found:

- Significant relationships between:
  - BI and SE
  - SA and PU
- Insignificant relationships between BI, PEOU, and SI

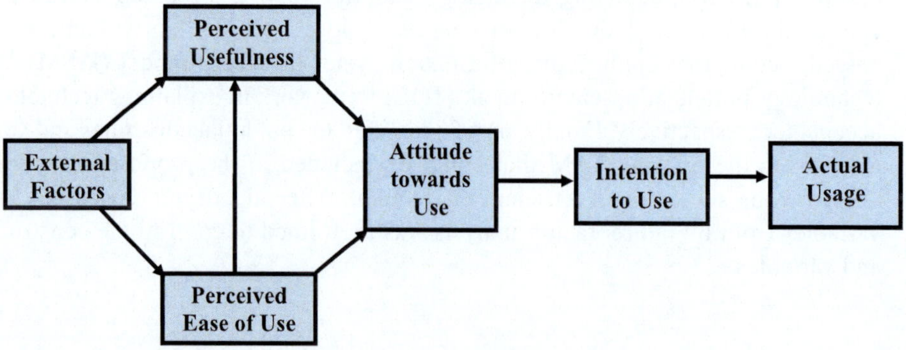


**Fig. 1** Technological acceptance model [7]

Based on the findings above, [11] recommended that universities' management organize trainings/seminars/workshops on technology use. The researcher's finding that "PEOU was insignificantly related to BI" contradicted the TAM model, and hence a further study on [11] findings, especially within the sub-Saharan African context, is needed. As a result, technological acceptance constructs that should be included in an expanded TAM in the sub-Saharan African context should include PEOU, PU, and BI and SE, SI, and SA. Chosen technological tools for co-teaching, co-researching, and co-learning in sub-Saharan Africa based on the proposed expanded TAM are more likely to give better results compared to the existing TAM.

King and He's (2006) meta-analysis of the technology acceptance model using 88 published studies provided sufficient data to be credible and makes a stronger case on why PU and BI should be incorporated into an expanded TAM model for the sub-Saharan African environment. King and He's (2006) meta-analysis concluded that the PU and BI were highly reliable and may be used in a variety of contexts. Hence, an expanded TAM in the context of sub-Saharan Africa should include PU and BI as the usefulness of a technological tool may influence behavior intention to adopt and use it.

Using TAM as theoretical background and collecting data from 400 university teachers in Ethiopian public universities for "empirical examination of e-learning acceptance and use," [2] found that PEOU and PU significantly determined behavioral intention of e-learning system use [2]. They also found that BI, user training, and management support e-learning system actual use [2] and also conclude that incentives for e-learning system actual use were insignificant to BI [2]. This supports the notion that *knowledge/training* and *support* are important variables to include in an expanded TAM in the context of sub-Saharan Africa as *knowledge/training* on a tool and *support* on how to use it for co-teaching, co-researching, and co-learning may be necessary for successful co-teaching, co-researching, and co-learning.

In the researchers' study to assess the adoption of MOOCs in developing countries, [19] collected data from 827 participants and performed hierarchical regression analysis to develop the model. Results of the study indicated that, among the technological characteristics, perceived usefulness significantly influenced behavior intention to adopt the most, followed by accessibility challenges and value of performance to cost [19]. Study results further revealed that *self-regulation* influenced MOOC adoption most significantly and also that female learners were more likely to adopt MOOCs. Finally, among *social environmental* variables, *tradition and social norms* and *lack of information* about MOOCs were negatively associated with adoption of MOOCs [19], thus confirming [15] finding of lack of understanding of a technology as having a negative impact on adoption. These results indicate the need to include PU and perceived performance (*accessibility and performance*) as constructs in an expanded TAM for the sub-Saharan African environment as usefulness and performance of a technological tool is necessary for its adoption and use for co-teaching, co-researching, and co-learning.

### 2.3 Concepts of the Original TAM to Include in an Expanded TAM in the Sub-Saharan African Context

Adopting TAM for their studies, [18] found the variables that influence the use of social media marketing in the tourism industry in South Africa. Applying a quantitative approach using questionnaires, data gathered from 150 travel agencies and tour operators were analyzed, and the results indicated that *managers' educational level* and *managerial support* influenced attitude towards social media marketing [18]. The study also found that external factors that influence social media marketing include PEOU, *competitors' pressure*, and perceived benefits [18]. Further, it was found that *technical knowledge* moderated the relationship between attitude towards use and social media marketing [18]. This re-emphasizes the importance of the variables of *knowledge/training* and *support* and the constructs PU and PEOU, hence suggesting their inclusion in an expanded TAM in the sub-Saharan African context [27]. Indicated in his study on determinants of PEOU, indicated this as well. Hence, co-teaching, co-researching, and co-learning technological tools' adoption and use may depend on their PU, PEOU, information technology staff *support* (better referred to as *service quality* in the ISSM model), and user *knowledge/training* on them.

While the previous studies discussed above were conducted in sub-Saharan Africa, the study by [3] was conducted in the Netherlands. In order to examine user-perceived e-procurement *quality* dimensions as key antecedent of employees to PU and PEOU, [3] collected and analyzed data from 139 participants at a university in the Netherlands. Results of the analyses showed that the following influenced individual employee e-procurement adoption.

- Processing
- Usability
- Professionalism

Since *processing*, *usability*, and *professionalism* influence *system quality* towards PEOU, *ease of use* towards *information quality*, and *culture* towards perceived performance, respectively, *system quality*, *information quality*, and *culture* variables need to be included in the expanded TAM for the sub-Saharan African environment. Hence, the system quality, information quality, and cultural variables (e.g., peer/important personnel influence, individualism/collectivism, language, power distance, uncertainty avoidance, long-term/short-term orientation, and masculinity/femininity) of co-teaching, co-researching, and co-learning tools are likely to play an important role in its adoption and use.

The discussed studies above found the existing TAM to be suitable for use in determining variables and constructs that influence technology acceptance in the context of sub-Saharan Africa: hence, all constructs of the existing TAM should be included in an expanded TAM for the sub-Saharan African context. Thus, technological tools for co-teaching, co-researching, and co-learning in the sub-Saharan African environment should be perceived useful and easy to use to enable their adoption and use. However, this is not conclusive enough to conclude that conditions for adopting and using technological tools for co-teaching, co-researching, and co-learning in the sub-Saharan African environment are determined by perceived usefulness and ease of use only, thus prompting further research and strengthening the case for an expanded TAM (going beyond perceived usefulness and ease of use) to include other possible factors (if any).

## *2.4 Research That Applied TAM with Modifications or Extensions*

Apart from the use of the original TAM in previous studies [2, 3, 11, 18], the modified or extended TAMs were also used in previous studies. Previous studies that used modified or extended TAMs include the studies of [4, 9, 22, 30].

Applying a modified technology acceptance model (TAM) as in Fig. 2, [4] examined variables and constructs that influenced consumers' intention and actual behavior to use sports brand applications. Using convenience sampling, data were collected from 261 Korean sports brand application consumers and analyzed by "partial least squares structural equation modeling (PLS-SEM)" [4]. The study results indicated that "*enjoyment level* positively affected PEOU and that PEOU also positively influenced PU" [4].

On constructs that influenced BI to use sports brand applications, perceived enjoyment was found to be the most influential. It was followed by PU and thereafter PEOU. The study further found that "Behavioral Intention positively affected Actual Behavior: however, differences were found between three age groups" (20s, 30s, and 40+) [4]. The variable of enjoyment may be directly due to the applications being sports brand related and hence may not be included in an expanded TAM for the sub-Saharan African environment. For co-teaching, co-researching, and co-

**Fig. 2** Modified TAM [4]

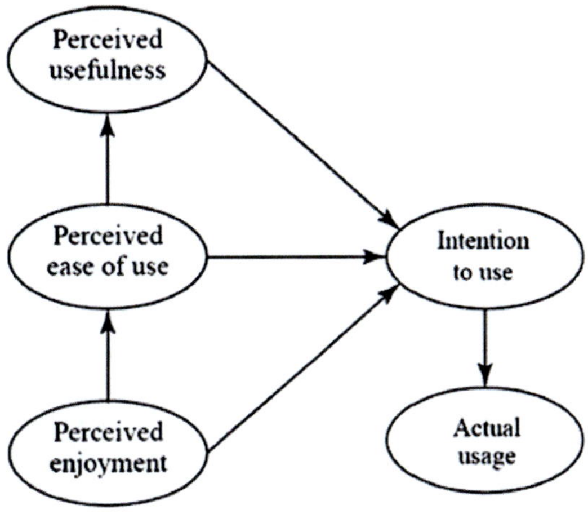

learning tools' adoption and use, the variable, enjoyment, may be insignificant as well.

In their bid to "understand why emerging information technology is adopted and factors that enhance their implementation," [30] proposed an "emerging information technology acceptance model" (EITAM) [30]. Using structural equation modeling (SEM) based on data from an open-ended questionnaire survey, hypotheses were tested, and the research identified factors that affect emerging information technology acceptance from engineering construction technology and innovation professionals. Creating a modified "technology acceptance model" (TAM) with "structural equation modeling" (SEM) and testing it using the statistical package SPSS Amos 20, [9] identified the variables that influence "hospital information system's (HIS) applications" most. Using questionnaires to collect data from several hospitals in Sleman, Daerah Istimewa Yogyakarta Province, Indonesia, [9] found that seven constructs influenced users in using an HIS. Among these constructs are subjective norm, PU, PEOU, user satisfaction (US), BI, attitude towards using, and actual system usage. This finding is an indication that adoption of co-teaching, co-researching, and co-learning technological tools may depend on the tools' perceived usefulness, ease of use, satisfactoriness, and peer/important personnel acceptance (subjective norm). As a result, the seven identified constructs in [9] study should be considered in an expanded TAM in the sub-Saharan African context.

[22] examined "factors that determined students' video usage and their learning satisfaction." Using educational videos, available on a learning management system environment for a Mathematics course, with a theoretical background from an "extended technology acceptance model" (TAM), questionnaires were used to collect data from 89 students and analyzed to examine the research model constructs. The results of the examination found that (1) PU, attitude, and Internet self-efficacy directly influenced the video usage, (2) learner-learner interaction, PEOU, and learning performance directly influenced learning satisfaction, and (3) the video usage had a significant effect on learning performance and learning satisfaction. Considering that attitude and learner-learner interaction are more likely to stem from culture which may influence frequency of use, experience, and Internet self-efficacy, it is suggested that an expanded TAM in the sub-Saharan African context includes *culture*, *frequency of use*, and *experience* as variables.

In their meta-analysis on the impact of trust on technology acceptance, focusing on moderating influence of subject and context type, [28] analyzed 126 studies on PEOU and PU of TAM and found that PEOU and PU influence attitude which in turn influences BI. Findings of [28] confirmed findings of King and He (2006) and hence add further reason why PEOU and PU should be included as constructs in the expanded model. In addition to PEOU and PU, [24] "analysis of affecting factors of technology acceptance in the application of knowledge management for small/medium enterprises in creative industry" concluded that *user willingness* to use and *knowledge management* influenced technology acceptance. As *user willingness* to use and *knowledge management* match f*requency of use* and *experience*, respectively, [24] findings give further reasons why *frequency of use* and *experience* should be included as variables in an expanded TAM in the sub-Saharan African context.

Constructs identified from the modified or extended TAM as above, which are suitable for an expanded TAM for the sub-Saharan African context, are PEOU, PU, PP, BI, and TA. The variables influencing the constructs are *culture*, *frequency of use*, and *experience*, subjective norm (*influence by peers* and *influence by important persons*), *accessibility*, and *performance.* This shows that adoption and use of co-teaching, co-researching, and co-learning technological tools depend on the constructs PEOU, PU, PP, BI, and TA and the variables culture, frequency of use, and experience, subjective norm (influence by peers and influence by important persons), accessibility, and performance.

## 2.5 Research That Applied Information Systems Success Model

The original and updated ISSM models as in Figs. 3 and 4 were found in very recent studies still to be an acceptable tool for predicting technological acceptance [16, 21, 25, 32].

To understand a user perspective organization's information systems, [21] "re-specified and validated" [5] ISSM model by "proposing a social exchange theory". Gathering data from 102 participants, [21] developed a model for testing information systems success. From the model, [21] concluded that "strong relationships existed between the following variables:

- *Perceived quality*
- *Perceived value*
- *User satisfaction*
- *Net benefits*"

*Further*, [21] found that *system quality* and *information quality* affect *service quality*. They further found that *information quality*, *system quality*, and *service quality* caused the formation of *perceived value* and *perceived value* causes the formation of *user satisfaction* and *net benefits*. Based on these findings, an expanded

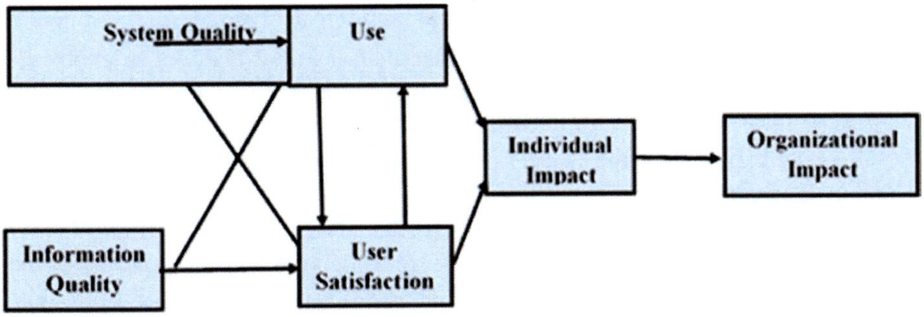


**Fig. 3** Information systems success model [6]

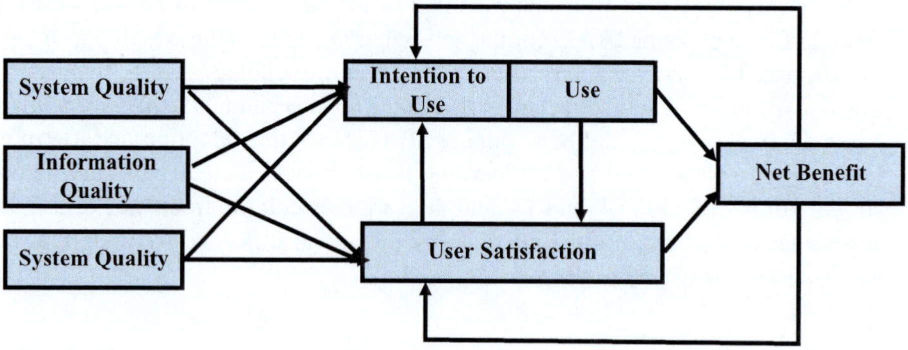


**Fig. 4** Updated information systems success model [6]

TAM in the sub-Saharan African context should include the variables *information quality*, *system quality*, *service quality*, and *satisfaction*. This also emphasizes the need that, for co-teaching, co-researching, and co-learning technological tools to be successful adopted and used, then they have to be chosen based on their information quality, system quality, service quality, and satisfaction.

Leveraging on the ISSM model (2003) and using questionnaires to collect data, [23] developed an "evaluation framework for health information systems." Considerations made in developing the framework included *collaboration*, *satisfaction*, *safety*, *procedures*, and *system quality* [23]. *Collaboration*, *satisfaction*, and *system quality* being part of considerations for [23] in developing a framework for evaluating health information systems reinforced the inclusion of *collaboration*, *system quality*, and *satisfaction* as possible variables for the expanded model. Another study that made use of *satisfaction* variable is that of [8], which developed an extended TAM with process satisfaction and outcome satisfaction as constructs.

"Use" as a construct in ISSM directly influences satisfaction and benefit and indirectly influences intention to use [5]. To examine this further, [16] reviewed relevant literature, including "information systems success model," and thereafter performed "user tests and interviews" with 12 participants. Results from the study revealed that "high organic search engine visibility is not the reason for ecommerce system use, but rather use is a result of high visibility" [16]. This result reinforces the inclusion of *frequency of use* and *satisfaction* as variables in the expanded model, as they influence technology acceptance (TA). The study also found that "there were more than just a few variables that users evaluate when choosing an e-commerce system to use" [16]. This finding informs an expanded model to have the variables *frequency of use* and *satisfaction* as variables.

To establish whether the "success of information systems affects job satisfaction and commitment among employees," [32] gathered data from 50 people in the "higher education" industry. Results of the analyzed data show that "there are dimensions associated with information systems that influence the success of the IS and this success in turn affects job satisfaction and commitment, and through the latter organizational performance" [32]. Dimensions associated with information systems that influence its success [32] may include system quality, information quality, service quality, and culture. By including *system quality*, *information quality*, *service quality*, and *culture* as variables in the expanded model, it becomes possible to verify the influence of information systems (IS) dimensions on IS success in the sub-Saharan African context.

In some organizations, information systems may not exist or exist but is not used as employees lack trust in it. To investigate users' (employees') trust in IS and preconditions for trust, [25] collected data from 30 professionals on occupational incidents related to IS trust. Content analysis of the data revealed 12 predictors of IS trust and distrust. Analysis of data further gathered from 179 professionals to validate the initial findings indicated that IS system quality and information quality are the most predictors of IS trust and distrust. In relation to job satisfaction and commitment, the study found well-being and performance as rating highest in trust events and experienced strain as rating highest in distrust events.

## *2.6 Research That Applied Technology-Organization-Environment (TOE) Framework*

The original TEO model as in Fig. 5 was found in very recent studies to still be an acceptable tool for predicting technological acceptance [7, 10, 12, 15, 19].

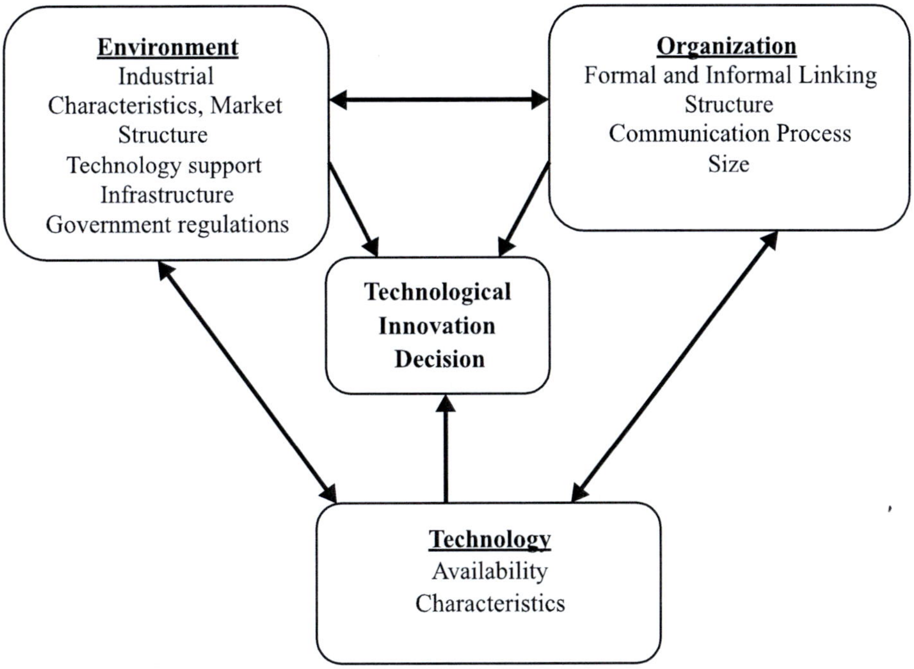


**Fig. 5** TOE framework [36]

[15] study to "investigate the drawbacks of transitioning from on premise system to a cloud computing service in relation to adoption" reviewed the TOE framework and thereafter conducted interviews to collect data from participants. Analyses of the data found that "*security*, *perceived loss of control*, and *lack of cloud computing understanding*" were found as drawbacks, and hence the researcher recommended that "further research on cloud computing comprehension" be undertaken [15]. As *security*, *perceived loss of control* and *lack of cloud computing understanding* were as a result of the cloud computing (system) quality and the researcher [15] recommended further study on it.

In [12] study in Malaysia in which they applied the TOE framework to examine existing information systems adoption empirical studies at the organizational level, [12] developed a model. The use of the TOE by [12] in developing a technological adoption model indicates the possibility of using some of TOE's constructs and variables in an expanded TAM in the sub-Saharan African context, hence supporting the use of *system quality*, *information quality*, and *service quality* (TOE technological dimensions) and *culture* (a TOE environment dimension).

In their study on factors that affect BI to use B2B e-commerce in Jordan using the TOE, [10] employed self-administrated questionnaire to collect quantitative data from 168 participants in Jordan and thereafter analyzed the data using "partial least squares structural equation modeling technique." The results indicated that "*relative advantage*, *top management support*, and *information intensity* have a significant indirect effect on behavior intention to continue using B2B e-commerce, through perceived usefulness (PU)" [10]. PU mediates the relationship of "*relative advantage* and *information intensity* with the behavior intention to continue using B2B e-commerce" [10]. Since *information intensity* and *top management* are linked to *information quality* and *service quality*, the results of [10] study indicate the need for *information quality* and *service quality* and PU to be included as variables, respectively, in an expanded TAM for the sub-Saharan African environment which will be useful when choosing co-teaching, co-researching, and co-learning tools for the sub-Saharan African environment.

In [7] study in the healthcare logistics setting using a multiple case study, it was found that a set of 17 factors influence technology adoption and the most significant relates to *quality*, *employee work conditions*, and *employee engagement* [7]. [7] pointing to *quality* as an important factor reinforce the inclusion of *information quality*, *system quality*, and *service quality* as variables in the expanded TAM and in technological tools for supporting co-teaching, co-researching, and co-learning tools for the sub-Saharan African environment.

The examples of recent studies as discussed above confirm that the modified or extended TAMs, ISSM, and TOE are still relevant in examining and identifying concepts that influence technology acceptance and hence a suitable theoretical basis for an expanded TAM for the sub-Saharan African environment will be more suitable in aiding the selection of tools for supporting co-teaching, co-researching, and co-learning technological tools for the sub-Saharan African environment.

## 3 Methodology

As part of the study, a review of existing literature was undertaken and confirmatory factor analysis undertaken. As part of the review, five technological acceptance-related thematic areas were taken into consideration. The six thematic areas include research that applied original TAM, existing TAM constructs to be included in an expanded TAM in the sub-Saharan African context, variables of the original TAM constructs to include in an expanded TAM in the sub-Saharan African context, research that applied TAM with modifications or extensions, research that applied information systems success model (ISSM), and research that applied technology-organization-environment (TOE) framework. Four literature materials were reviewed on the theme on research that applied original TAM, four on the theme on existing TAM constructs to be included in an expanded TAM in the sub-Saharan African context, three on the theme on variables of the original TAM constructs to include in an expanded TAM in the sub-Saharan African context, four on the theme on research that applied TAM with modifications or extensions, five on the theme on research that applied ISSM, and five on the theme on research that applied TOE framework. Various constructs and variables were discussed. The chosen literature for review was based on their authority, relevance, and recency.

Constructs and variables obtained from the reviewed literature based on the theoretical underpinnings were used in formulating questionnaires that were used to collect data from 308 students in 5 universities in the 5 regions of sub-Saharan Africa. Each university was chosen from one region and was using an e-learning system. Through confirmatory factor analysis (CFA), the data were analyzed and the eigenvalue and explained variance (EV), KMO and Bartlett's test, factor loadings, and mean and standard deviation values computed. The data collection stopped on the 308 student when the researchers realized that a point of saturation had been reached.

## 4 Data Analysis and Discussion

Table 1 presents a summary of the statistics (eigenvalue and explained variance (EV), KMO and Bartlett's test, factor loadings, mean and standard deviation values) for the quantitative data, gathered through completed questionnaire from the research respondents. Data in Table 1 enabled verification of the correlations between the constructs in the sub-Saharan African context. The results were generated from Cronbach's α measure of variables that formed the questionnaire. As given by the formula below, Cronbach's α of the questionnaire is how the questionnaire measures what it should measure. Scoring a high reliability means that it measures very close to what it should measure, and scoring a low reliability means it measures something else or possibly nothing at all. For this chapter, Cronbach's alpha value of 0.7 or above was accepted as high reliable value.

**Table 1** Constructs, Cronbach's α, KMO, factor loading, mean, and average values

| Construct | Variable | N | Cronbach's alpha | Eigenvalue, explained variance | KMO, Bartlett's test | Factor loading | Average mean, SD |
|---|---|---|---|---|---|---|---|
| Perceived usefulness | Ease of use | 308 | 0.78 | 4.3, 2.9 30 20 | 0.718 P < 0.0001 | 0.8 | 1.88 0.79 |
| | Usefulness | | | | | 0.45 | |
| | Frequency of use | | | | | 0.75 | |
| | Information quality | | | | | 0.82 | |
| | Culture | | | | | 0.54 | |
| | Satisfaction | | | | | 0.49 | |
| Perceived ease of use | Experience | 308 | 0.84 | 2.8, 2.7 18, 19 | 0.74 P < 0.0001 | 0.9 | 2.59 1.27 |
| | Student or user centeredness | | | | | 0.73 | |
| | Training | | | | | 0.88 | |
| | System quality | | | | | 0.67 | |
| Perceived benefit | Voluntary use | 308 | 0.77 | 1.3 2.4 10 17 | 0.72 P < 0.0001 | 0.7 | 1.95 0.78 |
| | Influence (peers) | | | | | 0.83 | |
| | Influence (important persons) | | | | | 0.67 | |
| | Satisfaction | | | | | 0.6 | |
| Perceived performance | System quality | 308 | 0.72 | 1.2 1.4 8 10 | 0.71 P < 0.0001 | 0.8 | 2.06 0.89 |
| | Influence (important persons) | | | | | 0.7 | |
| | Culture | | | | | 0.6 | |

Source: Field data (2018)

Cronbach's alpha formula: $\alpha = \dfrac{N \cdot \bar{c}}{\bar{v} + (N-1) \cdot \bar{c}}$ where:

- α = Cronbach's alpha
- N = the number of items
- $\bar{c}$ = average covariance between item-pairs
- $\bar{v}$ = average variance

Cronbach's alpha values for this chapter were generated by the use of SPSS software in which the data collected was entered and run for analysis.

In Table 1, the mean and standard deviation values were computed, and the summary of the direct, indirect, and total effects for the relationship between the variables and constructs was made. As in Table 1, of the 13 variables in the research instrument, 6 associated closely with a Cronbach's alpha of 0.78, to measure the PU construct (a construct in the existing). Four variables associated closely with a Cronbach's alpha of 0.84 to measure the PEOU construct. Since the six variables correlated to measure the PU construct (a construct of the existing TAM) and four variables correlated to measure the PEOU construct (a construct of the existing TAM), then it confirms the applicability of the two main constructs (PU and PEOU) in the existing TAM to the proposed constructs for the said environment in context.

With a Cronbach's alpha value of 0.77 as in Table 1, four variables further correlated to measure the PB construct (a new proposed construct). Furthermore, three variables associated closely with a Cronbach's alpha of 0.72 to measure the PP construct (a new proposed construct). Comparatively, the existing TAM model does not have the constructs perceived benefit (PB) and perceived performance (PP). While PB has three variables (culture, peers' influence, and important persons' influence) influencing it, PP has the variable culture, which would directly measure the sub-Saharan African context influence, hence adding the constructs PB and PP for the sub-Saharan African context. Although this may not be unique to the sub-Saharan African environment only as it may apply to other environments, the lack of these constructs in the said environmental context could negatively influence on results when measuring technology adoption in the said environment.

With a standard deviation score of 1.27 and the highest average mean of 2.58 as in Table 1, PEOU was rated as the construct with the most influence. Since the variables associated with PEOU in a technology were experience on use prior to using it, training on use, system quality, and student/user centeredness, then the result indicates that these variables enhance intention to accept and use a technology in sub-Saharan Africa.

The interview questions for the PB construct were reverse-phrased, and therefore the result must be interpreted with this in mind. With an average mean of 1.95 and standard deviation of 0.78, the construct PB rated lower than the PP construct. Since the variables associated with PP were influenced by peers and important persons (referred to as subjective norm in literature), voluntary use, and satisfaction, this result indicates that these variables account for intention to accept and use technology in sub-Saharan Africa.

PU was rated the lowest with an average mean of 1.87 and a standard deviation score of 0.78. Since the variables associated with PU included frequency of use, information quality, and culture, then these results show that these variables account for technology acceptance in sub-Saharan Africa.

The construct PP was rated as the construct with the second highest influence, with an average mean score of 2.06 and a standard deviation score of 0.89. This indicates that within the sub-Saharan African settings, culture, system quality, and influence by important persons enhance intention to accept and use technology.

Ease of use, frequency of use, and usefulness are variables that enhanced acceptance and use of technology. The construct PU from which they emanate was rated with an average mean of 1.88 and standard deviation of 0.79, as in Table 1.

## 5 Conclusion

To examine the concepts of the existing TAM in order to identify the concepts that may apply in the sub-Saharan African context for the adoption of technological tools for co-teaching, co-researching, and co-learning, two main issues need to be considered: firstly examination of concepts that address technological acceptance within the said environmental context above and secondly examination of the concepts that have the potential to predict and explain acceptance and use of technology within the said environmental context above. This chapter therefore discussed the theoretical underpinnings for an expanded TAM for the said environmental context above by analyzing the existing technological acceptance concepts in order to identify the technological acceptance concepts that apply to the sub-Saharan African environment and those that can potentially predict or explain technological tools' acceptance for supporting co-teaching, co-researching, and co-learning and later aid in the development of an expanded TAM for the sub-Saharan African environment.

Two (2) main constructs (consisting of 13 variables) were identified from literature and two (2) more generated from the factors analysis. The identified constructs from literature are PEOU and PU, and the two constructs generated from the factor analysis are perceived performance (PP) and perceived benefit (PB). Within these constructs, 13 variables were identified. The identified 13 variables in literature include *ease of use*, *usefulness*, *frequency of use*, *system quality*, *culture*, *satisfaction*, *experience*, *student or user centeredness*, *training*, *information quality*, *use mode*, *influence by peers*, and *influence by important persons.* Considering *external factors* as suggested by many studies including that of [26, 27, 35] studies, as a variable, would mean 13 variables in all.

Hence, when choosing technological tools (such as Vocaroo, WeVideo, Safe YouTube, Symbaloo, etc.) to support co-teaching, co-researching, and co-learning in sub-Saharan Africa, the said 4 constructs and 13 variables (as above) from existing technological acceptance concepts should be considered as they may have implications on adoption and use of the technological tools for co-teaching, co-researching, and co-learning. Previous studies on TAM focused on eight of the said variables and only handled partly or neglected the remaining constructs and variables. The partly handled or neglected variables are *use mode*, *culture*, *student or user centeredness*, *frequency of use*, and *training*. Further study is therefore needed to understand deeply these partly handled or neglected variables and constructs. Based on the comprehension of the neglected variables and constructs and the findings of these studies (4 constructs and 13 variables from existing technological acceptance concepts), an expanded TAM in the sub-Saharan African context should be developed.

## References


1. F. Abdullah, R. Ward, Developing a general extended technology acceptance model for E-learning (GETAMEL) by analysing commonly used external factors. Comput. Hum. Behav. **56**, 238–256 (2016)
2. A.A. Ayele, W.K. Birhanie, Acceptance and use of E-learning systems: The case of teachers in Technology Institutes of Ethiopian Universities. Appl. Inform. **5**(1), 1–11 (2018)
3. A. Brandon-Jones, K. Kauppi, Examining the antecedents of the technology acceptance model within E-procurement. Int. J. Oper. Prod. Manag. **38**(1), 22–42 (2018)
4. H. Byun, W. Chiu, J. Bae, Exploring the adoption of sports brand apps: An application of the modified technology acceptance model. Int. J. Asian Bus. Inf. Manag. **9**(1), 52–65 (2018)
5. W.H. Delone, E.R. Mclean, Measuring e-commerce success: reapplying the DeLone and McLean information systems success model. Int. J. Electron. Commer. **9**(1), 31–47 (2003)
6. W.H. DeLone, E.R. McLean, Information systems success: The quest for the dependent variable. Inf. Syst. Res. **3**(1), 60–69 (1992)
7. F.D. Davis, R.P. Bagozzi, P.R. Warshaw, Extrinsic and intrinsic motivation to use computers in the workplace. J. Appl. Soc. Psychol. **22**(14), 1109–1130 (1992)
8. D.C. Feibert, P. Jacobsen, Factors impacting technology adoption in hospital bed logistics. Int. J. Logist. Manag. **30**(1), 195–230 (2019). https://doi.org/10.1108/IJLM-02-2017-0043
9. R. Fayada, D. Paperb, *The technology acceptance model e-commerce extension: A conceptual framework*, in 4th World conference on business, economics and management, Utah State, USA (2015)
10. S.A. Guzmán, P.F. Fóster, P. Ramírez-Correa, E.E. Grandón, J. Alfaro-Perez, Information systems and their effect on organizational performance: an inquiry into job satisfaction and commitment in higher education institutions. J. Inf. Syst. Eng. Manag. **3**(4), 26 (2018)
11. V.N. Heliam, V.I. Asri, E. Kusrini, S. Miranda, *Modified technology acceptance model for hospital information system evaluation – A case study*, in MATEC web of conferences (2018), p. 154
12. A.L. Hussein, S.A. Baharudin, K. Jayarama, S. Kiumarsi, B2B E-commerce technology factors with mediating effect perceived usefulness in Jordanian manufacturing SMEs. J. Eng. Sci. Technol. **14**(1), 411–429 (2019)
13. T.A. Ibrahim, *The role of technology acceptance model in explaining university academics' acceptance and behavioural intention to use technology in education,* in the 4th International research conference on higher education, KnE Social Sciences (2018), pp. 1162–1172.14
14. N.S.W.W. Ismail, A.A. Azwadi, Conceptual model for examining the factors that influence the likelihood of Computerised Accounting Information System (CAIS) adoption among Malaysian SMEs. Int. J. Inf. Technol. Bus. Manag. **15**(1), 122–15115 (2016)
15. W.R. King, J. He, A meta-analysis of the technology acceptance model. Inf. Manag. **43**, 740–755 (2006)
16. T. Kilström, Factors limiting adoption of new technology: A study of drawbacks affecting transition from on-premise systems to cloud computing. KTH Industrial Engineering and Management (2016)
17. V. Laakkonen, Expanding the e-commerce system success theory: The impact of online visibility – Case Google. University of Jyvaskyla (2018)
18. Y. Lee, K.A. Kozar, K.R.T. Larsen, The technology acceptance model: Past, present, and future. Commun. Assoc. Inf. Syst. **12**(50), 752–780 (2003)
19. R. Matikiti, M. Mpinganjira, M. Roberts-Lombard, Application of the technology acceptance model and the technology–Organisation–environment model to examine social media marketing use in the South African tourism industry, South African. J. Inf. Manag. **20**(1) (2018)
20. L. Ma, S.C. Lee, Investigating the adoption of MOOCs: A technology–user–environment perspective. J. Comput. Assist. Learn. **35**, 89–98 (2018)

21. S. McCoy, D.F. Galletta, W.R. King, Applying TAM across cultures: The need for caution. Eur. J. Inf. Syst. **16**(1), 81–90 (2007)
22. Y. Nugroho, A. Prasetyo, Assessing information systems success: A re-specification of the DeLone and McLean model to integrating the perceived quality. Probl. Perspect. Manag. **16**(1), 348–360 (2018)
23. J.T. Nagy, Evaluation of online video usage and learning satisfaction: An extension of the technology acceptance model. Int. Rev. Res. Open Dist. Learn. **19**(1) (2018)
24. A. Stylianides, J. Mantas, Z. Roupa, N.E. Yamasaki, Development of an evaluation framework for health information systems (DIPSA). Acta Inform. Med. **26**(4), 230–234 (2018)
25. J.J. Siregar, A.W.R.A. Puspokusumo, A. Rahayu, *Analysis of affecting factors technology acceptance model in the application of knowledge management for small medium enterprises in industry creative*, in the 2nd International conference on computer science and computational intelligence, Bali, Indonesia (2017)
26. T.M. Thielsch, M.S. Meeßen, G. Hertel, Trust and distrust in information systems at the workplace. PeerJ (2018)
27. L.G. Tornatzky, M. Fleischer, *Process of Technological Innovation* (Lexington Books, Massachusetts/Toronto, 1990)
28. V. Venkatesh, H. Bala, Technology acceptance model 3 and a research agenda on interventions. Decis. Sci. **39**(2), 273–315 (2008)
29. V. Venkatesh, *Determinants of Perceived Ease of Use: Integrating Control, Intrinsic Motivation, and Emotion into the Technology Acceptance Model* (University of Maryland, Maryland, 2000)
30. K. Wua, Y. Zhaoa, Q. Zhua, X. Tana, H. Zhengb, A meta-analysis of the impact of trust on technology acceptance model: Investigation of moderating influence of subject and context type. Int. J. Inf. Manag. **31**, 572–581 (2011)
31. N. Wagner, K. Hassanein, M. Head, Who is responsible for E-learning success in higher education? A stakeholders' analysis. Educ. Technol. Soc. **11**(3), 26–36 (2008)
32. Z. Yang, Y. Wang, C. Sun, Emerging information technology acceptance model for the development of smart construction system. J. Civ. Eng. Manag. **24**(6), 457–468 (2018)
33. R. Woraporn, K. Seung, *A framework to study technology use: Alternatives to technology acceptance model*, in the Americas conference on information systems, proceedings 127 (2002), pp 883–889. Retrieved from: http://aisel.aisnet.org/amcis2002/127

# Prototype Design of Android App for Blood Donation

**Anay Dombe, Ganesh Bhutkar, and Aditya Dongre**

**Abstract** It is expected that the blood of required blood group will be available on demand. However, during certain situations such as accident or surgeries, this demand may be at the peak. Therefore, it may not be possible to make required blood components available to the patient in that crucial time frame. It may happen due to several vital problems such as insufficient blood stocks at the blood bank; delay in receiving required blood; lack of storage, separation, or preservation facilities; and even myths about blood donation. This paper discusses about a paper prototype design of the proposed application (app) for blood donation. A paper prototype for this app is designed based on correlated literature survey as well as peer Android app review. This prototype design includes several facilities such as requesting blood, donating blood, searching the donor, and tracking the donor efficiently. An important concern of this Android app is to spread awareness about blood donation addressing various myths and minimize the time for blood arrival as per user requirement.

**Keywords** Blood donation · Android app review · Paper prototype · Android app · Blood bank

## 1 Introduction

According to *The Times of India*, about 150,000 cardiac surgeries are performed in a year, in India [1]. The major surgeries related with heart or cancer ailments may involve significant blood loss. An article in *Business Standard* states that there was a shortage of 1900,000 units of blood in 2016–2017, in India, that could have aided more than 320,000 heart surgeries [2]. This data shows that a large amount of blood is still required to cover such shortage. One of the best ways to overcome such shortage is through the process of blood donation. Blood acquired through blood donation can be stored and used as per the need. However, statistics given by *India*

A. Dombe (✉) · G. Bhutkar · A. Dongre
Centre of Excellence in Human-Computer Interaction, Vishwakarma Institute of Technology (VIT), Pune, India
e-mail: anay.dombe15@vit.edu; ganesh.bhutkar@vit.edu



E. Eilu et al. (eds.), *Digital Literacy and Socio-Cultural Acceptance of ICT in Developing Countries*, https://doi.org/10.1007/978-3-030-61089-0_16

**Table 1** Data about blood discarded in 2016–2017 [6]

| Blood component details | Blood discarded (in units) |
|---|---|
| Whole blood | 150,567 |
| Packed cells | 174,848 |
| Plasma | 332,061 |
| Platelet concentrate | 524,378 |

*Today* reveal that only 8 out of 1000 Indians are voluntary donors [3] and 84 Indian districts lack access to blood transfusion facilities [4]. This shows that there is an exigency to demonstrate the importance of blood donation and thus to urge people for blood donation. Other contributing factors leading to the death of people due to blood shortage are:

- Unavailability of blood of appropriate blood group
- Delay in receiving required blood
- Lack of various facilities for separation and preservation of blood components
- Prevalent myths about blood donation

A large amount of blood is wasted due to lack of coordination between the hospitals and blood banks. Blood, plasma, and platelets are disposed of due to lack of usage before the expiry date. It is important to note that although we face a shortage of blood, a large amount of blood is wasted either due to lack of coordination between the hospitals and blood banks or due to lack of usage before the expiry date. According to *The Economic Times*, in 2016–2017 alone, over 650,000 units of blood and its products were discarded in India and 50% of the wasted units were plasma having a shelf life of 1 year [5]. Table 1 shows blood discarded in 2016–2017 in units, where 1 unit is roughly 525 milliliters [6].

The mission of this book chapter is to review the literature available as well as existing Android apps supporting blood donation and suggest a room for improvisation for motivating the users to donate blood and addressing the myths about blood donation. We are assertive that our research-based approach and related app prototype will lead to design of effective app for blood donation to minimize the time for blood arrival as per user requirement.

## 2 Literature Review

It is extremely essential to know if blood is available in nearby blood banks or with volunteers in case of an emergency. Rh++ is an integrated information system proposed by Sultan Turhan, and its aim has been to manage the blood donation and blood supply chain. The proposed system is used by the patients and/or relatives of the patients to notify their blood requirements and by the living donors to be aware of these requirements [7]. When the blood request is defined into the system, the system checks the stock of the blood banks in the neighborhood. If the blood needed is not available there, the system sends request to living donors of the same region, via the mobile app installed in donors' smartphones, and interprets the response coming from the donors. If there is any affirmative answer from the living donors, the system informs the related healthcare provider. Rh++ seeks to provide services to different kinds of users such as healthcare institutions, blood banks, volunteer donors, patients, or their relatives. This provides an efficient way to perform the blood donation activity and will be more useful in case of an emergency. **A similar approach to manage blood donation and blood supply chain will be a part of our proposed Android app to help in case of an emergency.**

There are several research studies focusing on different aspects of blood donation. Ringwald summarizes key elements to motivate blood donation such as explaining certain myths about blood donation, asking temporarily deferred blood donors to return, reducing blood donor's anxiety, and working on enhancing the reputation of blood donation service [8]. Nilsson et al. proposed that while it is essential to remove fears about blood donation, donors should be explained its advantages [9]. A study by Lemmens et al. among non-donors has found that campaigns to recruit blood donors should try to improve attitude and personal norms regarding blood donations [10]. Steele et al. have observed that while high levels of altruism, empathy, and social responsibility motivate people to donate blood, these indicators are not related to donor frequency [11]. They proposed that although these parameters will urge an individual to donate blood, certain factors, such as convenience of blood donations, safety, and the benefits of blood donation, should also be addressed to an individual.

Although majority of the apps are already live, there are various reasons because of which people do not donate blood. Getachew Arage et al. conducted a survey in Ethiopia [12]. They have listed various reasons for not donating blood. They found that 21.42% of people did not donate blood because of the fear of blood donation, 19.21% did not donate blood since they never thought of blood donation, 18.92% did not donate blood since they were physically unfit, 17.4% did not donate blood since no one never asked them about blood donation, and 16.1% did not donate blood due to medical problems. A similar research has been conducted by Anne Wilkinson et al. to find out the reasons for not donating blood among youths [13]. It states that one of the reasons for not donating blood is fear of blood transfusion needle. Some did not donate blood since their family did not allow them to do so; some did not know where to donate blood, and some were suffering from anemia. Some of the Android apps have included blood facts and guidelines in order to remove myths about blood donation. Once the user is aware of the benefits of blood donation and its myths, he/she is likely to contribute and promote blood donation. **Blood donation facts and guidelines will be included in our proposed app.**

Only some apps have the facility to provide credit to the donor after blood donation. This is an important aspect for promotion of blood donation. Anne Wilkinson et al. have presented the responses of the youth who donated blood. These statistics are important since they can be used to find out the most frequent donors. Some of the reasons stated by the donors for donating blood were noble cause, emergency for friend/family, need for money, and awareness about blood donation [13].

## 3 Android App Review

A systematic review has been conducted for eight Android apps during this research. The eight apps were randomly selected from the list of highly rated apps on Google Play Store. Two of the three researchers (the first author and the third author) reviewed all the eight apps independently based on different functionalities and features. Each researcher prepared his own observations, and the observations were later discussed with a usability expert (second author) to shortlist important app features and functionalities. The apps were then reviewed again based on the finalized features and functionalities, and the results were then finalized by discussing with the usability expert to avoid any bias. These results have been presented in Tables 2 and 3.

Figure 1 depicts screenshots of one of the selected apps named Blood Cure showing facts related with blood, user blood group, and blood donation history. The first screenshot of Fig. 1 states some important facts such as the units of blood a car accident victim might require, how frequently some or the other person needs blood, and the number of blood donations that are required every day. The second screenshot of Fig. 1 shows the profile information of the logged user. This includes his username, blood group, location, gender, and availability status and the number of donations made by the user till date.

The next two sub-sections discuss about the two types of reviews focusing on app functionalities and features.

### *3.1 Review of App Functionalities*

Table 2 shows the summary of app functionalities for all eight selected apps, namely, Simply Blood [14], Friends2support.org [15], Blood Donors [16], Blood Donor [17], Mblood [18], Telangana Blood Donors [19], Blood Cure, and Blood Donation

**Table 2** Summary of app functionalities

| App name | Profile creation (receptor) | Profile deletion | Donor listing | Donor tracking | Raising blood request to your contacts | Post blood request for another user |
|---|---|---|---|---|---|---|
| Simply Blood | No | No | Yes | Yes | Yes | No |
| Friends2support.org | No | Yes | No | Yes | No | No |
| Blood Cure | No | No | No | Yes | No | Yes |
| Blood Donors | No | No | No | Yes | No | No |
| Blood Donation app – INDIA | No | No | Yes | Yes | No | No |
| Blood Donor | Yes | No | Yes | Yes | No | No |
| Mblood | No | No | No | No | No | No |
| Telangana Blood Donors | No | Yes | Yes | Yes | No | No |

**Table 3** Summary of app features

| App name | Number of available blood groups | Geographical scope of use | Donor tracking range | Blood donation myths and facts | Blood group information | Promoting blood donation | Blood donation history of user |
|---|---|---|---|---|---|---|---|
| Simply Blood | 20 | Global | Within 5 km of the location | No | No | Yes | Yes |
| Friends2 support.org | 17 | Asia | Within your area | Yes | No | No | Yes |
| Blood Cure | 16 | Global | Within the state | Yes | No | No | No |
| Blood Donors | 8 | Global | Within the city | No | Yes | No | No |
| Blood Donation app – INDIA | 16 | India | Within the state | Yes | No | Yes | No |
| Blood Donor | 18 | America | Within the city | Yes | No | Yes | Yes |
| Mblood | 16 | India | Within the area | Yes | No | Yes | No |
| Telangana Blood Donors | No specific number | Telangana | Within the city | No | No | No | No |

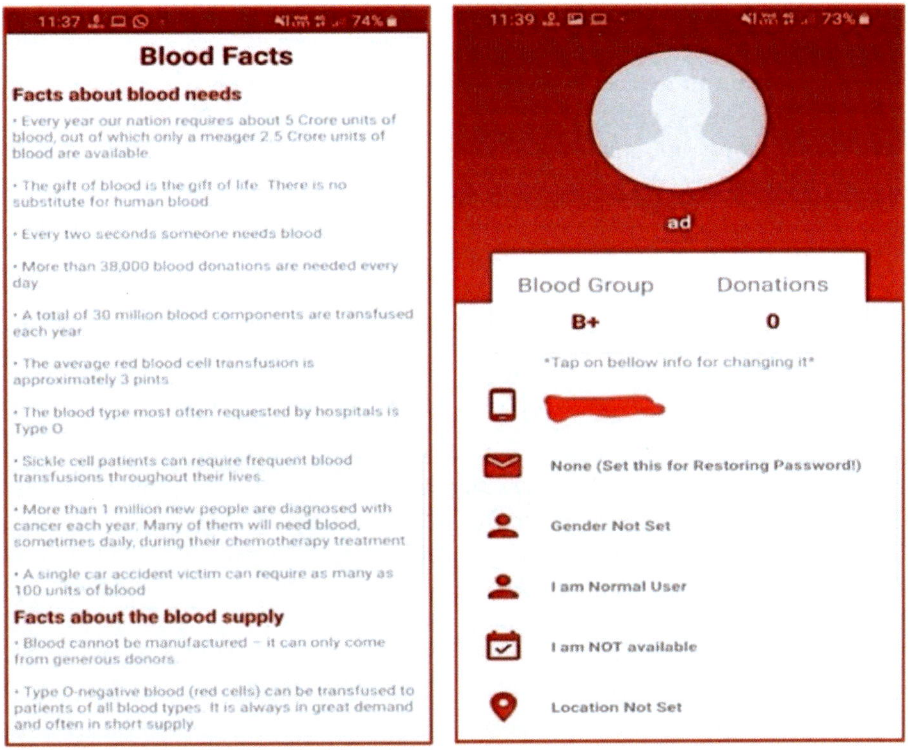


(a) (b)

**Fig. 1** Screenshots for Blood Cure app. (**a**) Facts and (**b**) profile information app – INDIA. Major app functionalities include profile creation of the receptor, profile deletion of the user (both donor and receptor), listing of the available donors, donor tracking, raising the blood request to the contacts, and posting a blood request on behalf of another user. Profile creation of the receptor or asking the receptor to create profile to receive blood will waste a lot of time. The receptor should have open access to receive blood. Doing so will make the entire process quite simple [20]. Similarly, the user should have the provision to delete the profile. The most important aspect while using any blood donation-related app is to track the donor. Since most of the apps list the contact details of the user, it may happen that the first user in the list is unavailable. In such cases, it would be beneficial for the receptor if the donor forwards the request to his contacts that are in the vicinity. Often it happens that the user is not registered; however, there is an exigency of blood. In such cases, it helps if the registered user has the provision to ask for blood by sharing the contact of the desired user. This can save a large amount of time. It is also essential to check if the required blood is available in the blood bank/hospital. Although no app listed here has such a provision, this can be done using RH++ [7]. The observations from entries in Table 2 are as follows:

- **Most of the apps (seven out of eight apps) provide vital facility for donor tracking.**
- **Most of the apps (seven out of eight apps) have facility to direct the receptor to the donor without creating the profile of the receptor.**
- **Some of the apps (four out of eight apps) list the contact details of the donors without asking the receptor to sign in.**
- **A provision to delete the user (donor/receptor) profile is available in only two apps (out of eight apps).**
- **There is only one app (out of eight apps) which has a facility to raise the blood request.**
- **Only one app (out of eight apps) has a facility to post blood request for another user.**

## *3.2 Review of App Features*

Table 3 depicts the summary of features related to different Android apps. Major app features include **number of available blood groups**, **geographical scope of use**, **donor tracking range**, **blood donation myths and facts**, **blood group information**, **ways to promote blood donation**, and **to record the blood donation history of the user**. Every app reviewed asks the user to select his/her blood group. The number of different blood groups varies app by app. For example, "Bombay Blood Group" is present only in few apps. It is important to include maximum number of blood groups since during emergency, blood requirement may arise for any blood group. There are some apps which cannot be used beyond a particular location. Emergency may arise from any location. As a result, it is essential to increase the geographical scope. Whenever blood is requested by the user, the app finds a donor; however, different apps have a different range. During emergency, the range plays a key role. The app should find a donor by taking into account parameters such as distance from the receptor and time to reach receptor's location or hospital. Finding a donor beyond a particular range limit is inefficient. However, this range may increase depending upon the availability of the blood donors in the nearby area.

There are majority of the users, who do not donate blood due to certain myths and fear [13]. These include fear of blood transfusion needle, fear of catching a disease [21], and fear of being weak after donating blood [21]. The inclusion of blood facts not only eliminates myths but also helps to promote blood donation. There are also other ways to promote blood donation such as adding reward points and showing the number of lives saved whenever any donor donates blood. The observations from Table 3 are as follows:

- **Most** of the apps (6 out of 8 apps) have more than 15 blood groups to select at the time of blood donation. A1+, A1−, A1B+, A1B−, A2+, A2−, A2B+, A2B−, A2−, A2B+, A2B−, Bombay Blood Group, INRA are some of the new blood groups.
- **Most of the apps (five out of eight apps) have included some myths and facts about blood donation.**
- **Some of the apps (four out of eight apps) endorse blood donation by offering reward points.**
- **Few apps (three out of eight apps) are globally available, and other apps are geographically confined to a particular area.**
- **Only one app (out of eight apps) finds a donor within 5 km (smaller range) of the requested area. Other apps find a donor within the city or state.**

## 4 Research Methodology

Based on the summary of app functionalities and features, it is evident that not all the apps have all these features and functionalities mentioned above. For instance, apps "Friends2support.org" and "Telangana Blood Donors" have profile deletion as one of its functionalities; however, these apps are limited to the geographical scope of use. Previous research conducted by Ouhbi, Sofia et al. [20] explains various reasons for less number of blood donation apps developed in Africa and how it would be useful if an app with majority of the features and functionalities will be made available to people in zones confronted by war and poverty. This explains why the geographical scope of use of the app is pivotal along with other features. On the other hand, apps "Blood Donors" and "Blood Cure" which are globally available do not have the provision to promote blood donation. Sharing blood donation information in social networks will not only help to spread awareness about blood donation but will also encourage more volunteers to donate blood. This will not only help WHO achieve its target [22] but it will also attract young donors [23]. Thus, the app should not only be available **globally** but it should also **promote blood donation**. The app "Simply Blood" which is available globally and which promotes blood donation, however, does not provide any blood donation myths and facts. Anne Wilkinson et al. [24] in their study have mentioned how important it is to educate people with the **myths and facts about blood donation** and how doing so can promote blood donation. Myths and facts about blood donation should therefore be included in the app.

The above review also shows that all the apps have the facility to find a donor. This is especially important in disaster situations since doing so will provide blood banks with additional support [25]. This will also help save a large number of lives. There is only one app ("Simply Blood") which finds a donor within a smaller range (5 km) of the requested area. However, if the donor is not available within such a smaller range, it might jeopardize the life of the requestor. The remaining apps find a donor within a larger range. Although this might increase the chances of getting a donor, it might be inefficient for the donor since the donor might have to make a long journey to donate blood. Previous research has shown how accessibility to clinics can positively affect donor turnout [26] and frequency of donations [27]. It is therefore important to locate a donor within a smaller range and, if the donor is not available within the smaller range, increase the range until the donor is located. It will also be helpful if the donor can divert the request to some other user if the donor is unable to accept it.

Based on the above analysis, it is clear that there is a need to design one single app with all the important functionalities and features. These functionalities and features will be discussed in the subsequent section.

## 5 Paper Prototype Design for Mobile App

A prototype is an experimental model of an idea [24]. A paper prototype is a low-fidelity option in mobile app design [28]. A paper prototype has also been proposed for eye-free Android app for visually impaired users by Sagale et al. [29]. Here, we are proposing a paper prototype design for blood donation app based on the requirement analysis as discussed in the above section.

The proposed prototype will start with the login process. If the donor is not registered, then he/she can register as a New User or choose the option of logging using an OTP (one-time password). If the donor decides to register as a New User, then he/she will redirect to the registration screen where the user will be asked to provide certain credentials. **Registration of donors is essential since it helps to maintain the history of all the donors.** By doing so, even the donor can maintain a track of the number of donations made and the number of lives saved by his donation, once the donor is logged in. This will enable the donor to share the donation history on social media, thereby motivating more users to donate blood [23]. **It will also be easier to track the donor once he/she is registered on the app.** If the user is a receptor, then he/she will have access to the contact details of the registered donors and, thus, can request blood without logging in. The receptor will most likely be a hospital, a blood bank, or any other organization requesting blood. This process of providing direct access to the app to request blood, after registering as a receptor, will help in saving a large amount of time which will be crucial in case of an emergency. However, once the process of blood donation is over, even the receptor will have to mention the details on the app. These may include details such as quantity of blood donated in units, name of the receptor, and contact details. It is important that we maintain a record of such details since it can help to identify whether any blood was discarded or wasted which is considered to be one of the important factors that leads to blood shortage [5].

**A user will be able to select a blood group from more than 15 different blood groups available.** This includes the new blood groups as well which most of the existing apps already have. Adding these new blood groups would ensure that the donor is not restricted from donating blood due to a wide range of available blood groups. **The app also has an additional facility to auto-search the donor without explicitly asking the receptor to contact each donor.** Previous research has shown that searching a donor using an app, especially in disaster situations, has helped laboratories and blood banks with additional support [25]. There is no separate option for auto-search. The app will search the donors within the area of the receptor as soon as the receptor requests blood. The nearest donor will be found by calculating the distance using an optimized approach as mentioned by Sultan Turhan [7]. If no single donor is found within the nearby area, then the app will increase the search limit by 5 km up to a particular range which will be the maximum search limit. As of now, in the first version of our app, the maximum search limit will be limited to the city. In the future, we will be using an algorithm to find the maximum search limit which can further maximize the chances of finding a donor. This additional facility of auto-search has been implemented in order to reduce the time required to find a donor and to maximize the chances of finding a donor.

A donor can receive multiple blood requests. For each blood request, a donor can see the details of the receptor (name, blood group, address) and the deadline to donate blood. Figure 2 (a) shows different blood requests received. **A donor can either accept the request or raise the request.** The "Raise" option allows the donor to forward the blood request to another registered user he knows. This option is useful in case the donor has an emergency and cannot donate blood. The donor can confirm the request by selecting the Accept option. Once the donor accepts the request, the contact details of the donor are shared with the receptor.

Figure 2 (b) shows different donors who are available to donate blood. The receptor has access to donor contact details and blood group. The receptor can also see how far the donor is located from his/her area. Once the donor accepts the request, the receptor can track the donor.

According to WHO, it is essential for each country to set its own blood donation recommendations and guidelines [22]. For instance, there is no upper age limit to donate blood in countries such as the USA, the UK, and Canada [20]. It is really important for the donor to know if he/she is eligible to donate blood or not based on the different guidelines since it will help to save the time, effort, and frustration of a

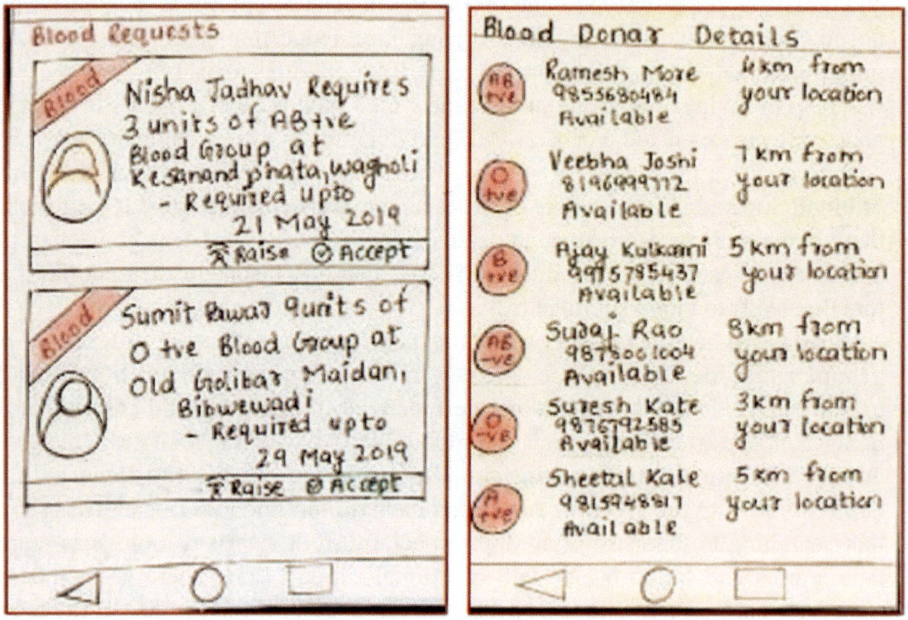


(a) (b)

**Fig. 2** Screenshots for (**a**) blood requests and (**b**) blood donor details

willing donor. It might demotivate the donor if such guidelines are not mentioned and if the donor, upon reaching the destination, discovers that he/she is not allowed to donate due to some non-medical reasons. As a result, **we have included the blood donation guidelines in our app.** These guidelines will differ from country to country and a user will be able to see the guidelines of any country. Thus, before accepting the blood donation request, a donor should check for such guidelines depending on where he/she is donating blood.

An inclusion of blood donation facts is critical since that not only helps to spread awareness about blood donation but it also helps to eliminate related myths. **A list of blood donation facts is available in our app.** It is essential that the donors and the receptors are aware of the benefits of blood donation. This awareness will urge the users to donate blood frequently and will also promote blood donation. Figure 3 (a) shows different blood donation facts, and Fig. 3 (b) shows explanation of one such fact, displayed in the app. **In order to increase the geographical scope, this app will be available globally.**

## 6 Future Research Directions and Conclusion

The proposed app for blood donation is quite interesting and useful. The app prototype has been designed by reviewing several peer apps as well as related literature that is available. Thus, spreading awareness on blood donation, eliminating all

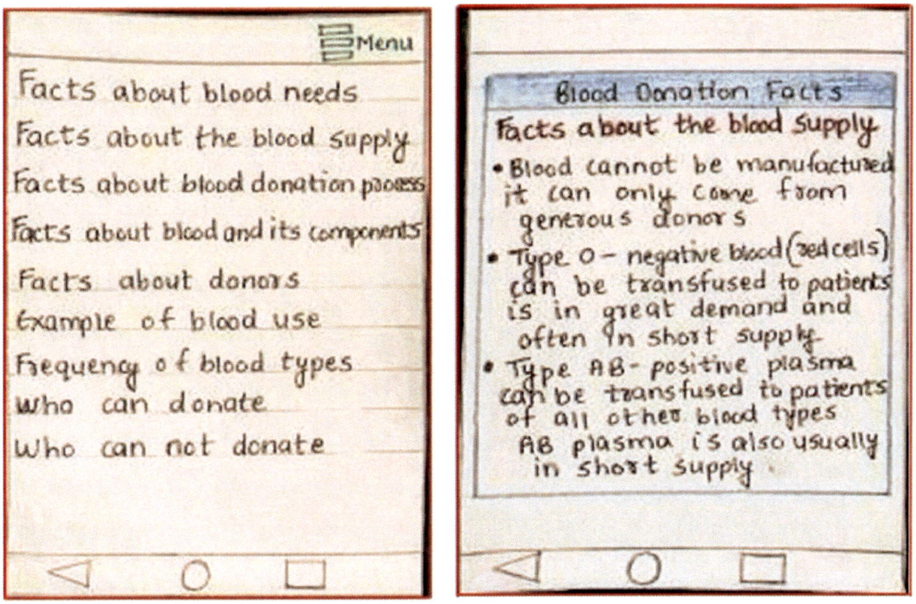


(a) (b)

**Fig. 3** Screenshots for blood donation facts

myths about blood donation, and locating suitable donors are the key aspects of proposed app designed in this research work.

In the future, we will launch the first version of our Android app. The app will be currently launched only in English. However, we are planning to include some other languages during its next version. In the existing app, there is a provision for the user to track the number of lives saved by his donations; however, we are not offering any reward points for this noble cause. In the next version, we plan to include this feature as well, so that the users can showcase such reward points and urge people to donate blood. Due to the coronavirus pandemic of 2020, we were unable to evaluate our prototype and plan to do it in the future.

## References

1. Cardiac surgeries performed. https://timesofindia.indiatimes.com/city/mumbai/32m-patients-but-cardiac-care-heartwarming/articleshow/16596237.cms. Last accessed 2019/05/04
2. Blood shortage in 2016–17. https://www.business-standard.com/article/current-affairs/blood-banks-gasp-for-oxygen-india-sees-shortage-of-1-9-mn-units-in-2016-17-118061400174_1.html. Last accessed 2019/05/04
3. Indian voluntary donors. https://www.indiatoday.in/fyi/story/blood-banks-india-wastage blood-donors-973478-2017-04-25. Last accessed 2019/05/04
4. Access to Safe blood. http://naco.gov.in/access-safe-blood. Last accessed 2019/05/04

5. Units of blood discarded. https://economictimes.indiatimes.com/industry/healthcare/biotech/healthcare/6-lakh-litres-of-blood-wasted-in-five-years-due-to-lack-of-coordination-between-blood-banks-and-hospitals/articleshow/58335139.cms?from=mdr. Last accessed 2019/05/04
6. Units of blood discarded in last 5 years. http://www.mymedicalmantra.com/in-last-five-years-28-lakh-unit-of-blood-disposed/. Last accessed 2019/05/04
7. S. Turhan, An android application for volunteer blood donors. Comput. Sci. Inf. Technol. **5**, 23–30 (2015). https://doi.org/10.5121/csit.2015.51103
8. J. Ringwald, R. Zimmermann, R. Eckstein, Keys to open the door for blood donors to return. Transfus. Med. Rev. **24**, 295–304 (2010)
9. B. Nilsson Sojka, P. Sojka, The blood-donation experience: Perceived physical, psychological and social impact of blood donation on the donor. Vox Sanguinis **84**, 120–128 (2003)
10. K. Lemmens, C. Abraham, T. Hoekstra, R. Ruiter, W. De Kort, J. Brug, H. Schaalma, Why don't young people volunteer to give blood? An investigation of the correlates of donation intentions among young nondonors. Transfusion **45**, 945–955 (2005)
11. W. Steele, G. Schreiber, A. Guiltinan, C. Nass, S. Glynn, D. Wright, D. Kessler, K. Schlumpf, Y. Tu, J. Smith, G. Garratty, Retrovirus epidemiology donor: The role of altruistic behavior, empathetic concern, and social responsibility motivation in blood donation behavior. Transfusion **48**, 45–54 (2008)
12. G. Arage, S. Ibrahim, A. Endeshaw, Blood donation practice and its associated factors among health professionals of University of Gondar Hospital, Northwest Ethiopia: a cross sectional study
13. A. Wilkinson, R. Gupta, Perceptions of blood donation amongst the youth
14. Simply Blood. https://play.google.com/store/apps/details?id=com.simplyblood
15. Friends2Support.org. https://play.google.com/store/apps/details?id=com.alen
16. BloodDonors. https://play.google.com/store/apps/details?id=com.virtuepoint.infotech.blooddonors
17. Blood Donor. https://play.google.com/store/apps/details?id=com.cube.arc.blood
18. Mblood. https://play.google.com/store/apps/details?id=com.jbsoftsystem.theblood
19. Telangana Blood Donors. https://play.google.com/store/apps/details?id=telangana.blood.donors
20. S. Ouhbi, J. Fernández-Alemán, A. Toval, A. Idri, J. Rivera, Free blood donation mobile applications. J. Med. Syst. (2015)
21. Fear of blood donation. https://www.giveapint.org/library/what-are-the-top-10-excuses-for-not-donating-blood. Last accessed 2019/05/04
22. World Health Organization: Blood donor selection: Guidelines on assessing donor suitability for blood donation (2012)
23. J. Mart'ın-Santanam, A. Beerli-Palacio, Achieving donor repetition and motivation by block leaders among current blood donors. Transf. Apheres. Sci. **47**(3), 337–343 (2012)
24. A. Patil, G. Bhutkar, M. Pendse, A. Tawade, A. Bodhke, S. Deshmukh, Prototype design of android app for mothers of preterm infants: fifth IFIP WG 13.6 working conference, HWID 2018, Espoo, Finland, August 20–21, 2018, Revised selected papers. https://doi.org/10.1007/978-3-030-05297-3_1.
25. M. Mart'ınez, Contingency planning for natural disasters. ISBT Sci. Ser. **6**(1), 212–215 (2011)
26. A. Paez, J. Esita, K. Newbold, N. Heddle, J. Blake, Exploring resource allocation and alternate clinic accessibility landscapes for improved blood donor turnout. Appl. Geogr. **45**, 89–97 (2013)
27. K. Cimaroli, A. Pez, K. Newbold, N. Heddle, Individual and contextual determinants of blood donation frequency with a 52 Page 20 of 20 J Med Syst (2015) 39: 52 focus on clinic accessibility: a case study of Toronto, Canada. Health Place **18**(2), 424–433 (2012)
28. Steele, Prototyping 101 the difference. https://theblog.adobe.com/prototyping-difference-lowfidelity-high fidelity-prototypes-use. Last accessed 2019/04/12
29. U. Sagale, G. Bhutkar, M. Karad, N. Jathar, An eye-free android application for visually impaired users, in *Ergonomics in Caring for People*, ed. by G. Ray, R. Iqbal, A. Ganguli, V. Khanzode, (Springer, Singapore, 2018), pp. 291–297

# Usability Evaluation of Cultural Heritage Crowdsourcing System (CHCS)

**Dinesh Katre**

**Abstract** The crowdsourcing systems are extremely complex as they require us to involve multidisciplinary knowledge domains for its design and development. This research article presents a detailed heuristic evaluation of Cultural Heritage Crowdsourcing System (CHCS) by using the existing crowdsourcing community heuristics. This evaluation is also combined with a usability test followed by an online survey conducted over the test participants. The crowdsourcing community heuristics applied in this project are helpful in identifying the missed out functional requirements. Combination of design heuristics with the community heuristics that focus more on functional requirements can help in customization of evaluation process. This will allow greater flexibility to designers and provide more realistic outcomes that can tally with the findings of usability tests.

**Keywords** Crowdsourcing · Heuristics · Usability · Metadata · Cultural heritage · Social computing

## 1 Introduction

### *1.1 Cultural Heritage Crowdsourcing System (CHCS)*

Our Cultural Heritage Crowdsourcing System (CHCS) aims to integrate and provide access to heterogeneous data repositories to the public. The system has to integrate diverse types of metadata standards like Dublin Core, MARC21, customized domain-specific metadata, and a wide variety of digital contents like rare books, e-books, manuscripts, museum objects, audio and video, research papers, and archived records. The volume of this data is amounting to millions of records. However, during the data collection phase, it was observed that majority of digital collections are poorly catalogued. The cataloging metadata fields are either empty

D. Katre (✉)
Human-Centred Design & Computing Group, Centre for Development of Advanced Computing (C-DAC), Pune, India
e-mail: dinesh@cdac.in



E. Eilu et al. (eds.), *Digital Literacy and Socio-Cultural Acceptance of ICT in Developing Countries*, https://doi.org/10.1007/978-3-030-61089-0_17

or containing information with a lot of typographic errors. The metadata associated with digital objects is very critical for classification based on various attributes and effective search and retrieval for proper understanding of the record by the public. The project has inadequate resources to engage domain experts and operators, which are necessary for correcting and curating millions of metadata records. Multilinguality is another major challenge in this task as significant volume of metadata records are available in different Indian languages that are spoken in different geographic regions of India. It is difficult to find domain experts with proficiency in multiple Indian languages in one location. Therefore, our team has resolved to design and develop a comprehensive metadata crowdsourcing system to encourage voluntary participation by library science professionals in this mega co-curation activity, which is aimed at correcting and enriching the cataloging metadata associated with digital contents in the repositories.

The functional prototype of the crowdsourcing system has been developed and designed in such a way that people should be able to view the digital objects, understand the metadata parameters, and then input the descriptive information in various metadata fields. Refer to Fig. 1 for the basic architecture of CHCS system. The system also provides detailed mechanisms for allocation of digital collections to curators and librarians, metadata version management, review, finalization, and approval of metadata by the experts. The system also supports folksonomy features and inputting of metadata in various Indian languages. A reward mechanism is also designed to recognize and motivate the contributors.

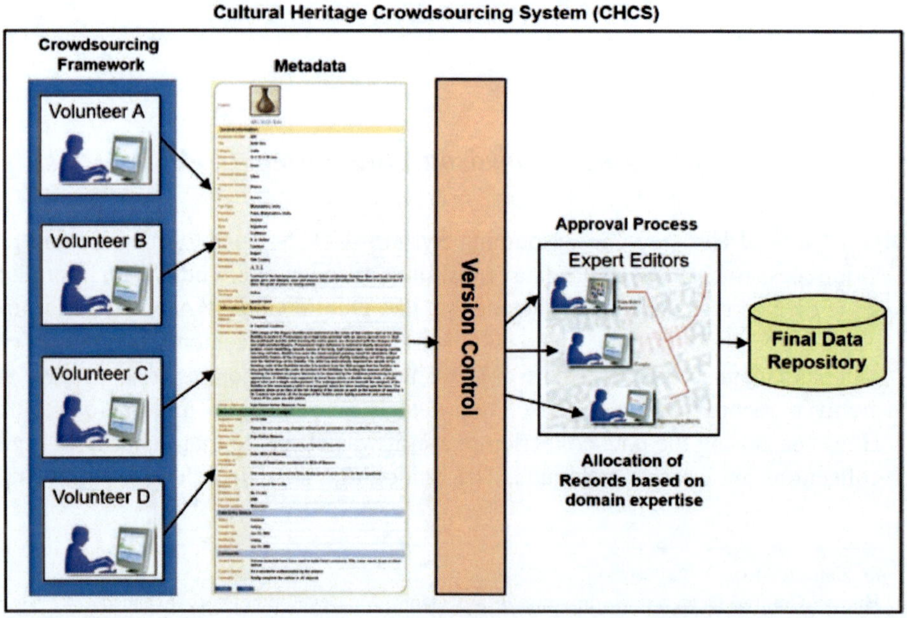


**Fig. 1** Co-curation/crowdsourcing framework of CHCS

Therefore, in order to proceed further with development, it is decided to first evaluate the usefulness and usability of the "Cultural Heritage Crowdsourcing System." The intention behind this work is to find opportunities for enhancing the design of system. Therefore, the objectives of this exercise are defined as follows:

- Conduct self-evaluation of the CHCS based on available crowdsourcing heuristic principles.
- Conduct a usability test to get feedback from potential users of the CHCS.

## *1.2 Related Work*

Crowdsourcing and open innovation are becoming buzz words for swarm intelligence (SI) or peer production [1]. Crowdsourcing encourages involvement of professional communities and the public in completing certain tasks collectively in a distributed way through the Internet. Crowdsourcing is a multidisciplinary subject requiring deeper studies in information systems, psychology, gamification, machine learning, and human-computer interaction in addition to the domain in which the crowdsourcing is being used. In this section, we briefly review the research focusing on multidisciplinary aspects of crowdsourcing.

- **Crowdsourcing as an information system (IS)**

Zhao et al. [2] have presented the theoretical conceptual framework for crowdsourcing in terms of who (crowd, specific group), how (competition, collaboration), why (motivation), and what (public or private goods) for defining it as an information system (IS). Geiger et al. [14] have defined the components of crowdsourcing system design and the typology in terms of crowd rating, crowd processing, crowd creation, and crowd solving. Daniel et al. have carried out a survey to identify the quality attributes, assessment measures, and assurance actions in crowdsourcing. This work identifies people, software, and data as the main constituents of quality in crowdsourcing [3]. Researchers have tried to identify the common processes and components from a variety of crowdsourcing systems developed in different domains while defining it as an information system (IS).

- **Role of psychology in crowdsourcing**

Crowdsourcing researchers have used theories from psychology to understand the motivational aspects of the crowd. Self-determination theory (SDT) has been applied [2, 4] to illustrate the motivation spectrum in terms of external, introjected, identified, and intrinsic types of motivations in crowdsourcing context. Experiments are carried out in which the contributions of paid participants and unpaid volunteers are compared to study the impact of motivational incentives [5]. The contest theory is also applied as an incentive strategy for attracting the volunteers for crowdsourcing. A study shows that network individuals are unwilling to participate in crowdsourcing projects for free and that they expect some compensation or reward for their contributions [6]. Afentoulidis et al. [12] have used social gamification as an implicit method of crowdsourcing in enterprise domain by involving the employees, as against the explicit calls for crowd contributions. Mobile gaming is also explored as a motivation for crowdsourcing, but it requires specialized and research question-specific games to be produced that can be costly to design [13]. Motivation of the crowd is identified as a key psychological factor involved in crowdsourcing.

- **Role crowdsourcing in supervised machine learning**

Rahmanian et al. have used human intelligence tasks (HITs) for ranking, sorting, and rating of user interface design for crowdsourcing system [18]. We also find that crowdsourcing of HITs is applied for ranking of books for improving the results of search and retrieval in digital library [21]. The Stanford Question Answering Dataset (SQuAD) project employed crowdworkers with over 1000 HITs through Amazon Mechanical Turk for creating questions [13]. Such crowdsourcing contributions can help in advancing the quality of machine learning [23].

- **Need of HCI research in crowdsourcing**

Limited research focusing on the HCI aspects of crowdsourcing is published. Pan et al. [7] have briefly covered crowdsourcing for social networks, academic purposes, enterprise applications, and mobile crowdsourcing to define the agenda for human-computer interaction. Brandtner et al. [8] have applied Malone's motivation principles for designing enjoyable use interfaces as the heuristic evaluation criteria for participatory evaluation of crowdsourcing platforms. Matias et al. [9] have presented their work in progress on reader sourcing to reduce gender disparities in Wikipedia by expanding the articles on women. Samdaria et al. [10] have evaluated inputting techniques for crowdsourced mobile micro-tasking on sample video clips to analyze the recognition of product brands and vehicles in city traffic. Balamurugan et al. [11] have explored crowdsourcing of form digitization activity and highlighted the need to widen the spectrum of HCI for crowdsourcing. Tinati et al. have stated that online citizen science projects involving crowdsourcing remain unpredictable in terms of the results even when sound user-centric design approach is followed by an experienced UX team [16]. There is insufficient understanding available on how the citizen science crowdsourcing systems are to be designed [17]. à Campo et al. [15] have formulated community heuristics in terms of purpose, moderation, members, common ground, contribution, and platform for the evaluation of crowdsourcing system design.

- **Need of crowdsourcing in cultural heritage**

Chhatwal et al. [20] have stated that the libraries have a massive probability to connect to the volunteers. They have enlisted various requirements for crowdsourcing and highlighted the fact that crowdsourcing can help in overcoming financial and staff crunch. In the library domain, crowdsourcing has been used for various applications such as distributed proofreading, participatory OCR correction, manuscript transcription, and folksonomy [22]. Similarly, in cultural heritage domain, creation of descriptive metadata about the artifacts is considered as a major challenge. In this regard, a report is available that presents 21 key design principles for crowdsourcing of cultural heritage [19].

## 2 Methodology

It is important to recognize that functional prototype of the Cultural Heritage Crowdsourcing System (CHCS) has been developed to be able to perform usability evaluation. It is being developed using agile software development model, which allows the designer of the system to perform evaluations and user testing over intermediate functional prototypes for incremental design enhancements in terms of identification of new requirements, new features, and improving upon the problems faced by users. Therefore, a two-step usability evaluation methodology is adopted in this study. It is defined as follows:

**Step 1. Self-evaluation of the CHCS system based on existing crowdsourcing heuristic principles by the designer of system**

It is a common practice, wherein a designer can evaluate his product or system against a set of heuristic principles or a checklist. It is a cost-effective and time-saving evaluation method during the design phase. This approach helps the desire in improving the system during iterative development life cycle when changes can be easily incorporated. Self-evaluation based on heuristic principles by the designer does not require the participation of experts or users.

**Step 2. Usability testing by involving potential users and feedback collection**

This usability test is meant to ensure whether the functional prototype is working towards addressing the basic purpose of the system, which is metadata crowdsourcing. This test involves a set of tasks that need to be performed by library and information science professionals/students using the functional prototype of CHCS. The test is completed by collecting feedback from the test participants in survey form which includes questions associated with the tasks performed by the users.

The insights found through self-evaluation and usability testing both together will help in confirming new requirements and changes in design that are necessary for the betterment of CHCS system.

## 3 Selection of Crowdsourcing Heuristic Principles

As per the proposed methodology, when we began to identify suitable heuristics for design evaluation of CHCS, it was found that there is very limited research available on crowdsourcing heuristics. Many have applied general usability heuristics, e.g., Malone's criteria for designing enjoyable user interfaces is adopted for evaluation of crowdsourcing platform [8]. However, it is observed that the original purpose of Malone's criteria is user interface design. We wanted to select the heuristics, which are specially defined for evaluation of crowdsourcing. The design principles for crowdsourcing cultural heritage by McKinley [19] present general usability best practices such as provision of clear and concise instructions, minimization of user efforts and errors, simplicity and prioritization of tasks, provision of task options, attractiveness of design, and conveying a sense of fun. This work provides specific guidelines for crowdsourcing of cultural heritage such as recognizing contribution, displaying project progress, and conveying a sense of community. But this report does not provide guidelines on how to evaluate or measure the success of crowdsourcing system design. Therefore, we needed a specialized set of heuristics along with evaluation method specially meant for crowdsourcing systems.

The community heuristics proposed by à Campo et al. [15] already draw upon the previous works in this domain and consolidate the design advice relevant to crowdsourcing and provide a well-defined marking system; hence, we choose this for the self-evaluation by the designer of CHCS. During this evaluation process, we also aim to consider if any new additions can be made to the existing set of heuristics or provide feedback for its enhancement.

## 4 Heuristic Evaluation

The heuristic evaluation is done on the scale of 0–2 (0 absence, 1 partially addressed, 2 fully addressed). We also discuss the modules and features of the CHCS as we evaluate each heuristic principle. The crowdsourcing heuristic principles, namely, purpose, moderation, members, common ground, contribution, and platform, are mentioned as subsection labels. The subcategories under the heuristic principles are given in tabular format along with the observation of evidences and ratings. Also, the relevant screenshots are included to show the evidence and for discussion on applicability of heuristics to CHCS.

### 4.1 Purpose

| Subcategories | Observation of evidence | Rating |
|---|---|---|
| *Clarity* | Video-based tutorials explaining the purpose of crowdsourcing and guidance on how the task is to be performed are prepared and linked on crowdsourcing page | 2 |
| *Visibility* | Current system provides a "crowdsourcing" link above the display window of every record, which is visible enough. This link acts as the initial trigger for the volunteers to engage in metadata co-curation activity | 2 |
| *Idealism* | How the voluntary contributions to enrichment of the metadata can benefit the society and professional community needs to be described in the crowdsourcing module. We have assumed that library and information science professionals are already aware of the requirements. Therefore, the idealism factor has not been addressed in the system | 0 |

Therefore, the CHCS scores 4 out of 6 based on "purpose" heuristic.

As shown in Fig. 2, online tutorials are prepared to explain the purpose and method of crowdsourcing the metadata by referring actual digital object, which is to be described

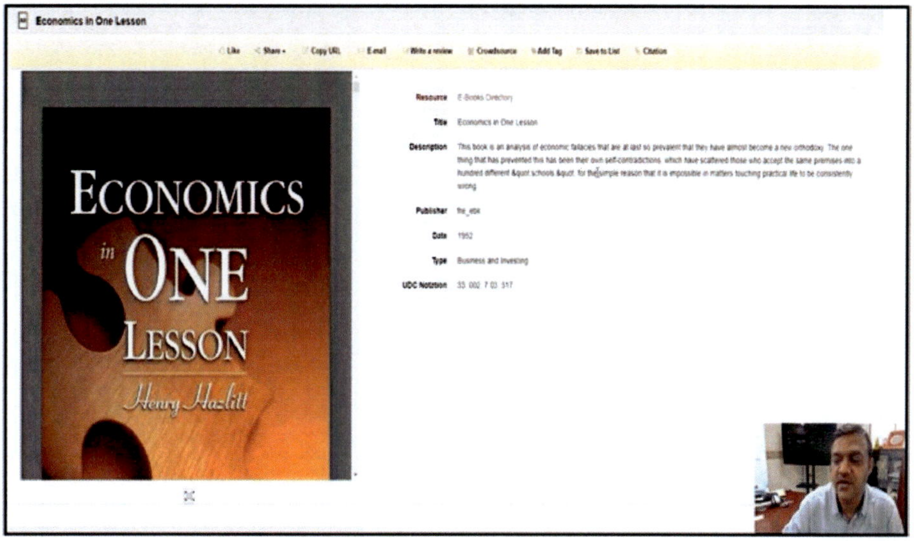


**Fig. 2** Sample e-book along with metadata

by the volunteers. The folksonomy features provided above the record include a link labeled as "crowdsource," clicking on which the crowdsourcing module gets loaded.

The idealistic values of metadata enrichment are assumed to be known by the volunteers, and hence they are not covered in the present scope of design. However, based on this heuristic evaluation, the need of communicating the ideals to the community is well recognized. Therefore, the principle of "purpose" as represented in the system scores 4 out 6 marks.

## 4.2 *Moderation*

| Subcategories | Observation of evidence | Rating |
|---|---|---|
| *Monitoring* | The CHCS supports thematic collection-wise experts who can act as reviewers for the metadata corrections and additional information received from volunteers. They can merge the inputs received from multiple volunteers; edit, refine, and delete unwanted information; and approve the final modifications<br>There is no feature for reporting misuse or undesired behavior by volunteers. However, it should be noted that the contributions by volunteers become visible to the public only after they are approved by the expert | 1.5 |
| *Regulations* | Regulations and policies regarding CHCS need to be documented and published | 0 |
| *Moderators* | There is a provision for appointing experts to review and finalize the crowdsourced metadata records. Perhaps, we need to provide an online discussion forum wherein the volunteers and curators are able to discuss their difficulties and more finer issues involved in the enrichment of the metadata | 1 |

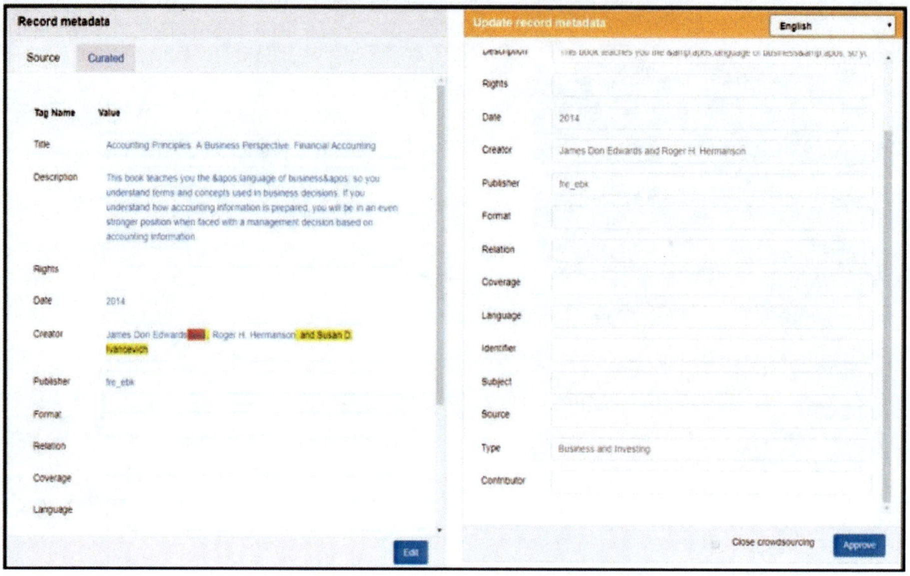


**Fig. 3** Experts can review multiple versions curated by volunteers or curators

Therefore, the CHCS scores 2.5 out of 6 based on "moderation" heuristic. CHCS enables to appoint the experts, who can review the inputs received from volunteers as well as curators. As shown in Fig. 3, an expert can log in to review the modified information, which is visible in track change mode, and approve the final version. The expert can also compare the modified version with source record. They can fine-tune the record by referring all the inputs received through crowdsourcing and then approve the final version, if acceptable. The expert can reject the edits received from volunteers if they are inaccurate.

It will be useful to block some volunteers if they are adding junk or malicious information through crowdsourcing. Also, an online discussion forum could help in discussing and knowing the finer issues involved in metadata enrichment, e.g., people get confused while understanding some metadata parameters like "coverage" in Dublin Core metadata elements. They do not know what values should be provided and in what manner. In such discussions, the role of moderators becomes very critical in explaining and getting correct information inputs from the volunteers. However, such functionalities need to be developed in CHCS.

## 4.3 Members

| Subcategories | Observation of evidence | Rating |
|---|---|---|
| *Self-presentation* | The CHCS allows the users to create their own profile by registering their name, geographical identity, uploading of photo, interests, professional background, languages known, etc. The "saved lists" and "watch lists" of repository contents prepared by users get maintained in the personal account of users which also get added in their profile information | 2 |

| Subcategories | Observation of evidence | Rating |
|---|---|---|
| *Deep profiling* | The logs of user behavior in terms of the records liked, tagged, and commented, records shared in social media, records edited by the user, number of edits, and number of accepted edits by the user are maintained by the system | 2 |
| *Life cycle* | There is a provision to publish, certify, and recognize the active contributors in CHCS. Additional features for instructing new users, empowering the leaders, and honoring of elders need to be developed as expected in this heuristic principle | 1.5 |
| *Recruitment* | The CHCS allows admin user and crowdsourcing manager to invite experts and curators who are appointed for activities like metadata verification, approval, and enrichment. However, presently, the system does not allow general users to invite his friends to join | 1.5 |
| *Virtual co-presence* | The feature to support showcasing of the contributions by members from the community needs to be developed | 0 |

On the whole, the crowdsourcing heurist evaluation of parameter "members" reveals that CHCS requires to provide more features to support crowdsourcing community development. An independent observation regarding this particular heuristic can be recorded that user profiling and deep profiling can be separated from crowdsourcing community development parameters such as life cycle, recruitment, and virtual co-presence. User's behavior in his personal space and community-focused social behavior need to be observed and evaluated separately. Also, this would require different sets of features and functionalities. Therefore, a separate heuristic on "community development" will bring greater focus and emphasis on social behavior and contributions by users, e.g., a user helping another user by catering to his/her interest. Therefore, the CHCS scores 7 out of 10 based on "members" heuristic (Fig. 4).

## 4.4 Common Ground

| Subcategories | Observation of evidence | Rating |
|---|---|---|
| *Sub-groups* | The CHCS does not support creation of sub-groups. Based on this heuristic, it is possible to visualize sub-groups focusing on different types of digital repositories such as rare books, audio and video, museum collections, etc. as the metadata requirements vary depending on the type of repository. Also, the sub-groups may focus on different metadata schemes to discuss and evolve uniform understanding about the parameters of information | 0 |
| *Diversity* | The data repositories in CHCS and interests of the users are expected to be diverse. However, CHCS does not support creation of sub-groups by the users | 0 |
| *Events* | There is no provision for community events in CHCS | 0 |
| *Rituals* | There is no provision for community rituals in CHCS | 0 |

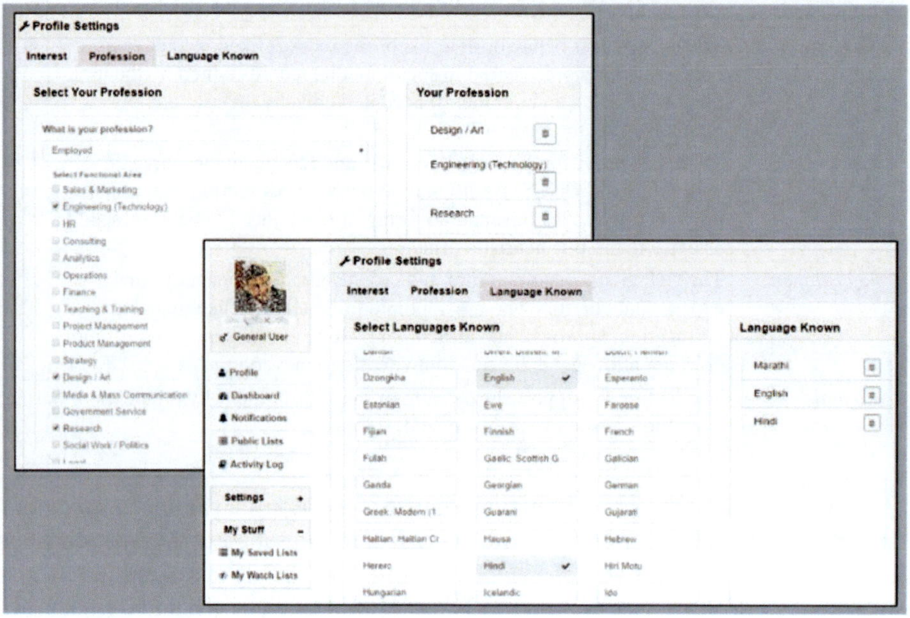


Fig. 4 User profiling in CHCS

The crowdsourcing heurists are assuming that it is a community-based activity. On the contrary, the CHCS is designed from the volunteer's individual perspective without any linkages or interactions between members with similar interests, and hence it scores poorly on the "common ground" heuristic. Therefore, the CHCS scores 0 out of 8 based on "common ground" heuristic.

## 4.5 Contribution

| Subcategories | Observation of evidence | Rating |
|---|---|---|
| *Threshold* | The users require minimum efforts to register. If desired, they can start contributing to metadata curation activity. Many shortcuts are provided for fetching accurate metadata inputs using the international protocols such as Z39.50 and Virtual International Authority File (VIAF) which provide standardized and accurate names of the authors. However, it is expected that they have prior knowledge of metadata standards like Dublin Core and MARC21. In case they are not aware of these standards, then it may take longer time for them to contribute | 1.5 |
| *Motivators* | The CHCS maintains a log of edits contributed by every user and the number of edits which have been accepted by the experts. The volunteers can earn points based the importance of tags and the usefulness of inputs provided. The contributions of volunteers are recognized by issuing a Certificate of Recognition in Bronze, Silver, Gold, Platinum, and Diamond levels. The certificate is meant to help librarians, curators, and archivists in proving their professional excellence and seeking jobs and promotions | 2 |

| Subcategories | Observation of evidence | Rating |
|---|---|---|
| *Comparative* | CHCS publishes the names and photographs of contributors along with the level of certification. This is intended to trigger competitive spirit among the volunteers and motivate them to contribute more for raising their level of certification from say Bronze to Silver and beyond | 2 |
| *Request list* | There is a need to provide such feature for enlisting the records in dire need of metadata enrichment. The experts can publish a priority list of such records | 0 |
| *Request description* | There is a need to describe the tasks in order to make them more meaningful for selected specialists. It will be possible to announce special metadata requests for the high-ranking volunteers say Gold-level certification and above | 0 |
| *Targeted requests* | We need to provide a feature for inviting specific individuals to contribute in selected records | 0 |

Therefore, the CHCS scores 5.5 out of 12 based on "contribution" heuristic.

CHCS allows the volunteers to extract metadata records from other repositories. As shown in Fig. 5, CHCS incorporates standard protocols and for correcting the author names as per the Virtual International Authority File (VIAF) for standardized names. This approach not only saves time but also ensures accuracy of information as per the international standards.

## 4.6 Platform

| Subcategories | Observation of evidence | Rating |
|---|---|---|
| *Reputation* | The CHCS platform has been recently made online. It is yet to be publicly announced and popularized. Therefore, presently, we do not have any endorsements and public reviews to assert its reputation | 0 |
| *Aesthetics* | The CHCS is designed with proper user experience, which will be evaluated through the usability tests conducted by us. The results of usability test are discussed in the subsequent section | 2 |
| *Uniqueness* | The CHCS is designed to integrate diverse data repositories and also supports popular metadata schemas with many Indian languages for metadata crowdsourcing. Thus, CHCS becomes a very unique crowdsourcing platform in cultural heritage domain. At least in India, such solution is not available | 1 |
| *Tools* | CHCS supports metadata curation activities through many tools and technologies like VIAF, Z30.59, ISBN search, metadata sourcing from open libraries, multilingual support, collaborative framework, and a variety of metadata standards | 2 |
| *Motive* | The CHCS also serves as a digital repository management system with search and retrieval and customizable portal homepage; as a result, crowdsourcing becomes one among the many activities. Therefore, the crowdsourcing motive needs to be highlighted in the overall system | 0 |

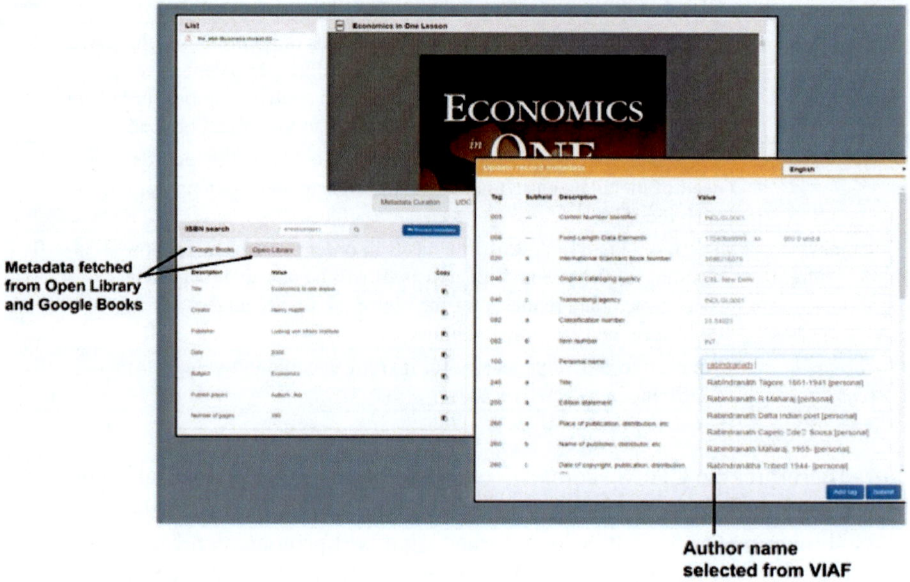


**Fig. 5** ISBN-based metadata fetching and integration of VIAF for accurate author names

Therefore, the CHCS scores 5 out of 10 based on "platform" heuristic.

## 4.7 *Aggregate Scores of Evaluation*

As per radar chart as shown in Fig. 6, the CHCS scores 27.34 out of 60 in total, i.e., 45.56% as per the crowdsourcing heuristics. Based on this exercise, we can observe that the crowdsourcing heuristics are basically helpful in identifying missing functional requirements of a given system. They are not adequate to ensure usability and successful acceptance of the crowdsourcing system by the users.

# 5 Usability Testing of CHCS

Subsequent to heuristic evaluation, which is a self-evaluation by the designer of the CHCS, the usability test was conducted in the computer lab at C-DAC, Pune, by involving 32 students of Library and Information Science program studying at Abasaheb Garware College of Arts and Science, Pune. The test participants comprised of 18 undergraduates, 13 post-graduates, and 1 PhD holder (a faculty member). During the introductory session, the students were first explained about the metadata crowdsourcing concept along with its need and significance to digital repositories. The CHCS was briefly demonstrated to all the students.

| Heuristics | Score out of 10 |
|---|---|
| Purpose | 6.6 |
| Moderation | 4.16 |
| Members | 7 |
| Common Ground | 0 |
| Contribution | 4.58 |
| Platform | 5 |
| Total | 27.34 |

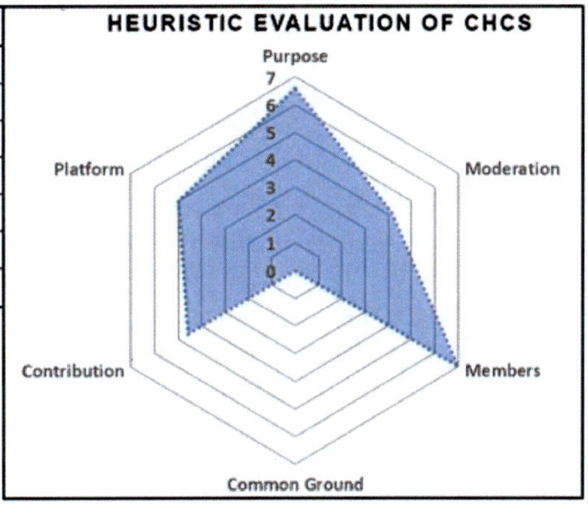


**Fig. 6** Radar chart of the aggregate scores of heuristic evaluation

## *5.1 The Scope of Usability Test*

We had many questions with regard to the usability of our system. Therefore, the following set of tasks were selected to be performed by the test participants, and later an online survey was conducted to consolidate the feedback from test participants.

- Can the participants effortlessly register and log into CHCS?
- Are they able to search and retrieve a desired record?
- Are they familiar with Dublin Core (DC) metadata standard, and can they edit DC tags using the crowdsourcing module?
- Are they familiar with MARC21 metadata standard, and can they edit MARC tags using the crowdsourcing module?
- Are they able to type in Indian languages using the phonetic keyboard supported in CHCS?
- Do they understand the meaning of various metadata tags?
- Can they add missing MARC21 tag along with relevant information inputs?
- Can they add a UDC tag?
- Are they able to fetch records using Z39.50 protocol from libraries?
- Are they able to fetch metadata with the help of ISBN search from open libraries?
- Can they use VIAF for getting standardized author name?
- Do they find the certification scheme motivating enough?
- Will they like to actively contribute to the metadata crowdsourcing?

A task list based on above questions was presented to the test participants.

## 5.2 The Outcomes of Usability Test

The students were given 60 minutes to perform the tasks. They were asked to identify the anomalies in the metadata, do the corrections, and enrich the metadata. In this usability test, we have focused mainly on operational aspects of the system and excluded the quality of inputs provided by the test participants, as the qualitative evaluation of metadata inputs requires involvement of subject experts, which can be undertaken separately. After all tasks were performed repeatedly for multiple records, their feedback was collected in an online survey form. The overall results are presented along with questions, parameters, and vote counts by the participants (Fig. 7).

| Sr. Nos. | Questions | Parameters | | |
|---|---|---|---|---|
| 1 | User registration | Easy | Difficult | Could not do it |
| | | 28 | 3 | 1 |
| 2 | Finding desired record | Easy | Difficult | Could not do it |
| | | 18 | 13 | 1 |
| 3 | Familiarity with Dublin Core | Basic | Expert | Need Guidance |
| | | 14 | 15 | 3 |
| 4 | Familiarity with MARC21 | Basic | Expert | Need Guidance |
| | | 18 | 9 | 5 |
| 5 | Use of Z39.50 Protocol | Very Useful | Could not get required metadata | Other method required |
| | | 25 | 6 | 1 |
| 6 | Addition of New MARC21 Tag | Easy | Difficult | Need Guidance |
| | | 22 | 7 | 3 |
| 7 | Typing in Indian Languages | Easy | Difficult | Could not do it |
| | | 30 | 2 | 0 |
| 8 | Use of VIAF | Easy | Difficult | Could not do it |
| | | 31 | 1 | 0 |
| 9 | Version management in metadata editing | Very Helpful | Needs improvement | Not useful |
| | | 32 | 0 | 0 |
| 10 | UDC Tagging | Easy | Difficult | Need Guidance |
| | | 20 | 9 | 3 |
| 11 | Certificate of Appreciation | Highly Motivating | Greater recognition needed | Not effective |
| | | 32 | 0 | 0 |
| 12 | Possible use of certification | Improves credibility | May help in career advancement | Not useful |
| | | 18 | 14 | 0 |
| 13 | Will you like to contribute? | Happy to contribute | May consider | Not interested |
| | | 28 | 4 | 0 |

**Fig. 7** The results of survey followed by the usability test

## 6 Discussion

- **Evaluation based on community heuristics**

As per the heuristic evaluation, the following ten new requirements are discovered that are not addressed by CHCS.

(i) **Idealism** – Highlight social benefits of crowdsourcing.
(ii) **Regulations** – Document the regulations and policies with regard to crowdsourcing.
(iii) **Virtual co-presence** – Showcase the contributions made by members of the community.
(iv) **Diversity** – Support creation of sub-groups depending the diverse interests of users.
(v) **Events** – There should be a provision to announce community events.
(vi) **Rituals** – Provide support for community rituals.
(vii) **Request list** – Experts should be able to publish a list of records that need metadata enrichment on priority.
(viii) **Request description** – It should be possible to describe the tasks to make it meaningful for the specialized volunteers.
(ix) **Targeted request** – It should be possible to invite high-ranking volunteers to enrich metadata for selected records.
(x) **Motive** – The social motive behind crowdsourcing needs to be highlighted.

The community heuristics tend more towards evaluation of functional aspects of the crowdsourcing system. It is helpful in identifying the missing requirements. Of course, the designer has the prerogative to incorporate the observations depending on the type of system and its usage context.

There is a need to formulate design heuristics, which would guide the design of community perception and social dimensions of crowdsourcing. Presently, the crowdsourcing systems are enabling individuals to contribute through the Internet, but they continue to function in isolation. The crowd perception is dependent on being able to influence and interact with other members of the group. Online discussion forums and community development features may not actually achieve the feeling of being a part of the crowd or a community. Therefore, it is not adequate to be able tick "online discussion forum" during the heuristic evaluation. It only states the availability of a particular functionality in the crowdsourcing system. It does not assure its usability for the specified user community.

The crowdsourcing system evaluation requires the functional heuristics to be combined with design heuristics. During the evaluation of CHCS, we have applied the crowdsourcing heuristics to identify the missing functional aspects of the system, whereas the usability test has confirmed some of the findings of heuristic evaluation and helped in ensuring overall usability of the system in its present form.

- **Usability testing feedback**

87.5% of the test participants found the user registration process to be quite easy. 93.75% of the test participants were comfortable with Indian language typing. Based on this percentage, it is possible to infer that the test participants demonstrate adequate familiarity with use of computer. On the contrary, only 56.25% of the participants found it easy to find the desired record. It indicates that the user interface design for search and retrieval and presentation of search results needs improvements.

43.75% of the test participants have basic knowledge of Dublin Core (DC) metadata standard, and 46.87% of the participants claim to have expertise in DC metadata. Only 9.3% of them require guidance on this metadata standard. 56.25% of the participants have basic familiarity with MARC21 metadata standard, and 28.12% of the participants claim to be experts in this topic. About 15.62% of the participants require guidance on MARC21.

96.87% of the test participants were extremely comfortable in using VIAF, and 78.12% of the participants found the use of Z39.50 protocol to be very helpful. Also, 28.12% of the participants found it difficult to use UDC tagging. This may be partly due to usability issues and partly due to unavailability of guidance.

100% of the test participants find the certificate of recognition to be highly motivating. They believe that such formal recognition of their contributions through crowdsourcing will improve their job prospects and career growth. 87.5% of the test participants have conveyed their happiness to contribute in contributing to crowdsourcing.

## 7 Conclusion

As per the initial heuristic evaluation, CHCS has scored 27.34 out of 60, i.e., 45.56%. This shows that more than 50% of the functional requirements remain to be addressed in the Cultural Heritage Crowdsourcing System (CHCS). We partially agree with the outcomes of heuristic evaluation, as the community heuristics are assuming a fixed template of functionalities, which must be available in a crowdsourcing system.

Ideally, the designers should be provided with the flexibility to select the required crowdsourcing functionalities and not forcefully incorporate all functionalities as implied in the community heuristics. In case of CHCS, we believe that online discussion forum is not a high-priority requirement. It should be possible to give weightages to various heuristic parameters depending on the application context.

The design heuristics should be combined with the community heuristics that focus on functional requirements to achieve proper balance in the evaluation process.

The results of actual usability test are contrary to the findings of the heuristic evaluation. As per the online survey conducted after the usability test, almost 87.5%

of the participants have expressed their willingness to contribute in co-curation activity.

Need of sufficient guidance on metadata standards happens to be the common finding of heuristic evaluation and the usability test.

**Acknowledgments** The author acknowledges the entire development team of Human-Centered Design and Computing Group at C-DAC, Pune, for their valuable contributions in the development of CHCS.

## Appendix: Definitions of Domain-Specific Terms

- Dublin Core: The Dublin Core Schema is a small set of vocabulary terms that can be used to describe digital resources (video, images, web pages, etc.).
- MARC21: MARC (Machine-Readable Cataloging) standards are a set of digital formats for the description of items catalogued by libraries, such as books.
- Z39.50: It is an international standard client-server, application layer communications protocol for searching and retrieving information from a database over a TCP/IP computer network.
- VIAF: The Virtual International Authority File (VIAF) is an international name authority file.
- UDC: The Universal Decimal Classification (UDC) is a bibliographic and library classification representing the systematic arrangement of all branches of human knowledge organized as a coherent system in which knowledge fields are related and interlinked.

## References

1. T. Bücheler, J.H. Sieg, Understanding science 2.0: Crowdsourcing and open innovation in the scientific method. Procedia Computer Science **7**, 327–329 (2011)
2. Y. Zhao, Q. Zhu, Evaluation on crowdsourcing research: Current status and future direction. Inf. Syst. Front. **16**(3), 417–434 (2014)
3. F. Daniel, P. Kucherbaev, C. Cappiello, B. Benatallah, M. Allahbakhsh, Quality control in crowdsourcing: A survey of quality attributes, assessment techniques, and assurance actions. ACM Computing Surveys (CSUR) **51**(1), 7 (2018) 7–40
4. R.M. Ryan, E.L. Deci, Self-determination theory and the facilitation of intrinsic motivation, social development, and well-being. Am. Psychol. **55**, 68–78 (2000)
5. Redi, J., Povoa, I.: Crowdsourcing for rating image aesthetic appeal: Better a paid or a volunteer crowd?. In Proceedings of the International ACM Workshop on Crowdsourcing for Multimedia (2014) 25–30
6. Xu, S., Liu, L., Cui, L., Zheng, Y.: Optimal crowds contest model for crowdsourcing. In ACM Proceedings of the 2nd International Conference on Crowd Science and Engineering (2017) 72–76

7. Pan, Y., & Blevis, E.: A survey of crowdsourcing as a means of collaboration and the implications of crowdsourcing for interaction design. In IEEE International Conference on Collaboration Technologies and Systems (CTS) (2011) 397–403
8. Brandtner, P., Auinger, A., & Helfert, M.: Principles of human computer interaction in crowdsourcing to foster motivation in the context of open innovation. In International Conference on HCI in Business. Springer, Cham. (2014) 585–596
9. Matias, J. N., Diehl, S., & Zuckerman, E.: Passing on: Reader-sourcing gender diversity in Wikipedia. In Proceedings of the 33rd Annual ACM Conference Extended Abstracts on Human Factors in Computing Systems (2015) 1073–1078
10. Samdaria, N., Sowndararajan, A., Vennelakanti, R., & Madhvanath, S.: Mobile interfaces for crowdsourced multimedia microtasks. In ACM Proceedings of the 7th International Conference on HCI, IndiaHCI (2015)
11. Balamurugan, C., & Roy, S.: Human computer interaction paradigm for business process task crowdsourcing. In ACM Proceedings of the 11th Asia Pacific Conference on Computer Human Interaction (2013) 264–273
12. Afentoulidis G, Szlávik Z, Yang J, Bozzon A.: Social gamification in enterprise crowdsourcing. In Proceedings of the 10th ACM Conference on Web Science (2018) 135–144
13. Dergousoff, Kristen, and Regan L. Mandryk.: Mobile gamification for crowdsourcing data collection: Leveraging the freemium model. In Proceedings of the 33rd Annual ACM Conference on Human Factors in Computing Systems. (2015) 1065–1074
14. Geiger D, Rosemann M, Fielt E, Schader M. Crowdsourcing information systems-definition typology, and design. In George, Joey F (Ed.) Proceedings of the 33rd International Conference on Information Systems, Association for Information Systems / AIS Electronic Library (AISeL), Orlando, Florida, The United States of America. (2012)
15. S. à Campo, V.J. Khan, K. Papangelis, P. Markopoulos, Community heuristics for user interface evaluation of crowdsourcing platforms. Futur. Gener. Comput. Syst. **95**, 775–789 (2019)
16. Tinati R, Van Kleek M, Simperl E, Luczak-Rösch M, Simpson R, Shadbolt N. Designing for citizen data analysis: A cross-sectional case study of a multi-domain citizen science platform. In Proceedings of the 33rd Annual ACM Conference on Human Factors in Computing Systems (2015) 4069–4078
17. Maher ML, Preece J, Yeh T, Boston C, Grace K, Pasupuleti A, Stangl A. NatureNet: A model for crowdsourcing the design of citizen science systems. In Proceedings of the Companion Publication of the 17th ACM Conference on Computer Supported Cooperative Work & Social Computing (2014) 201–204
18. Rahmanian B, Davis JG. User interface design for crowdsourcing systems. In Proceedings of the 2014 International Working Conference on Advanced Visual Interfaces (2014) 405–408
19. McKinley D. Design principles for crowdsourcing cultural heritage. Non-Profit Crowd. (2016)
20. A. Chhatwal, P. Mahajan, Crowdsourcing in libraries: An operational framework. International Journal of Digital Library Services **5**(3) (2015)
21. Kazai G, Kamps J, Koolen M, Milic-Frayling N. Crowdsourcing for book search evaluation: Impact of hit design on comparative system ranking. In Proceedings of the 34th International ACM SIGIR Conference on Research and Development in Information Retrieval (2011) 205–214
22. Andro M., Saleh I.: Digital Libraries and Crowdsourcing: A Review. Samuel Szoniecky; Nasreddine Bouhaï. Collective Intelligence and Digital Archives: Towards Knowledge Ecosystems, ISTE; Wiley, (2017) 135–162
23. Oomen J, Aroyo L. Crowdsourcing in the cultural heritage domain: Opportunities and challenges. In Proceedings of the 5th International Conference on Communities and Technologies (2011) 138–149

# Use and Challenges of ICT for Classroom Teaching

**Sunita Dhotre and Abhijit Banubakode**

**Abstract** As humans, we are not born knowing how to live. Knowledge is a continuous process which comes with learning. In this process, a learner and teacher play a very important role. To make the learning process joyful, teachers need to believe that "to teach is to learn twice." When students are trained, it is essential for the teacher to play the role of a facilitator and should leave the students alone, so that they can develop further and learn themselves. In the words of Swami Vivekananda, "Plants always remain small under a big tree." The world is changing very fast and there is a need for making learning equally rapid. There is a need for learners to develop skills that will empower them in modern society, and technology plays a very important role in the same. The educational institutions are putting in efforts to design a well-focused curriculum by integrating unique Information and Communication Technology (ICT) infrastructure into the delivery of educational programs, online curricula, and management systems for both students and teachers. Hence the role of ICT in the classroom is becoming increasingly prominent. The recent changed working scenario has generated the need for use of tools for learning. This chapter focuses on the role of ICT in classrooms and the analysis of data gathered.

**Keywords** Technology-aided learning · Passive · Active

## 1 Introduction

The use of computers has rapidly changed our lives in the past decade. Information and Communication Technology (ICT), a powerful tool transforming education, is a type of technology that creates, stores, displays, shares, or exchanges and conveys information by electronic means [1, 2]. To implement ICT, many tools are used such

Sunita Dhotre (✉)
Department of Computer Engineering, Bharati Vidyapeeth (Deemed to be University) College of Engineering, Pune, Maharashtra, India
e-mail: ssdhotre@bvucoep.edu.in

A. Banubakode
MET Institute of Computer Science, Mumbai, Maharashtra, India



E. Eilu et al. (eds.), *Digital Literacy and Socio-Cultural Acceptance of ICT in Developing Countries*, https://doi.org/10.1007/978-3-030-61089-0_18

as mobile phones, satellite systems, computers along with equipment and services associated with these technologies. The services include videoconferencing, e-mails, blogs, etc. which are also used for innovation. To manage the educational services and make them affordable and available anytime and anywhere, ICT can be used. It eases the information distribution, teaching, and managing of educational services. Today ICT is replacing the concept of whiteboards or blackboards, thus encouraging green computing [10]. It requires changes in the curriculum content, the instructional process, and above all changes in teacher-student relationships. It emphasizes that students should be produced as knowledge generators and active participants while teachers act as facilitators [5, 8, 15]. For this pattern of ICT, this chapter presents an in-depth analysis of the usage of ICT tools by teachers and students. To perform the analysis, the data is collected as to which medium is used the most.

### *1.1 Standardization of Teaching*

Education is undertaken to gain information and understand the facts. Professional education must prepare the would-be professionals for a lifetime of independent learning. Conventional professional education does not specifically cultivate an approach to life-long learning. Using the computer tools, learners can have a creative learning experience. Dropouts can be identified, and there can be a happy learning process and a nonthreatening evaluation. Various evaluation criteria used through the use of ICT are non-biased and helpful in efficient evaluation. To enhance the real dignity of human being, the education needs to be standardized. Different e-learning systems have been developed, studied, and validated such as technology adoption and gratification (TAG) model [3], contextualization [11], and intelligent e-learning systems using Web 3.0 [9].

### *1.2 Stages of ICT Development (Fig. 1)*

Four stages can be followed by the educational institutes while developing ICT:

1. Emerging
2. Applying
3. Infusing
4. Transforming

Emerge the new technologies, apply them, infuse or spread the technology or idea, and finally transform the education to the new level.

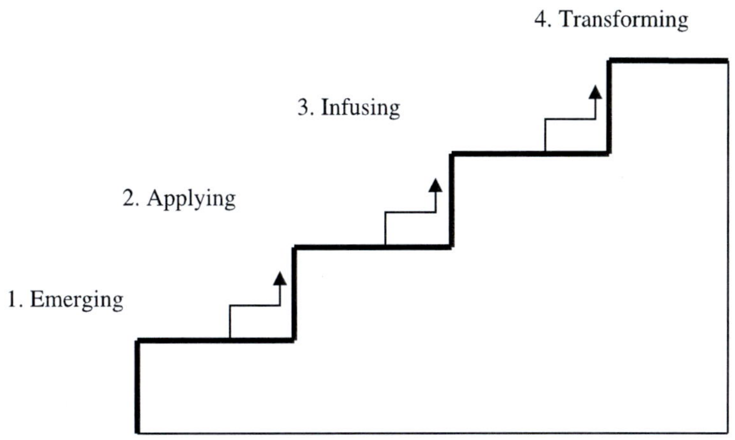


**Fig. 1** Different stages of ICT development

### *1.3 Learning: Teaching Style and ICT*

There are various learning styles and modes. Different people prefer different kinds of learning. While some prefer to watch and grasp the concept through a visual approach, others prefer reading or listening. Hence, it is necessary that learning opportunities must be provided through various means. It has been proven that when a preferred learning platform is provided the learners are motivated thus improving the end results.

There is a passive reception of knowledge when a teacher-centered learning approach is used whereas; learner-centered approach encourages a process of active inquiry. It turns out to be an active process as the responsibility lies on the learners themselves.

With the popularity of computers, a major shift in the learning paradigm [13] is experienced as the technologies are becoming more interactive encouraging active learning. Today, with the help of ICT, teachers can provide powerful learning opportunities wherein the learners can define their learning needs, search for information, and assess its value which ultimately leads to gain of knowledge.

## 2 Teachers and ICT

It is necessary that teachers have an optimistic mindset in applying and using ICT. As teachers begin to use ICT in classrooms, their roles have been changed. The concept of delivery of transfer of contents from teacher to student is vanishing away. Instead, critical thinking and problem-solving based on information accessed from a variety of resources such as television, phone, satellite systems, computers, and internet lead to learning and knowledge gain. The teacher's role is to design contents which can be used to study real-life examples. The contents should also be applied to solve existing problems [4, 11], thus featuring innovation.

To become confident, teachers must undergo training so that awareness and adaption to newer technologies [14] can be made. Teachers should be:

1. Provided access to professional development programs [3] that enables to have multiskills both in the use of technology and in task design
2. Able to develop good time management skills
3. Aware of various learning styles available and effectively use them by providing verbal instructions to accompany on-screen instructions
4. Able to make hard copies of the screen version of text available to learners
5. Able to develop the ability to interpret theoretical instructional models into active student-centered teaching practices

For example, teachers can put images into their work that will make the presentation better and will make it easier for learners to understand. Commercial simulation packages can be used for teaching difficult concepts. Pedagogical activities should be included to encourage and challenge the thinking of learners.

## *2.1 Students and ICT*

Students will have more interactive sessions while using ICT than normal teaching. A survey was made in which students told that now they enjoy attending classes. Students say that ICT helped them to achieve better grades through better presentation. There are interactive sessions between students and teachers. But they also felt that ICT was not fully responsible for this. They concluded that the use of technology does not lead to obtaining superior grades; it just makes learning look easier. Most of the students use internet website for searching educational contents.

## *2.2 ICT Herald a Paradigm Shift in Education*

In earlier days, traditional method of teaching was used which was hard. The process of transfer and delivery of the content was linear, that is the content was divided into small modules and each was delivered in a sequence. The last three decades have undergone a phenomenal change in education. The introduction of ICT has changed this concept. It has provided a new definition of learning and bridged the gap between teaching and learning rapidly. The learning is active, social, linear as well as non-linear, neutral, contextualized, and integrative based on the learning ability of students. There is a need for teachers to have competencies in ICT [8]. The focus has changed from traditional instruction to virtual learning environment. This environment involves the transformation of roles of both teachers [3] and learners creating more interactive and engaging roles. The Table 1 below indicates the

**Table 1** Changes in teachers' role [6]

| Shift from | To |
|---|---|
| Knowledge transmitter | Learner, co-learner, facilitator, collaborator, coach, knowledge navigator |
| Teacher controls and directs all aspects of learning | More options and responsibilities are given by teachers for students learning |

**Table 2** Changes in students' role [3]

| Shift from | To |
|---|---|
| Passive receiver of information | Actively participates in the learning process |
| Reproducing knowledge<br>Learning as a solitary activity | Producing knowledge<br>Collaborative learning and concept-based learning |

changing roles of teachers. A new way of thinking and understanding of the new learning process is required by the teachers. The teachers should also have access to infrastructure, variety of tools in order to improve access to learning opportunities and be digitally literate. Interaction which policy-makers, administrators should also take place at regular intervals.

Table 2 indicates the changes in role of students [3]. This new way involves the learners and makes them active participants in the learning process.

## 3 Engineering Education and ICT in India

In the past few years, a major change has been witnessed in engineering education. The demand for superior-quality and on-demand education is rising, which has led to the adoption of Information and Communication Technologies thus extending the outreach of education. Several technology-enabled initiatives taken up by the Government of India consist of the National Programme on Technology Enhanced Learning (NPTEL)), use of "virtual classrooms" and "virtual laboratories," and use of educational satellite called the EDUSAT. Blending collaborative learning with interactive technology-enhanced learning initiatives and finding ways of providing support for learners' queries are the approaches used in these initiatives. [12]

For this purpose, educational institutes should:

- Have a good infrastructure to support the facilities provided by ICT
- Have well-equipped classrooms with computers and projectors
- Have the ability to develop and maintain IT Infrastructure
- Provide opportunities for regular collaboration among teachers
- Provide electricity during working hours. Have a better security system
- Create new approaches [7]
- Should arrange regular workshops or seminars so that teachers get updated with technology

India is progressing in providing education through ICT. The objectives and goals of ICT in India are made clearer for education programs [12]. They are as follows:

- Use of technology-aided learning to enhance learning effectiveness
- Empower teachers and make learning joyful and student-centric by promoting a spirit of inquiry and critical thinking

Educational institutes in India have started building IT infrastructure and are provided with computers, networking facilities, printers, and scanners. Computer software is also provided with various tools, software, and Educational CDs. However, internet connection is limited to some cities and remote places. There are still many challenges and obstacles like software content, internet bandwidth availability, lack of skilled teachers, and lack of enthusiasm in all concerned workforce required to execute and maintain the whole system and infrastructure.

## 4 Aims of the Study

The objective of this study is to identify the availability of ICT tools, its usage, and the preferred mode of communication by students and teachers. The prior knowledge of computer applications and the opinion about the use of ICT in learning are studied and analyzed.

## 5 Research Methodologies

This area gives a glimpse of the methods used to achieve the objectives of this research study. The focal point of the research design is that this research is descriptive research. The data was collected through Survey (Online Questionnaires) and the source of data is either primary or secondary depending on the need of the research. Data is analyzed on the target population of the study, sampling techniques, sample size, conceptual framework, and the data collection is also done for the study.

A questionnaire was designed after conducting a comprehensive review of the related literature. An online survey was conducted at the academic institutions of Mumbai Educational Trust, Bandra, Mumbai and Bharati Vidyapeeth (Deemed to be University) College of Engineering, Pune. The online survey form was forwarded to more than 200 faculty members and staff; out of 200 we had received response of 175 faculties and staff. Data analyzed the target population of the study, sampling techniques, sample size, and conceptual framework.

**Research Design**

The research design explains the survey type of research. We have used three main types of research design:

*Data collection*: A survey link shared among faculty members and staff of the institution and collected data.

*Measurement*: Descriptive quantitative methods applied for data measurement.

*Analysis*: Extracted useful information from data and the decision based upon the data analysis.

### 5.1 Primary Data

In this research, the target population is academic staff and faculty members of the Mumbai Educational Trust (MET) and Bharati Vidyapeeth (Deemed to be University) College of Engineering (BVDUCOE). Considering the importance of Information and Communication Technology, we have collected data from teachers and students of both academic institution who are using ICT tools during their lecture, and because of this, there is a great impact on student's learning.

### 5.2 Sample Framework

The sample framework for this research are the various institutes of Mumbai Educational Trust (MET)) and Bharati Vidyapeeth (Deemed to be University) College of Engineering (BVDUCOE), Pune. The sample has been collected from around eight institutes of MET and BVDUCOE.

### 5.3 Sample Size

The size of the sample is 200 at MET and BVDUCOE. Respondents were 175 from whom data was collected. This sample is borrowed from the questionnaires survey.

## 6 Results and Discussion

We have created a rubrics shown in Fig. 2 and collected data from different MET institutes of MET league of colleges and BVDUCOE. We have marked the survey on a scale of 1 to 4 based on the need of ICT tools to relate with departmental functions [15]. The scale was marked as follows:

- None
- Very Little
- Some
- Very much

| *ICT- Tools* | *None* | *Very Little* | *Some* | *Very much* |
|---|---|---|---|---|
| Website | | | | ✓ |
| Audio | | ✓ | | |
| Computer | | | | ✓ |
| Mobile | | | | ✓ |
| ERP | | | ✓ | |
| PPT | | | ✓ | |
| e-Readers | ✓ | | | |

**Fig. 2** ICT rubrics

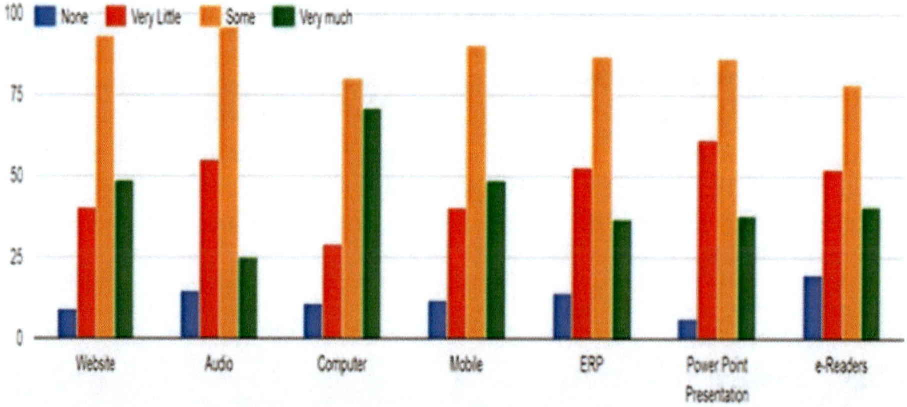


**Fig. 3** Implementation of ICT tools at an educational institution

- None
- Very little
- Some
- Very much

None represents that nobody uses ICT tools in academics. Very little represents that very few people use ICT tools in academics. Some represent that some of the faculty members and staff are using ICT tools. Very much represents most of the faculty members and staff are using ICT tools in academics. Figure 3 shows the graph of seven ICT tools, i.e., website, audio, computer, mobile, enterprise resource planning, e-reader power point presentation. These ICT tools are the digital infrastructures used in Information and Communication Technology among students to students, students-teacher interaction. Institute can upload various circulars, notice on website. Faculty members are using computer systems and audio systems for effective teaching-learning system and power point presentations for better understanding. Students can pay online fees using ERP system.

Table 3 exhibits the Information and Communication Technology survey of seven ICT tools taken across various departments under categories such as:

- None
- Very little
- Some
- Very much

After taking survey and collecting samples from various departments, the count of categories converted into percentage count.

The first row states that 57% of faculty members preferred none, 20% of faculty members preferred very little, 48% of faculty members preferred some, 26% of faculty members opted very much in website ICT Tool. This is followed for the entire six tools and the percentage for each module under each category is given in Table 3. Then we generalized the categories none, very little, some to percentage of faculty members who prefer ICT tools and the categories very much, required restructuring to percentage of faculty members who don't prefer ICT in the academic system.

Table 4 exhibits the final survey result on two specific categories for the seven ICT tools. The result of Table 4 is derived from Table 3 result. According to the result generated in Table 4, almost in all the modules, more than 70–80% of faculty members prefer use of ICT tools in academic system and 25–30% of faculty members don't prefer ICT tools in the academic system. More number of faculty members are using ICT because of following advantages:

- Provides a wide range of latest and ready material
- Enables effective learning through the use of images, text, video, audio, and animation

**Table 3** Implementation of ICT in the academic system

| Sr. no. | ICT tools | None | Very little | Some | Very much |
|---|---|---|---|---|---|
| 1. | Website | 57% | 20% | 48% | 26% |
| 2. | Audio | 2% | 17% | 30% | 39% |
| 3. | Computer | 10% | 11% | 29% | 37% |
| 4. | Mobile | 28% | 16% | 24% | 16% |
| 5. | ERP | 35% | 20% | 30% | 13% |
| 6. | PowerPoint presentation | 16% | 15% | 30% | 46% |
| 7. | e-readers | 26% | 12% | 28% | 15% |

**Table 4** Use of information and communication tool analysis in the academic system

| Sr. no. | ICT devices | Percentage of faculty members who prefer ICT tools in the academic system | Percentage of faculty members who don't prefer ICT tools in the academic system |
|---|---|---|---|
| 1. | Website | 75% | 25% |
| 2. | Audio | 72% | 28% |
| 3. | Computer | 70% | 30% |
| 4. | Mobile | 68% | 32% |
| 5. | ERP | 85% | 15% |
| 6. | PowerPoint Presentation | 81% | 19% |
| 7. | e-readers | 66% | 34% |

- Increases the chances of effective learning through interaction and collaboration
- Provides platform that can engage students

Since more number of faculty members prefer to use ICT tools in the academic system and they are happy using it, so our research finally concludes that in this new technology era use of ICT tools is a great benefit for faculty members as well as for students. In spite of the use of ICT tools, there are certain key challenges in providing ICT infrastructures in India. There is a lack of technology, people are not using the internet in rural areas and masses also lack of availability of qualified trainers.

**Future Enhancement** In this research, we have discussed about seven ICT tools in academic system. In future, many more ICT tools can be implemented such as Tablets, Interactive whiteboard or smartboard, flipped classrooms, etc. By doing this, all the information can be maintained online in the database and interaction will be more comfortable and at ease.

Several benefits from the implementation of ICT tools in an academic system are:

- Improved information supply and flow for planning and controlling processes of an institution
- Improved service for faculty members, students, and staff
- Reduced expenditures through increased process efficiency

## 7 Conclusion

Information and Communication Technology (ICT) has brought a change in education policy used by academic institutions. It has replaced the concept of whiteboards or blackboards. It has made the sessions between students and teachers more engaging and lucrative. ICT has changed the level of education in an academic institution. With the help of ICT, the use of internet has been increased; students and teachers are in touch with the latest technology or media. The use of Information and Communication Technology enables individualized instruction and collaborative learning. Keeping the importance of ICT, it is observed that teachers are using Information and Communication Technology tools during their lecture because of which there is a great impact on student's learning.

In our research work, we conclude that there is a significant number of teachers uses ERP system as ICT tool as ERP implemented in most of the academic institution. Because of the ease of use and technology awareness, most of the teachers are using power point presentation, website, and audio as second-largest ICT tools. The incorporation of ICT in education can bring significant changes in the teaching and learning process and overcome existing challenges. Further research can be carried out on more ICT tools.

## References


1. S.R. Das, S. Mohapatra, Social and public impact of ICT enabled education. *2008 International Conference on Information Technology*, Bhubaneswar, 2008, pp. 300–303
2. M. Vesisenaho, ICT education and computer science education for development – Impact and contextualization. *2010 IEEE Frontiers in Education Conference (FIE)*, Washington, DC, 2010, pp. F4J-1–F4J-6
3. I. Simonics, Use of ICT equipment by engineer teachers and mentors. *2017 IEEE Global Engineering Education Conference (EDUCON)*, Athens, 2017, pp. 527–535
4. D.J. Sujatha, R. Jobanputra, R.V. Joshi, A tutorial on enhancing the level-of-learning through ICT-Enabled Teaching In Engineering Education. *2011 IEEE International Conference on Technology for Education*, Chennai, Tamil Nadu, 2011, pp. 303–304
5. J. Chanwijit, W. Lomwongpaiboon, O. Dowjam, P. Tangworakitthaworn, Decision support system for targeting higher education. *2016 Fifth ICT International Student Project Conference (ICT-ISPC)*, Nakhon Pathom, 2016, pp. 154–157
6. P. Zahariev, N. Bencheva, G. Hristov, Y. Ruseva, ICT convergence challenges in education and their impact on both instructors and students. *2013 24th EAEEIE Annual Conference (EAEEIE 2013)*, Chania, 2013, pp. 193–197
7. L. Dhandabani, R. Sukumaran, Use of ICT in engineering education: A survey report. *2014 IEEE International Conference on Computational Intelligence and Computing Research*, Coimbatore, 2014, pp. 1–5
8. G.I. Mwandosya, C.S. Montero, Towards a mobile education tool for higher education teachers: A user requirements definition. *2017 IEEE AFRICON*, Cape Town, 2017, pp. 881–887
9. S. Dhotre, S. Patil, Intelligent e-learning systems using web 3.0. J. Eng. Res. Stud. **1**(2), 230–232 (2010)
10. T. Worthington, A green computing professional education course online: Designing and delivering a course in ICT sustainability using Internet and eBooks. *2012 7th International Conference on Computer Science & Education (ICCSE)*, Melbourne, VIC, 2012, pp. 263–266
11. A. Chhabra, Information and Communication Technology (ICT): A paradigm shift in teacher education. Sch. Res. J. Humanity English Language.I(III) (2014)
12. B. Bhattacharya, Centre for Educational Technology, IIT, Kaharagpur, Innovations in Education and Teaching International, India. 45(2), 93–101 (2008)
13. M. Belagra, C. Benachaiba, B. Guemid, Using ICT in higher education: Teachers of electrical engineering department at the University of Bechar: Case study. *Proceedings of the 2012 IEEE Global Engineering Education Conference (EDUCON)*, Marrakech, 2012, pp. 1–6
14. A.Y.M.A. Islam, M.M.C. Mok, X. Gu, J. Spector, C. Hai-Leng, ICT in higher education: An exploration of practices in Malaysian universities. IEEE Access **7**, 16892–16908 (2019)
15. Faculty of Education at IUB, Library Philosophy and Practice (e-journal). 677. (2011). https://digitalcommons.unl.edu/libphilprac/677

# Index





E. Eilu et al. (eds.), *Digital Literacy and Socio-Cultural Acceptance of ICT in Developing Countries*, https://doi.org/10.1007/978-3-030-61089-0

## K



## L



## M